Fuels and Petrochemicals Division 2017

Core Programming Area at the 2017 AIChE Spring Meeting and 13th Global Congress on Process Safety

San Antonio, Texas, USA
26 - 30 March 2017

ISBN: 978-1-5108-4147-5

Printed from e-media with permission by:

Curran Associates, Inc.
57 Morehouse Lane
Red Hook, NY 12571

Some format issues inherent in the e-media version may also appear in this print version.

Printed by Curran Associates, Inc. (2017)

For permission requests, please contact AIChE
at the address below.

AIChE
120 Wall Street, FL 23
New York, NY 10005-4020

Phone: (800) 242-4363
Fax: (203) 775-5177

www.aiche.org

Additional copies of this publication are available from:

Curran Associates, Inc.
57 Morehouse Lane
Red Hook, NY 12571 USA
Phone: 845-758-0400
Fax: 845-758-2633
Email: curran@proceedings.com
Web: www.proceedings.com

TABLE OF CONTENTS

482030 AGILE Award Keynote Address

Abstract:

Factors in Selecting AGILE Honoree

1. Overall sustained excellent industry leadership in businesses that employ chemical engineers

2. Recognized for efforts in innovation and creating strategies that ensure success

3. Value of the keynote to the meeting attendees and the members of the Institute Schedule

Event	Time	Location
Continental Breakfast	7:00AM - 8:00AM	Henry Gonzalez Convention Center, Lila Cockrell Theatre
AGILE Keynote Address	8:00AM - 9:15AM	Henry Gonzalez Convention Center, Lila Cockrell Theatre

The 2017 AIChE Spring Meeting and 13th GCPS AGILE Keynote Address will be presented by AGILE award recipient John Y. Televantos, Partner, Arsenal Capital Partners.

John Televantos is a Partner at Arsenal Capital Partners, a middle market private equity firm focused on the specialty industrial, healthcare and specialty financial service sectors. Before joining Arsenal in 2006, Dr. Televantos was President of Aqualon Company, owned by Hercules, Inc.; President and Founder of Helios Chemical Company; and Chief Executive Officer for Foamex International.

Dr. Televantos began his career with Union Carbide in 1977 in silicones research. There he developed two key processes for the manufacture of Methyl Chloride and Siloxane polymers that were commercialized. These processes not only improved efficiency and reduced cost but also dramatically reduced waste. He moved through a variety of management positions there and later at ARCO Chemical, where he became vice president of strategy and development.

At ARCO John co-invented and led a team that developed and commercialized a breakthrough technology in polyol manufacture (a key raw material for polyurethanes) eliminating a whole process step along with waste and emissions, while at the same time opening the door to the creation of new higher-performance polyurethanes. The technology has since been licensed by most of the major global polyol manufacturers with significant waste reduction worldwide.

At Arsenal Capital, Dr. Televantos serves as co-head of the specialty industrial practice. During his tenure, he has been part of several acquisitions and successful exits in the chemicals and materials sectors. In addition to serving on the boards of several Arsenal-owned companies, he is lead director on the board of Balchem Corp.

482218 Energy-Industry Trends and Insights from the Wellhead to Downstream Products

Abstract:

HSB Solomon Associates LLC (Solomon) has conducted Comparative Performance Analyses™ (CPA™), also known as the Solomon studies, for over 35 years. These studies benchmark the Refining, Petrochemicals, Midstream, Upstream, and Electric Power industries, with the earliest studies starting in 1980. As part of these studies, Solomon gathers and analyzes data from around the globe on all phases of operations. Key Performance Areas include yields and losses, energy consumption, reliability and maintenance, personnel, efficiency, and financial performance. These benchmarking studies give Solomon clients valuable insights into the key performance areas of their facilities as well as the overall industry trends. The studies provide reliable information based on proprietary, patented, and time-tested methodologies, to foster the continuous improvement needed for companies to remain competitive.

The first Solomon studies began in 1980. Over the years, the studies have documented all of the ups and downs of the oil-and-gas-related industry as it reacted to changes in prices, markets, technology, and government regulations. For this paper, Solomon consultants have examined the trends and shared insights learned from their data-driven CPA on all energy-related industries from the wellhead to the downstream derivative chemical manufacturing, and all stages in-between.

BURNS McDONNELL

Alkylation and Other Processes for Refinery Propylene

presented by

Ben Ketter
Dominic Varraveto

March 2017

Agenda

- Overview of Refinery Propylene Processing Options
- Propylene Alkylation of Isobutane
- Propylene Alkylation of Benzene

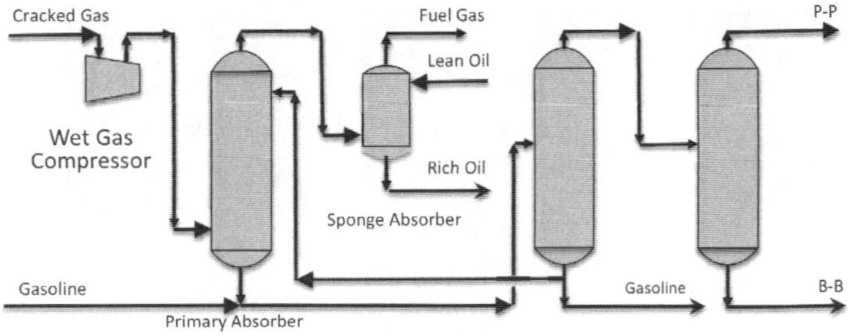

Overview of Refinery Propylene Processing

- Propylene Sources
  - Catalytic (FCC)
  - Thermal (DCU) Cracking
- Recovery in "unsats" Gas Plant of 90+%
- Separating Propane and Propylene
  - Ethane (BP = -128.2°F)
  - Propylene (BP = -53.7°F)
  - Propane (BP = -43.6°F)
  - Butane (BP = 30.2°F)

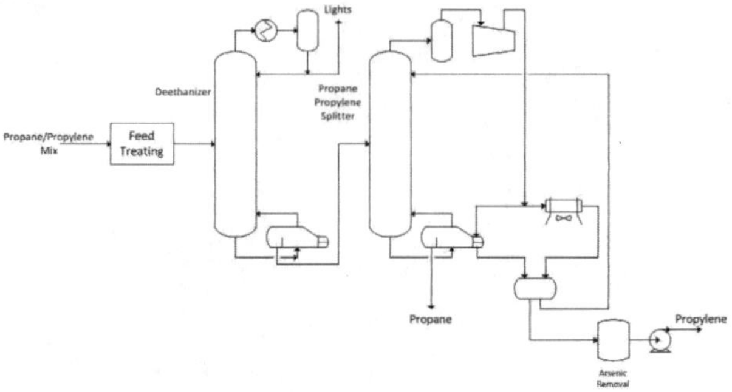

Propylene Purity Grades, %

- Refinery Grade, 70
- Chemical Grade 92-96
- Polymer Grade 99.5

Process Description of Refinery Alkylation

- Turn LPG into Alkylate, a relatively high octane gasoline blendstock
- Utilizes a strong liquid acid catalyst, HF or H_2SO_4, to react olefins including propylene, with isobutane.
- Reactions occur at moderate pressure and less than ambient temperature

Isobutane Feed

- Contaminants
  - Sulfur, caustic, water
- Source
  - Refinery straight run
    - iC4 ~25%
  - Refinery DIB
    - iC4 ~80-95%
  - Pipeline (NGL Fractionator)
    - iC4 ~95-99%

Sulfuric Acid Alkylation Block Flow & Feedstocks

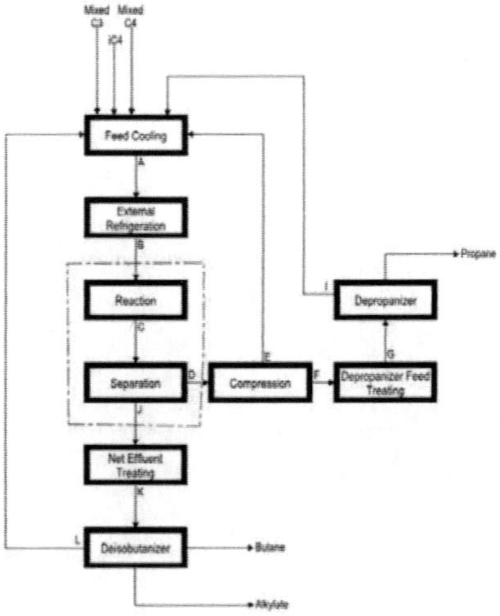

Feed Contaminants: Dienes, sulfur, aromatics, naphthenes, water

- Propylene, $C_3^=$
  - ~75% C3 / 25% C_3=
  - Typically mixed with $C_4^=$
  - Requires highest acid spend strength
  - Relatively low product octane
  - Net propane from Hydrogen transfer reaction
- Butylene, $C_4^=$,
  - ~35 iC4 / ~10% nC4
  - Most desirable olefin feedstock
  - Results in highest product octane
- Amylene $C_5^=$,
  - ~35 iC5/~5% nC5
  - Least severe operating conditions
  - Can be left in unsats gas plant debut bottoms aka cat gasoline
  - Least common alkylation feed
  - Net isopentane from Hydrogen transfer reaction

Alkylation Equipment Piping and Metallurgy

- Operating conditions
  - Low pressure (<300 psig), low temperature (0-300 °F), but not cryogenic, almost all hydrocarbons lines are LPG
- Equipment
  - Traditional heat exchangers, reboilers, and pumps
  - Large compressor (HP = alkylate * 0.60)
    - Centrifugal with low suction pressure (~15 psia and low discharge ~150 psia)
    - Single machines with no interstage cooling
- Piping
  - Large lines sizes and equipment (internal rate is ~6x alkylate rate) (6 - 10"), some large lines around the compressor (30 – 60")
- Controls
  - Some mass flow meters, Some very large control valves around compressor,
  - Compressor protection and antisurge system
- Materials
  - Mostly carbon steel, some alloy 20 and stainless

Alkylation Reactions

C5=														
	Selectivity													
Rxn 1	0.80	1	C5=	+	1	iC4	=	1	iC9					
Rxn 2	0.20	1	C5=	+	2	iC4	=			1	iC5	+	1	iC8
BBL (norm)		1.00	C5=	+	1.16	iC4	=	1.32	iC9 +	0.22	iC5	+	0.30	iC8
C4=														
Rxn 1	1.00	1	C4=	+	1	iC4	=	1	iC8					
BBL (norm)		1.00	C4=	+	1.10	iC4	=	1.71	iC8					
C3=														
	Selectivity													
Rxn 1	1.00	1	C3=	+	1	iC4	=	1	iC7					
Rxn 2	0.00	1	C3=	+	2	iC4	=			1	C3	+	1	iC8
BBL (norm)		1.00	C3=	+	1.28	iC4	=	1.77	iC7 +	0.00	C3	+	0.00	iC8

- Reaction #2 are side reactions are dependent on:
  - Catalyst
  - Temperature
  - iC5 content
  - iC4 content
- Reaction isobutane is kept at ~10:1 excess

Alkylation Products

- True Alkylate
  - Mixture of C_7, C_8, & C9 molecules
  - ~ 40 to 50 individual molecules
  - 90 - 94 Octane ~ 2 psia RVP
    - Gasoline Octane, (R+M)/2 = 84-91
    - Gasoline RPV, 7 – 15 psia (seasonal)
  - Contains no toxics (olefins or aromatics)
- Whole Alkylate
  - Alkylate including $C6^-$

Methyl heptane, (29)

Di-methyl hexane, (72)

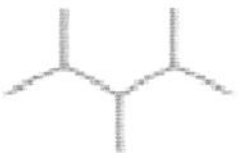

Tri-methyl pentane, (100)

MSAT II STRATEGY: POST FRACTIONATION & EXTRACTION

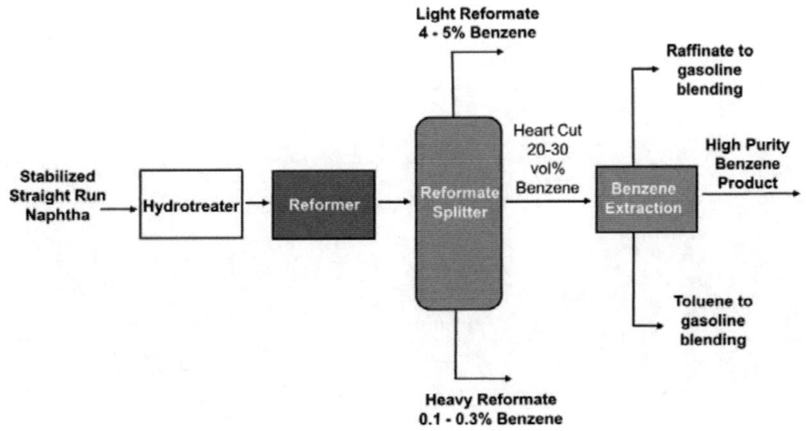

MSAT II STRATEGY: POST FRACTIONATION & SATURATION

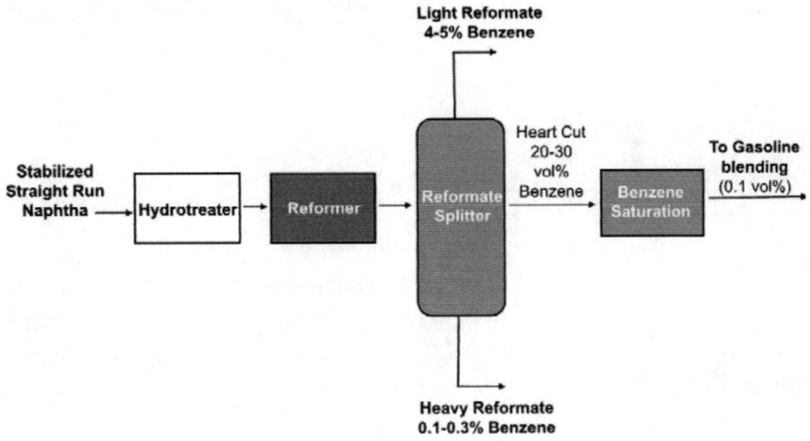

ALTERNATIVE: DILUTE BENZENE ALKYLATION

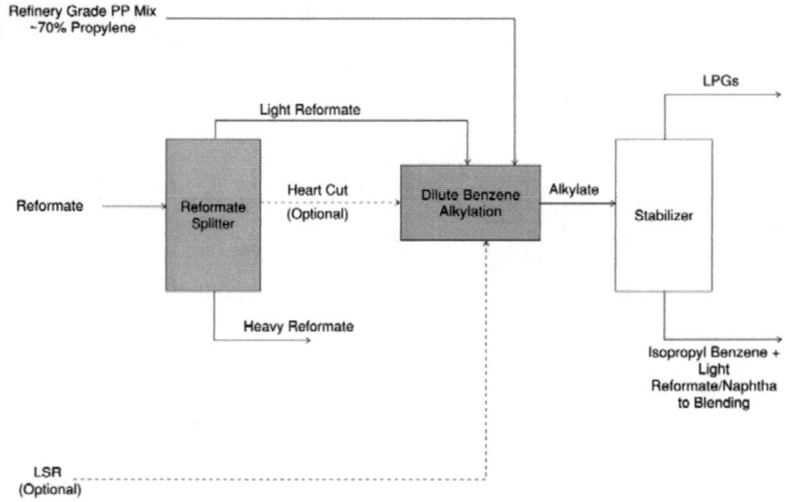

FEEDS

LIGHT REFORMATE

- C4-C6 Range
- 10-15 vol% Benzene

AND/OR

REFORMATE HEART CUT

- C6 Range
- 20-30 vol% Benzene

AND/OR

LSR

- C4-C6 Range
- 1-5 vol% Benzene

REFINERY GRADE PP

- 70% Propylene, 30% Propane

	BENZENE	ISOPROPYL BENZENE
RON	103	112
MON	105	102

REACTION

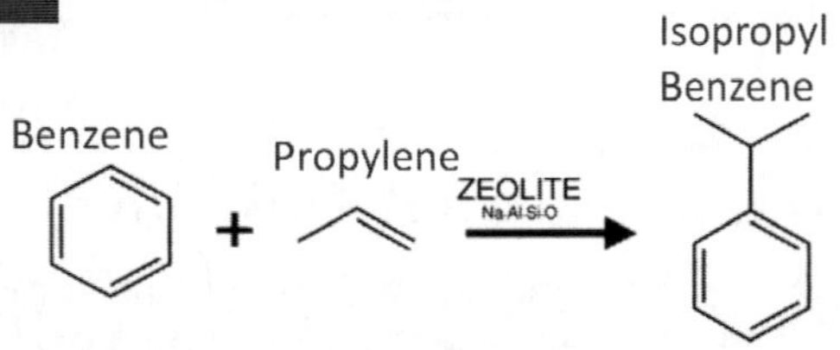

CATALYST

- Zeolite Catalyst (solid fixed bed)
- Example:
  ZSM-5 (Zeolite Socony Mobil-5) $Na_nAl_nSi_{96-n}O_{192}\cdot 16H_2O$ (0<n<27).
- Lewis Acid – Friedel-Crafts Alkylation Reaction.
- Regenerable.

CONDITIONS

- Liquid Phase Reaction.
- Reaction conditions: 350-450 °F @ 800-850 psig.
- Exothermic.

REACTION (CONT'D.)

SECONDARY/SIDE REACTIONS

- Main undesired product: Di-isopropyl Benzene.
- Propylene polymerization, tri-isopropyl benzene, other reactions much less common.
- Heavily alkylated single-ring aromatics have relatively high octane numbers (compared to other gasoline components) but octane numbers generally get lower as MW increases.

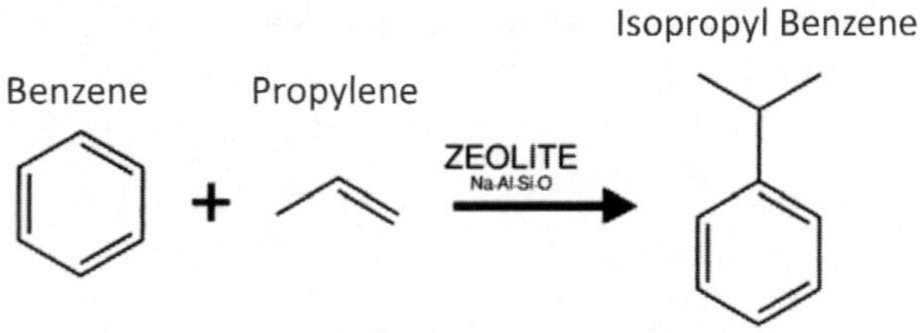

CAPITAL STRATEGIES

POSSIBLE TO CONVERT BENZENE SATURATION UNIT TO BENZENE ALKYLATION?

- X Reaction section and heat conservation exchangers likely occur in the vapor phase at a much lower pressure than the high-pressure liquid phase benzene alkylation (as currently developed technology).
- ✓ Stabilizer easily re-purposed for benzene alkylation.
- ✓ Mothball/repurpose hydrogen recycle compressor, heater, flash pots, etc. for another service.

BURNS & McDONNELL 15

CASE STUDY, ROUND NUMBERS

TYPICAL MID-SIZE 150 MBPD REFINERY

- FCC Capacity: 50 MBPD – 10 vol% C3 Yield ⟶ 5,000 BPD C3/C3= Mix
- Reformer: 30 MBPD – 5 vol% Benzene ⟶ 1,275 BPD BENZENE
- BENZENE LIMITING ⟶ Assume ~80% Conversion, 1,600 BPD Cumene
- Small volumetric swell on Reformate Gasoline Blendstock,
- Resultant 1 vol% Benzene in Reformate.

BURNS & McDONNELL 16

INDUSTRY EXAMPLES

CALUMET SUPERIOR

- 45,000 BPD refinery located in Superior WI.
- 5,500 BPD product - LSR/Light Reformate Feed.
- Engineered, Procured, Constructed by Burns and McDonnell.

Badger Congratulates Calumet on Start-Up of World's First BenzOUT™ Process Unit

March 15, 2012 09:18 AM Eastern Daylight Time

CAMBRIDGE, Mass.--(BUSINESS WIRE)--Badger Licensing LLC congratulates Calumet Specialty Products Partners L.P. on the successful completion and start-up of the world's first BenzOUT™ process unit for gasoline benzene reduction at Calumet's refinery in Superior, WI. The unit converts benzene contained in the refinery's gasoline blend-stock into high octane alkyl-aromatics by reacting benzene with refinery grade propylene.

INDUSTRY EXAMPLES (cont'd.)

WYOMING REFINING CO

- 14,000 BPD refinery located in Newcastle WY.
- 4,000 BPD unit.
- Light Reformate Feed – some olefin pretreatment/fractionation required.
- Engineered, Procured, Constructed by Burns and McDonnell.

Badger Licensing to Provide Proprietary BenzOUT™ Technology for Grassroots Plant in Newcastle, Wyoming

September 24, 2014 02:00 PM Eastern Daylight Time

BOSTON--(BUSINESS WIRE)--Badger Licensing LLC (Badger) today announced that Wyoming Refining Company has selected Badger to provide its proprietary BenzOUT™ technology for a 4,000 barrels per day grassroots plant to be built in Newcastle, Wyoming. The unit converts benzene contained in the refinery's gasoline pool into high octane blendstock by reacting the benzene with refinery grade propylene. The award includes technology license, process design, and start-up services.

INDUSTRY EXAMPLES (cont'd.)

Calumet Superior		Wyoming Refining Co.	
Small	✓	Small	✓
Geographically isolated	✓	Geographically isolated	✓

Advantages

- Isolation from chemicals markets makes benzene/propylene export costly and impractical.
- Refinery throughput not big enough to achieve economies of scale – PP splitter/Benzene extraction large capital investments.
- No hydrogen demand. No access to pipeline, and on-purpose hydrogen production is an additional investment.

CONCLUSION

ADDITIONAL ADVANTAGES

- Reduction of Benzene in Gasoline Pool.
- Octane upgrade of 2-3 numbers.
- Upgrade propylene to high-octane gasoline blendstock (from fuel gas).
- HD-5 quality propane stream is a product of stabilizer.
- Reformer/FCC feed and severity flexibility.
- Alkylation capital cost comparison:

BENZENE ALKYLATION	ISOBUTANE ALKYLATION
BASE, PER BARREL PRODUCT (ISOPROPYL BENZENE + LIGHT NAPHTHA)	~4 x BASE, PER BARREL ALKYLATE PRODUCT

WORKS CITED

- Ghosh, Prasenjeet, Karlton J. Hickey, and Stephen B. Jaffe. "Development of a detailed gasoline composition-based octane model." Industrial & engineering chemistry research 45.1 (2006): 337-345.
- Do, Phuong TM, et al. "Catalytic strategies for improving specific fuel properties." Catalysis 20 (2007): 33-64.
- BUSINESS WIRE. (2012, March 15). Badger Congratulates Calumet on Start-Up of World's First BenzOUT™ Process Unit. Retrieved March 10, 2017, from http://www.businesswire.com/news/home/20120315005052/en/Badger-Congratulates-Calumet-Start-Up-World%E2%80%99s-BenzOUT%E2%84%A2-Process
- BUSINESS WIRE. (2014, September 24). Badger Licensing to Provide Proprietary BenzOUT™ Technology for Grassroots Plant in Newcastle, Wyoming. Retrieved March 10, 2017, from http://www.businesswire.com/news/home/20140924005114/en/Badger-Licensing-Provide-Proprietary-BenzOUT%E2%84%A2-Technology-Grassroots
- Varraveto, D., & Phillips, B. (2012, February 7). Gasoline: from MSAT2 to Tier 3 [Tech Tuesday Presentation]. Burns and McDonnell, Kansas City, MO.

BURNS McDONNELL

Dave Smith
Honeywell UOP

Trace Contaminant Removal In Petroleum Refining Processes

Honeywell UOP

March 26-29, 2017 | 2017 AIChE Spring Mtg & 13th Global Congress on Process Safety | San Antonio, Texas

UOP 7828B-0

1

Overview of The Presentation

- **Section 1: Adsorption Fundamentals & Types Of Adsorbents Offered By UOP**
  - Physisorption versus chemisorption
  - Selectivity
  - Adsorbent reactivity issues
  - Adsorbents innovation
  - Rationale for adsorbent use

- **Section 2: Contaminant Removal In Petroleum Refining Processes**
  - Fluid catalytic cracking
  - Catalytic reforming
  - Isomerization
  - HF alkylation
  - Benzene saturation
  - Others

Honeywell UOP

UOP 7828B-1

Pore Size Distribution of Various Adsorbents

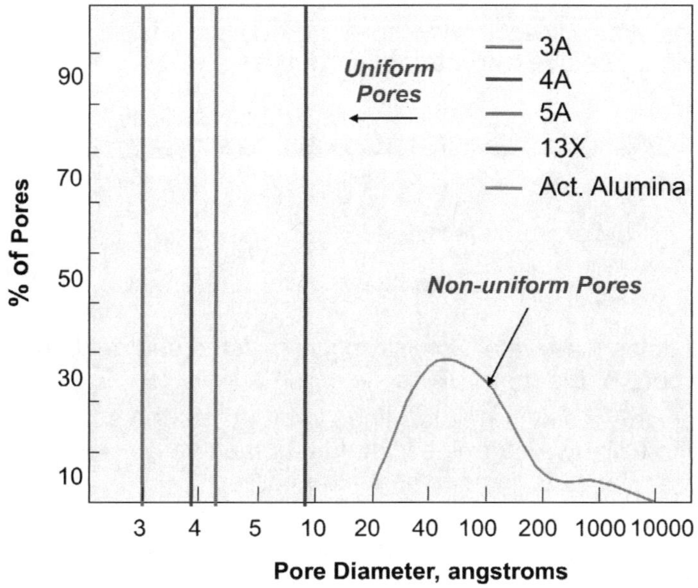

Molecular Sieving Effect

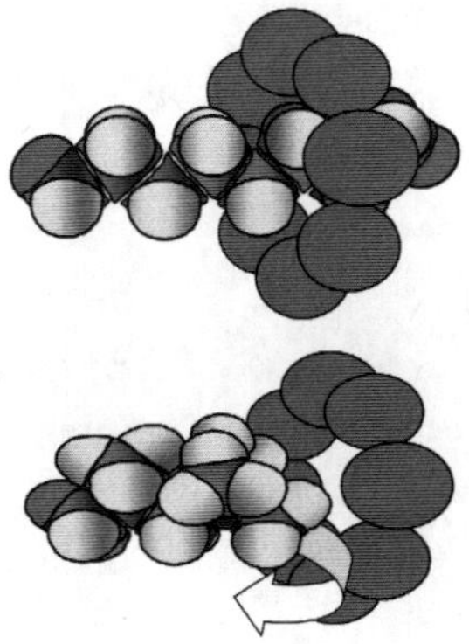

Honeywell UOP

UOP 7828B-2

Contaminant Removal Mechanism

a) Physical Forces (physisorption)

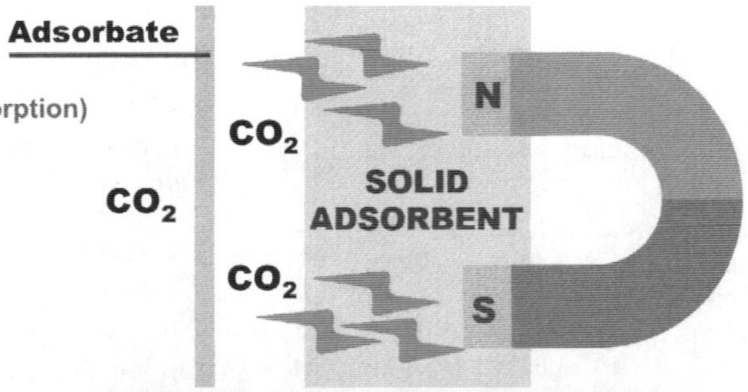

b) Chemical Reaction (chemisorption)

$H_2S + MO \rightarrow MS + H_2O$

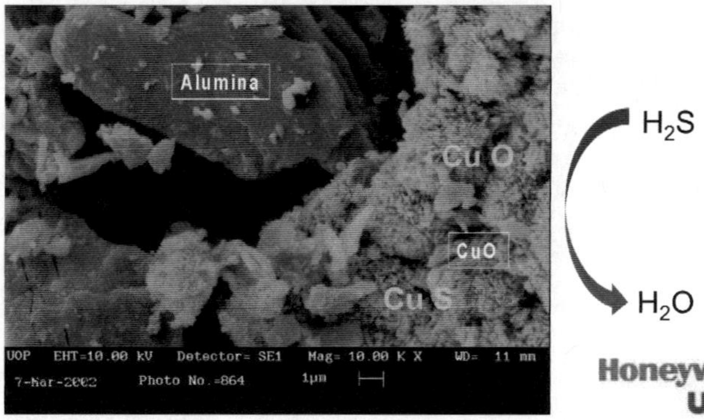

Honeywell UOP

UOP 7828B-3

Molecular Sieves Contaminant Removal Mechanism - Polarity

4

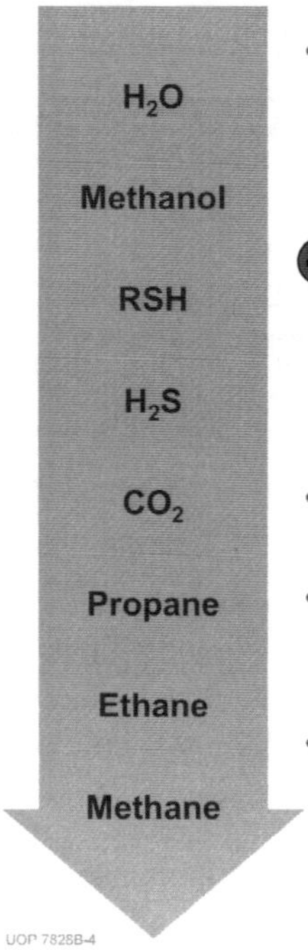

- Molecular sieves attract and adsorb molecules based on polarity (local positive and negative charges).

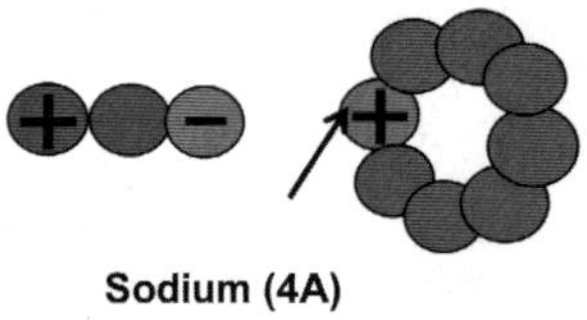

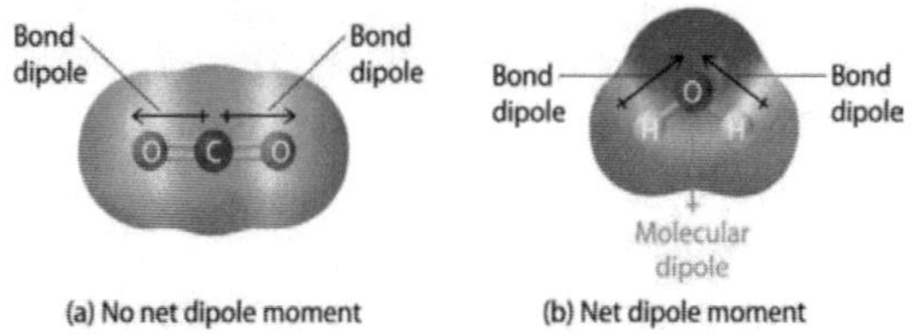

- Polar molecules (like water) are strongly attracted and held inside the pore of the molecular sieve.
- Size Exclusion (or sieving) determines what enters the pore opening and Polarity determines what is then actually adsorbed (tendency to remain inside the pore)
- The more strongly adsorbed molecule, the more heat releases when adsorbed on the adsorbent (H_2O generates more than 3 times the heat of H_2S)

Honeywell UOP

UOP 7828B-4

Adsorbent Options for Hydrocarbon Purification

5

1) Molecular Sieves:

- Use of a small pore zeolites enables exclusion of reactive species and is very effective for dehydration.
- Other contaminants may need a larger pore (such as a 4A, 5A, or 13X type) which may adsorb a reactive carrier stream thus a preload step may be required.
- Careful selection of base material helps control high surface energy and active sites to meet adsorption requirements in reactive streams.
- Special attention is required on the formulation of the binder system (in adsorbent structure) to make it robust enough to avoid particle break-up at severe operating conditions.

2) Promoted Aluminas

- Promoted activated aluminas are synthetic transitional phase aluminas with a tailored pore structure to facilitate rapid chemisorption, resulting in a short mass transfer zone.
- Low affinity for reactive hydrocarbons, thus a preload step not needed.
- Light acid gases such as CO_2, H_2S & COS are removed by chemisorption on promoted aluminas to very low effluent concentrations.

Honeywell UOP

UOP 7828B-5

Adsorbent Options for Hydrocarbon Purification

3) Hybrid Adsorbents

- Engineered homogenous combination of a promoted alumina and a large pore zeolite.
- Combines high selectivity for light acid gases with low olefin reactivity (low heat of adsorption of hydrocarbons) and high capacity for polar molecules thereby making these products ideally suited for purification of olefinic streams.
- Reducing the reactive nature of an adsorbent is a key to assuring trouble-free process operations and long effective adsorbent life.

4) Metal Oxides & Sulfides

- Critical applications for hydrocarbon streams where ultra low contaminant effluent specifications are required.
- Trace arsine and phosphine removal from hydrocarbon streams to protect sensitive noble metal or polymerization catalysts.
- Specialty applications - such as Hg & S compounds removal.
- Trace CO, O_2 scavenging from C_2H_4 and N_2.
- Non-regenerative guard bed applications.

Honeywell
UOP

UOP Adsorbents Innovation

- UOP continues to develop new technology solutions to meet industry demands
- Recently introduced UOP adsorbent products:

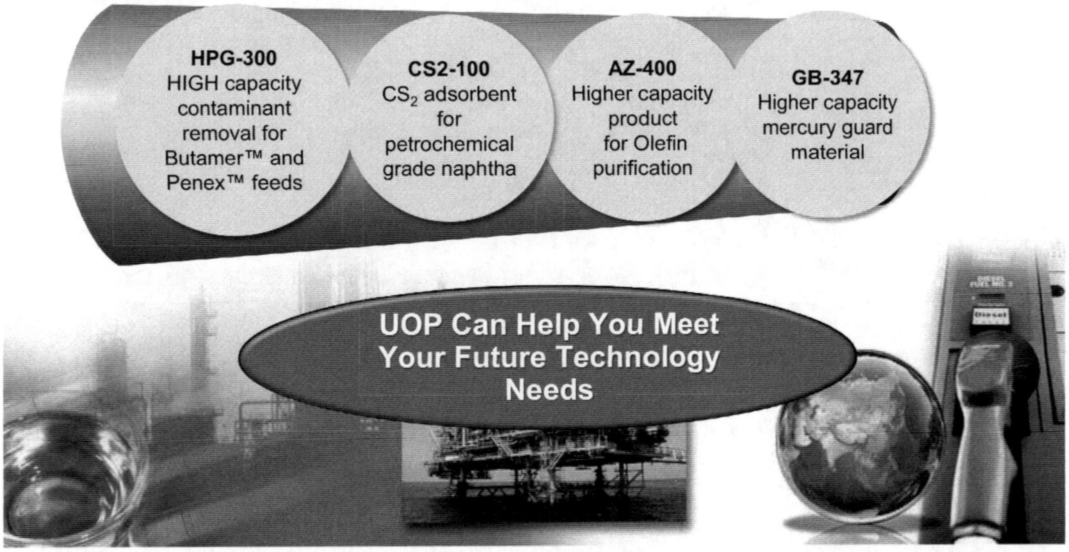

Reduce Potential Plant Safety Issues- A Recent Incident

Optimize the adsorbent potential, by minimizing

- Adsorbent reactions with process stream
  - Coking
  - Side reactions & new contaminants created
- Safety concerns due to exothermic side reactions

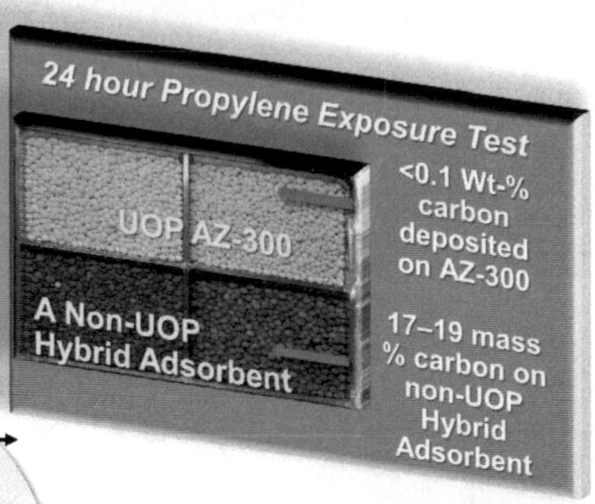

Lower Catalytic Reactivity: Key to Enhanced Performance & Reduced Plant Safety Issues

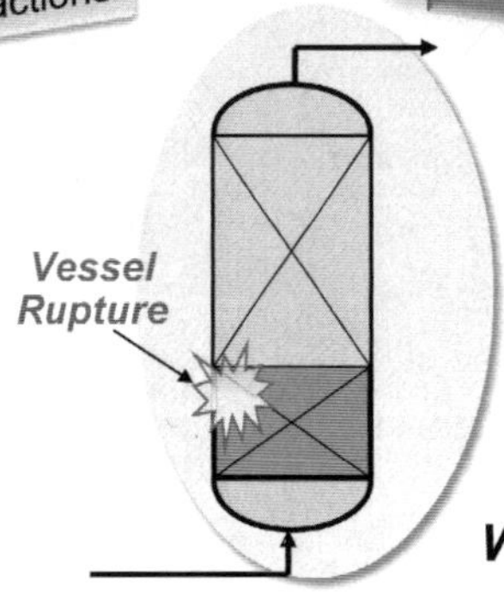

Some Adsorbent chunks found over 200 feet outside vessel

What caused the vessel rupture?

Adsorbent reactivity, often overlooked, is key to reliable operation

Accelerated Adsorbent Reactivity Testing with Iso-butylene

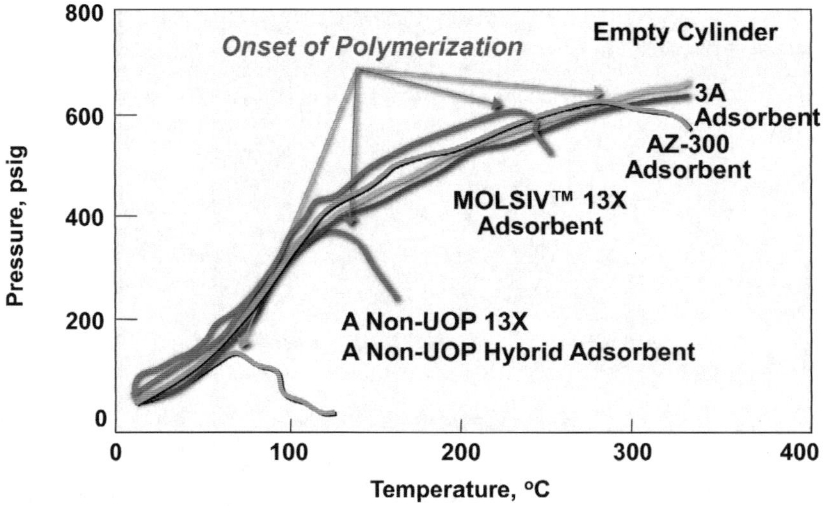

- **Coke build-up on adsorbents occurs due to heavy hydrocarbon deposits on reactive adsorbent surfaces**
- **Heavy hydrocarbon deposits on adsorbents, hinder their performance**

Honeywell UOP

UOP 7828B-8

Typical High Conversion Petroleum Refinery

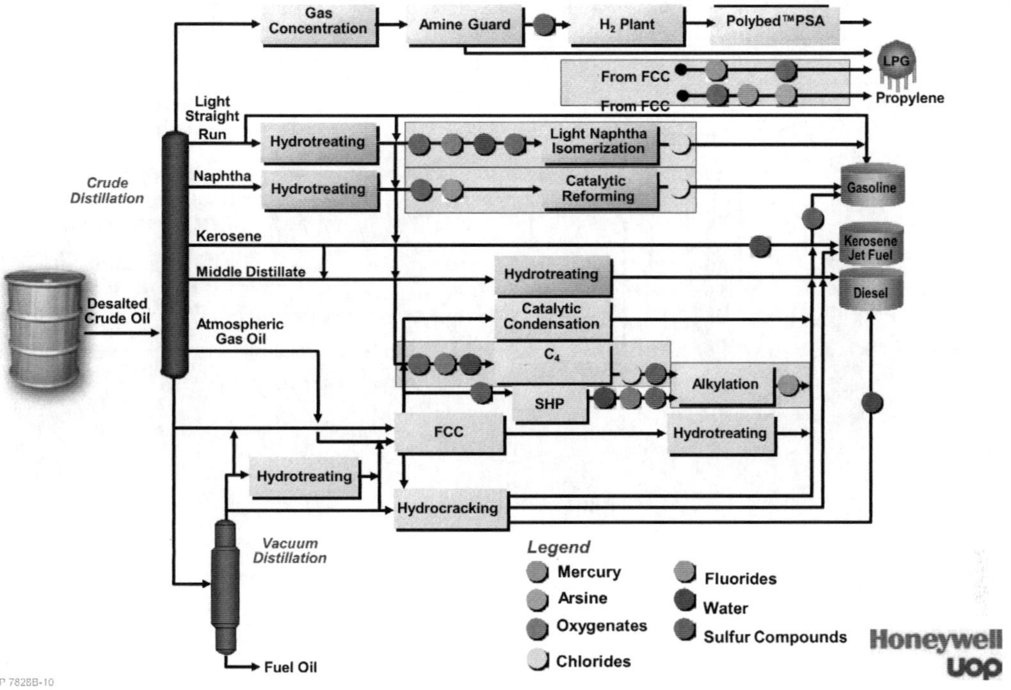

Honeywell UOP

UOP 7828B-10

Rationale For Adsorbent Use In Petroleum Refining Processes

Catalyst Deactivation

- Expensive and sensitive platinum group metal catalysts
- Typically 1 kg of O_2 (as oxygenates) kills 100 kg of Isom catalyst
- S, Hg, H_2O, AsH_3, Cl, Oxy

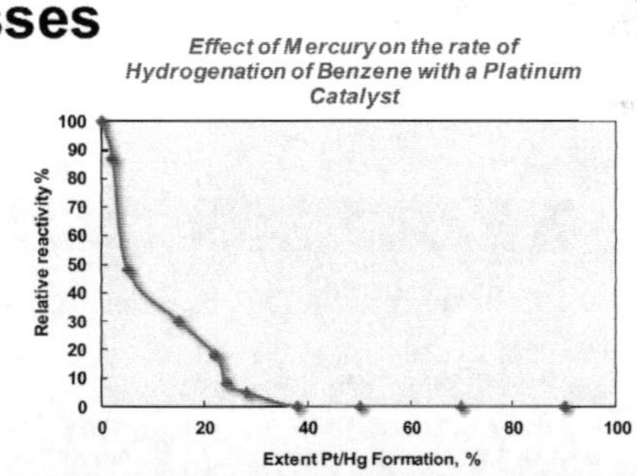

Equipment Fouling

- Corrosion
- Hg embrittlement in alumina heat exchangers
- Green oil formation
- Hg, HCl

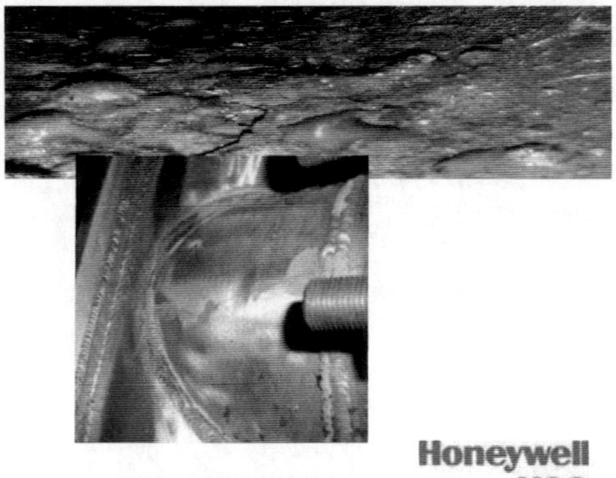

HS&E Issues

- Increased toxicity awareness
- Tighter TLV limits / emission controls
- S, F, Hg

Value Of Exported Products

- Product is offspec
- Hg, AsH_3, O_2

Honeywell UOP

UOP 7828B-11

Fluid Catalytic Cracking Process

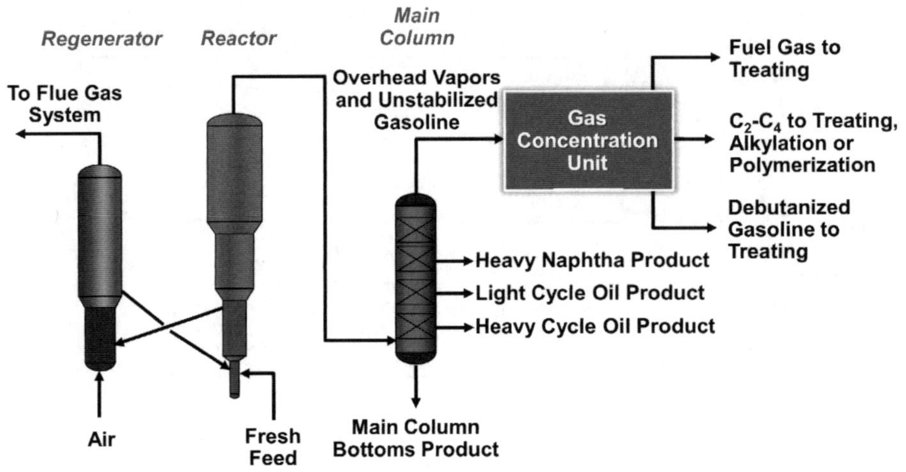

Honeywell UOP

UOP 7828B-12

Contaminants of Concern in $C_3^=$, $C_4^=$, LPG, and Off Gas Streams in FCCU's – Appropriate Adsorbents for Their Removal

Product Name & Type	Operating Mode*	Contaminants Removed
ODG-442 molecular sieve	R	H_2O
AZ-300, AZ-400 composite, higher capacity molecular sieves	R or NR	H_2O, oxygenates, H_2S, COS, CS_2, RSH, DMS, DMDS, NH_3,amines, nitriles, CO_2
SG-731 promoted activated alumina	R or NR	H_2S, COS, CO_2
GB-238, GB-417, GB- 420, ADS-120 metal oxides	NR	H_2S, COS, AsH_3, PH_3
GB-346S, GB-347S, GB-562S, GB-567S metal sulfides	NR	Hg

* : R = regenerative, NR = non-regenerative

Breadth of adsorbent products available to effectively remove a wide range of contaminants from $C_3^=$, $C_4^=$, and other streams

A Typical Propylene Recovery Unit's Purification System

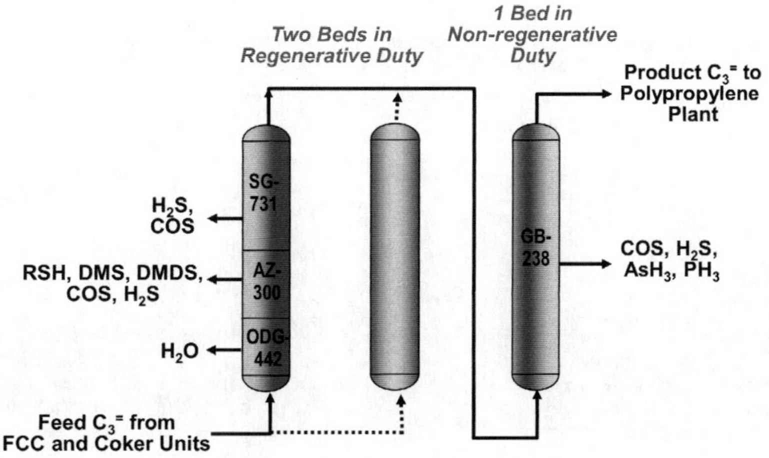

Combination adsorbent beds provide solution for removal of multiple contaminants

UOP CCR Platforming^{TM} Unit Scheme

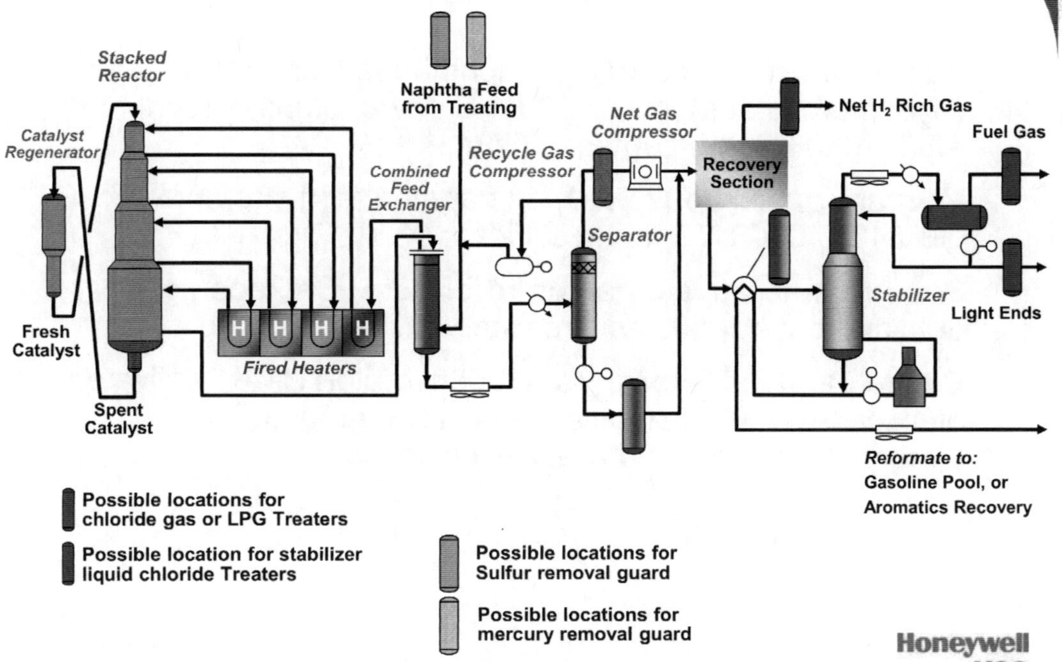

Honeywell UOP

UOP 7828B-15

UOP Chloride Adsorbents

Product	CLR-204	CLR-304	CLR-454	PCL-100/ PCL-200
Base Material	Activated Alumina	Mixed Metal Oxide	Molecular Sieve	Molecular Sieve
Service	Vapor Phase	Vapor Phase	Treat H_2 net make gas - Vapor Phase	Debutanizer Feed- Liquid Phase
Green oil Formation	nil	negligible	No green oil or organic chloride formation	none
Adsorption capacity	Optimum for HCl & good for organic chloride	Excellent capacity for HCl	Good for HCl & optimum for organic chloride	Good for HCl & optimum for organic chloride and ammonium chloride

UOP adsorbents have high chloride capacity, handle liquid carryover, and have low green oil formation

Additional Adsorbent Applications In Catalytic Reformers

- **Sulfur guard bed (SGB)** on naphtha feed stream: ADS-120 (H_2S, mercaptans), ADS-130 (H_2S, mercaptans, thiophenes), ADS-11L (H_2S, mercaptans, thiophenes)
- **Mercury guard bed (MRU)** on naphtha feed stream: GB-346S, GB-347S metal sulfides
- **Semi-regenerative catalytic reformers:** Cat regen gas dehydration: AW-500 acid resistant molecular sieve
- **Cyclic catalytic reformers:** H_2 dehydrator: D-201 activated alumina, AW-500 molecular sieve; H_2S removal from H_2 stream: AWX promoted activated alumina

Adsorbents for Isomerization Processes

HPG-250
HPG-429
HPG-300 Use in feed treaters to remove a broad range of impurities. Add to make-up H_2 dryers for both H_2O and removal of CO_2, H_2S or methanol.

PDG-418 For moisture removal in make-up H_2 dryers

ADS-106 For trace CO removal in make-up H_2 dryers

ADS-120 Cost effective non-regenerative sulfur removal adsorbent for treating Penex and Reforming feed streams

ADS-130 Non-regenerative sulfur removal adsorbent for treating BenSat feed streams which are more sensitive to water. No water generated at start-up

ADS-11L Custom product. Pre-reduced adsorbent for thiophene removal.

ADS-280 Zinc oxide based sulfur guard adsorbent for H_2S removal from naphtha stream in Parlsom™ process

Simplified Penex™ Flow Diagram

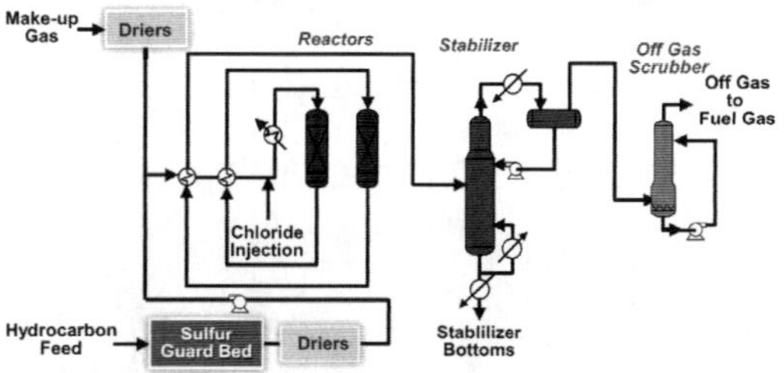

Simplified Butamer™ Flow Diagram

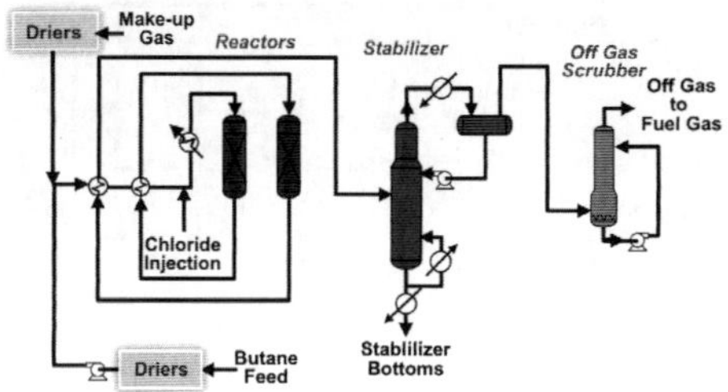

Honeywell
UOP

UOP 7828B-18

UOP HPG-300 MOLSIV™ Adsorbent for Improved Paraffin Isomerization Feed Treating

HPG-300 offers increased capacity for contaminants compared to current generation Butamer and Penex adsorbents. Key customer benefits include:

- Ability to increase throughput in existing feed driers by as much as 30% without contaminant breakthrough
- Ability to tolerate higher feedstock impurities and help olefin content
- Less frequent regenerations in existing vessels will save operating costs and result in longer adsorbent life

Acetone

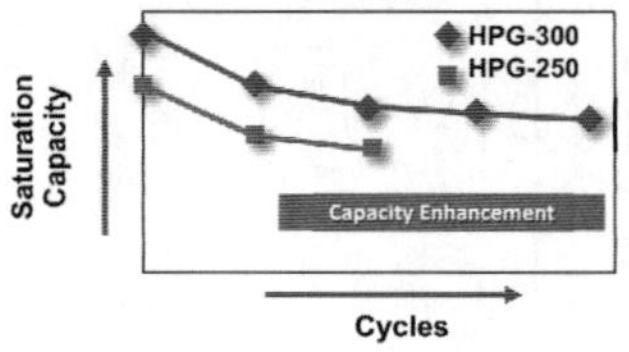

Acetonitrile

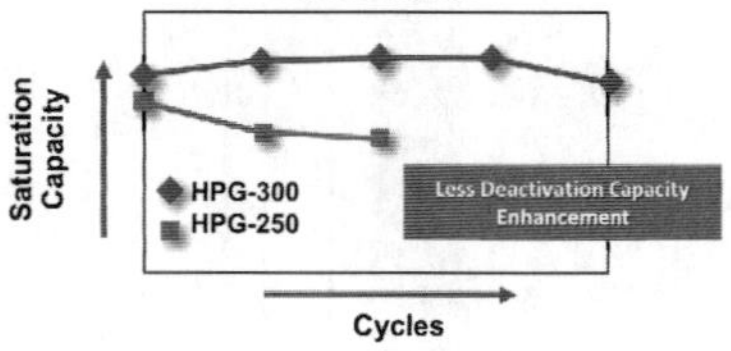

Dimethyl Sulfide

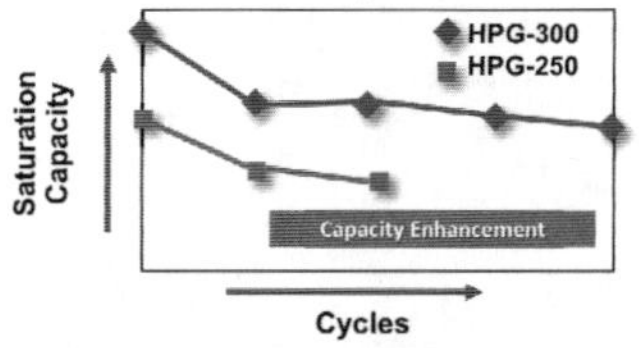

Ethyl Mercaptan

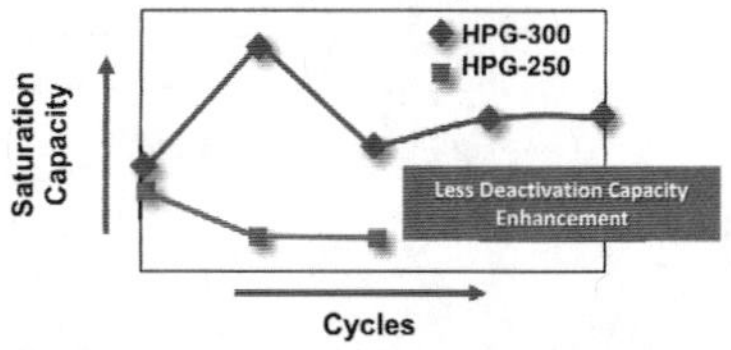

UOP HPG-300 Adsorbent Has Higher Contaminant Capacity & Ability to Tolerate Higher Feed Olefins

UOP HF Alkylation Process

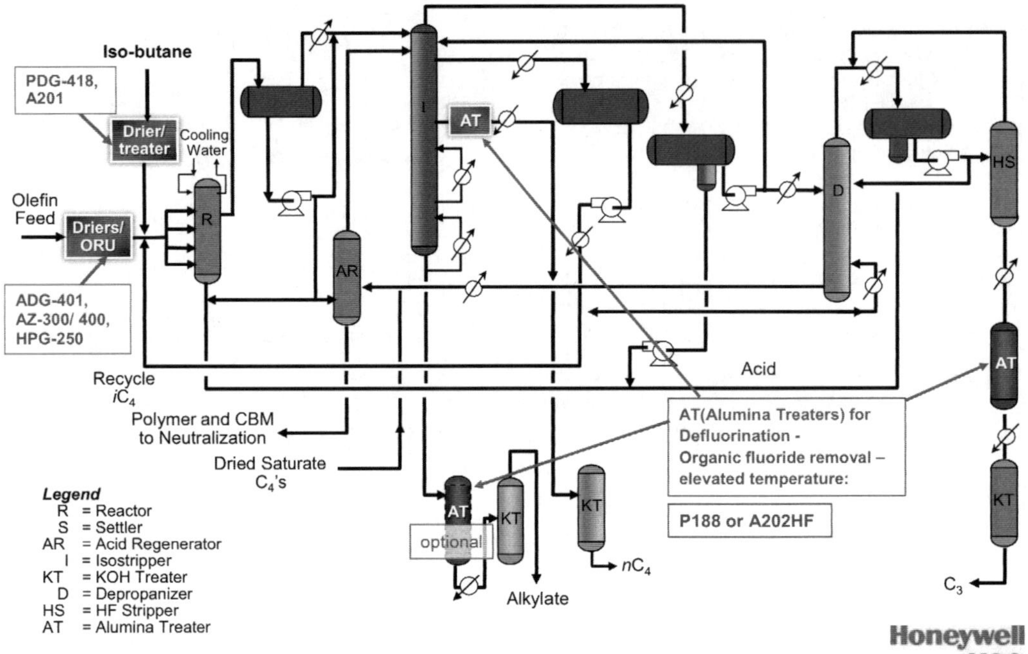

Honeywell UOP

UOP 7828B-20

UOP BenSat^{TM} Process Diagram

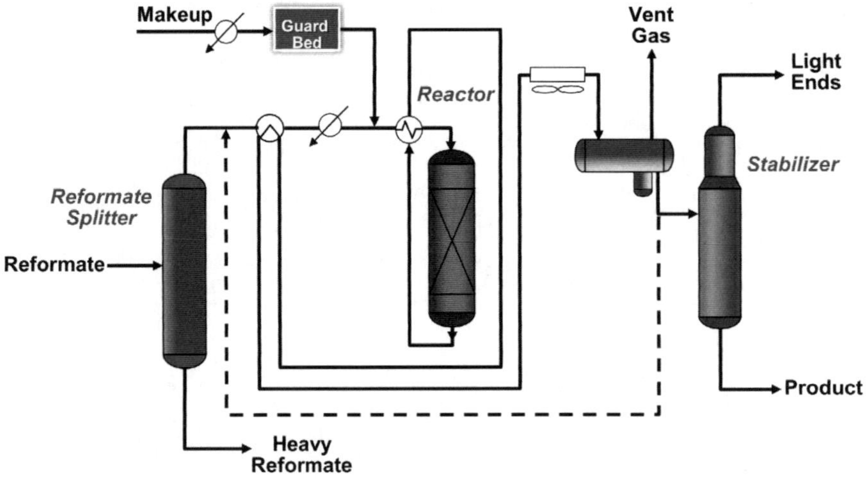

Honeywell UOP

UOP 7828B-21

Adsorbent Applications In Benzene Saturation And Other Petroleum Refining Processes

22

- **BenSat™** make-up H_2 treater: CLR-204 (at bed inlet for HCl removal), ADS-120/ 130 (at bed exit for H_2S removal); BenSat product treater: PCL-100/ 200 (for organic chlorides)
- **CSP (Complete Saturation Process):** C_3 feed Treater: Regen unit: ODG-442 (H_2O removal), AZ-300/ 400 (oxygenates & organic sulfur compound removal); SG-731 (COS & H_2S removal), Guard bed: GB-238 (for trace H_2S, COS, AsH_3 removal), GB-346S/ 347S: Hg removal
- **COS hydrolysis units on C_3's from FCCU and coker unit:** A-201, A-204-1 activated aluminas
- **CS_2 removal from petrochemical grade naphtha:** CS_2-100 molecular sieve
- **Process/ instrument air dryers:** D-201, A-201 activated alumina, 4A-DG, 13X molecular sieves

Honeywell UOP

UOP 7828B-22

Additional Adsorbent Applications In Other Petroleum Refining Processes

23

- **Etherification (MTBE, ETBE):** ORG-E and ORG-HP molecular sieve (removal of oxygenates in ORU), NRU-429 and NRU-300 molecular sieve (removal of nitriles in NRU)
- **Steam reformer H_2 plant:** GB-280 (zinc oxide based for SGB on natural gas or naphtha feed stream)
- **Safe Cat TIP™ unit:** GB-280, SCA-1 (for H_2S removal)
- **HydroIsom: Feed C_4 treater:** GB-238 (H_2S, COS, AsH_3 removal), GB-346S/ 347S (Hg removal)
- **Trim to Merox™ unit (to meet ultra low sulfur spec on LPG):** AZ-300/ 400 (removal of trace organic sulfur contaminants such as mercaptans, DMDS)

Honeywell UOP

UOP 7828B-23

24
Q&A
Honeywell
UOP
UOP 7828B-24
UOP 7116-24

Ese Jeroro
Robert J. Schmidt

New Cumene Alkylation Technology

Honeywell UOP

March 27th, 2017 | AIChE Spring Meeting | San Antonio, TX

UOP 7200-0

1

Agenda

- Highlights
- Alkylation and Transalkylation Chemistry
- Process Flow-scheme
- QZ-2500H™ Catalyst Performance
- QZ-2500H™ Commercial Experience
- Summary

Honeywell UOP

Q-Max™ Process for Cumene Production

- Zeolite-based technology commercialized in 1996
- QZ-2500H™ catalyst introduced in 2012
  - 4th generation Alkylation catalyst
  - Improved Cumene Selectivity over previous generation catalyst
  - Superior product quality and yields
- 22 Q-Max units licensed to date
  - Total capacity of over 6.1 million MTA of cumene

Alkylation Chemistry

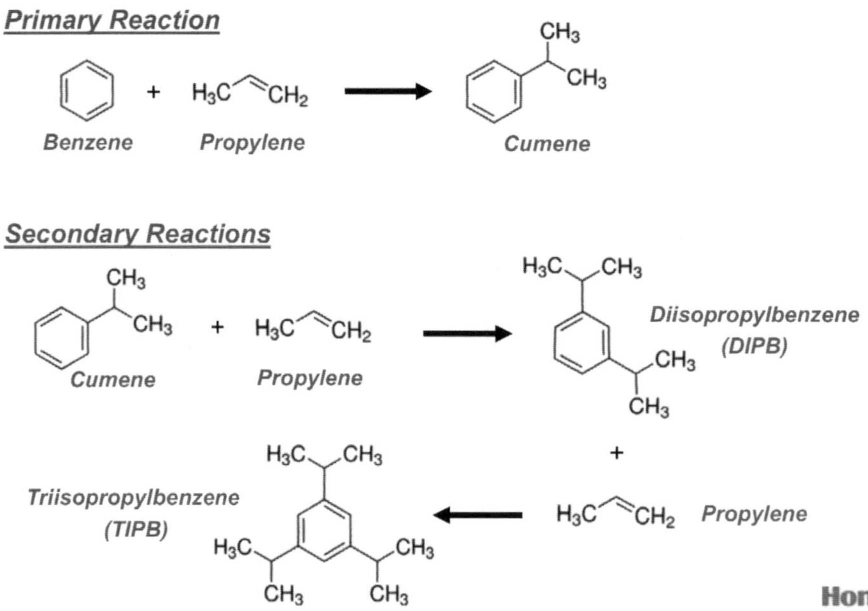

Transalkylation Chemistry

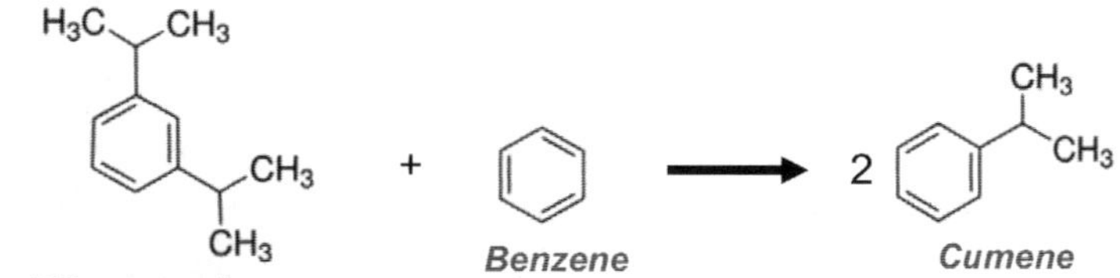

Diisopropylbenzene (DIPB) *Benzene* *Cumene*

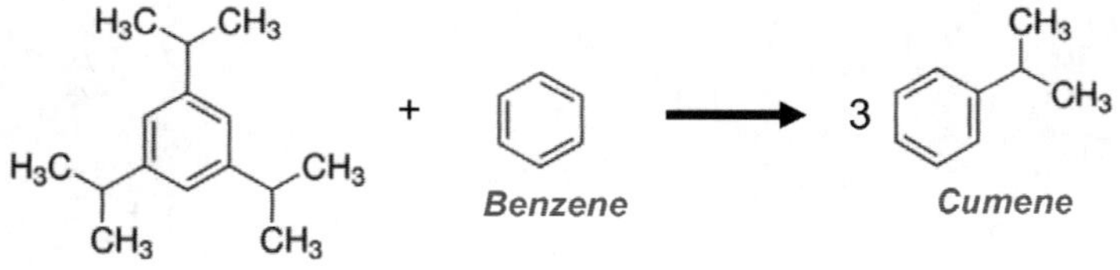

Triisopropylbenzene (TIPB) *Benzene* *Cumene*

Honeywell
UOP

Q-Max™ Process

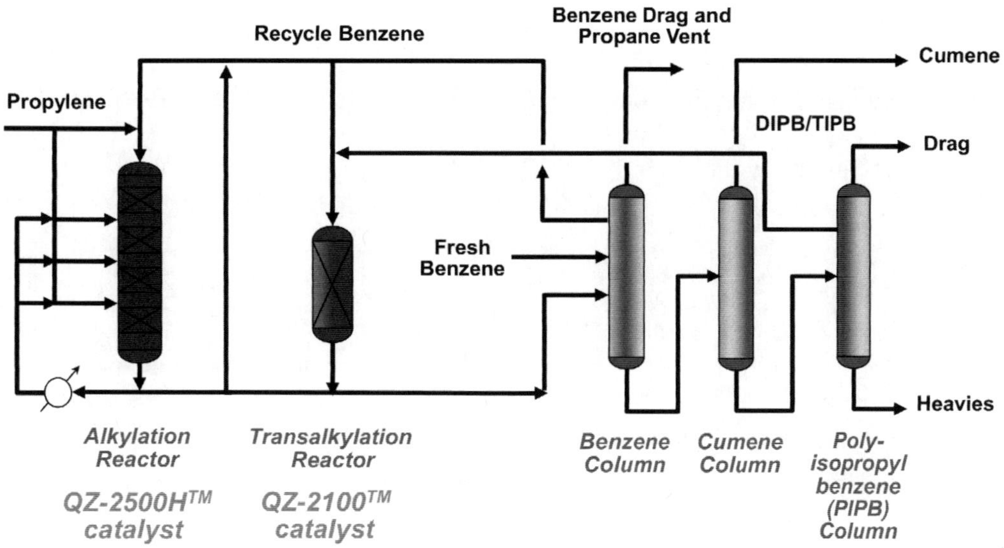

Simplified flow-scheme with QZ-2500H™ and QZ-2100™ catalysts

UOP Q-Max Catalysts

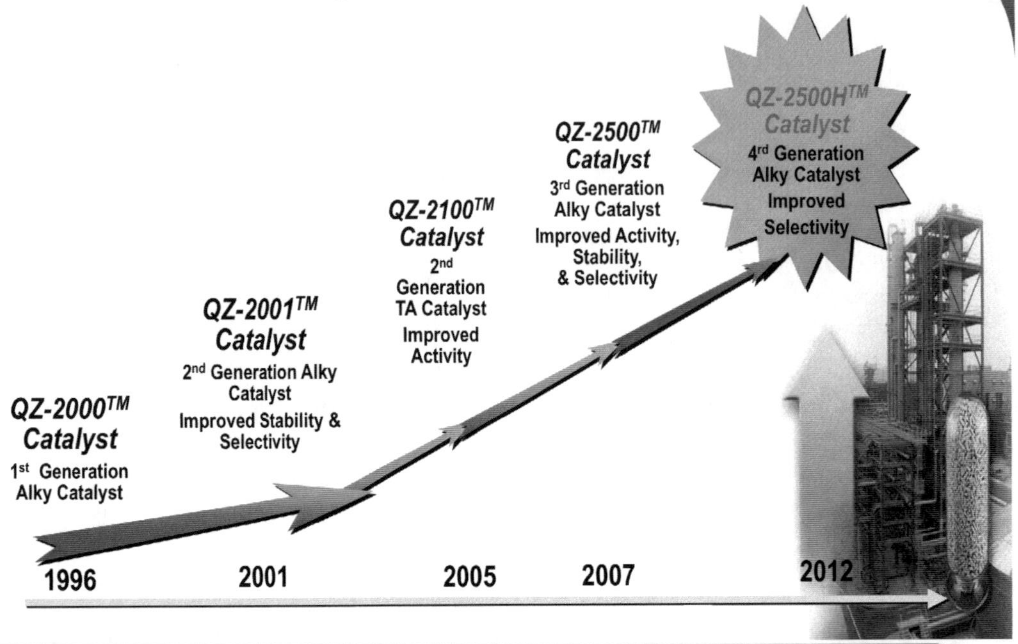

Lengthy history of steady catalyst improvement

QZ-2500H™ Alkylation Catalyst

- Features
  - Based on zeolite platform developed by UOP
  - Improved Cumene selectivity: ~6 mol.% more selective than QZ-2500
    - Alkylation Cumene Selectivity, mol.% = Cumene/(Cumene + DIPB + TIPB)*100
  - Same activity and stability as QZ-2500
  - Improved Cumene yields: 0.1-0.2% higher yields than QZ-2500

- Benefits
  - Lower feed consumption and fewer byproducts formed
  - Better product quality: less Ethylbenzene, *N*-Propylbenzene in Cumene product
  - Reduced load on TA section
  - Lower utility costs

Higher yields, improved product quality at lower operating costs

QZ-2500H™ Pilot Plant Cumene Selectivity

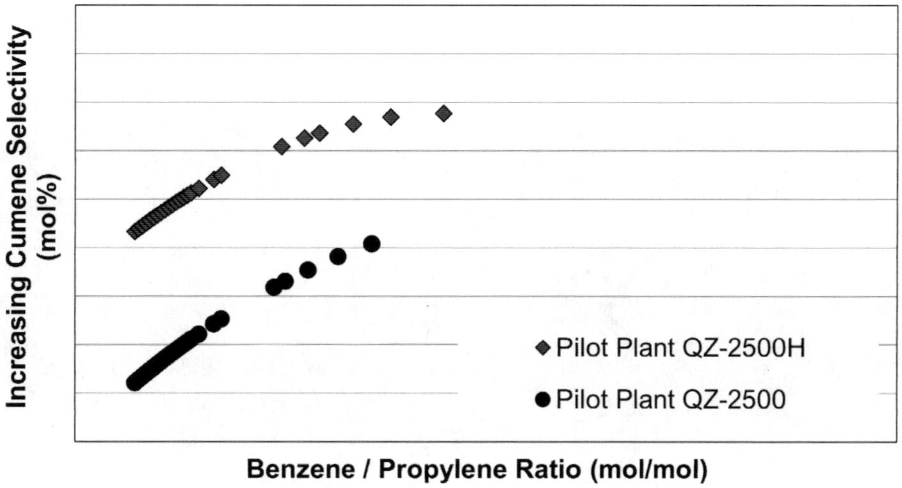

Superior selectivity to QZ-2500™

QZ-2500H™ Pilot Plant vs. Commercial Cumene Selectivity

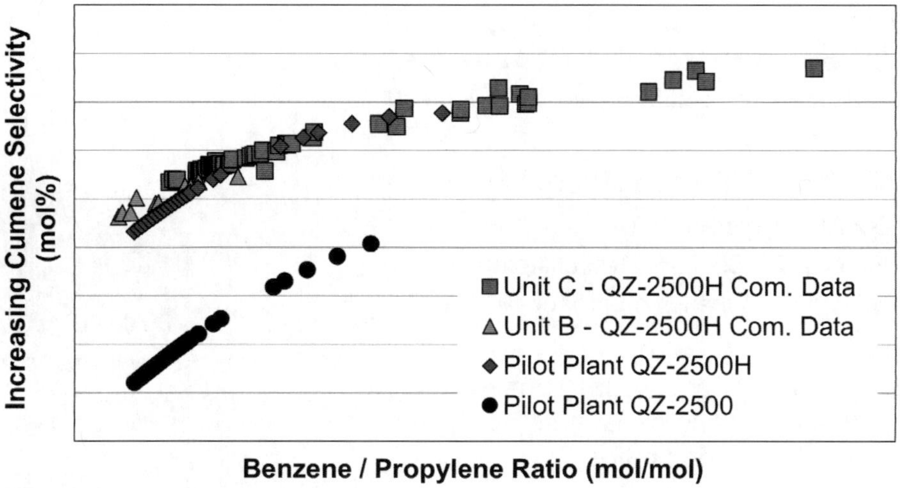

Selectivity improvement validated commercially

10

Lower Propylene and Benzene Consumption with QZ-2500H™/QZ-2100™ system

Benzene Consumption/Cumene
wt/wt
0.651

Propylene Consumption/Cumene
wt/wt
0.352

1.003 Total Feed/Cumene, wt/wt
or 99.7 wt% Yield

Achieves near stoichiometric production of Cumene

11

Some Examples of QZ-2500H™ Commercial Experience

- 'Unit B: SPA Cumene Unit revamped to Q-Max
  - Started up with QZ-2500™ / QZ-2100™ system
  - Upgraded to QZ-2500H catalyst in June 2013
- 'Unit C': New Q-Max unit QZ-2500H™ and QZ-2100™ system
  - First QZ-2500H™ catalyst customer
  - Started up in the 2nd quarter of 2013
- Both units have consistently achieved high Cumene purity and excellent stability at low Benzene/Propylene ratios

Honeywell
UOP

12

Unit C Data – Alkylation Benzene/Propylene Ratio

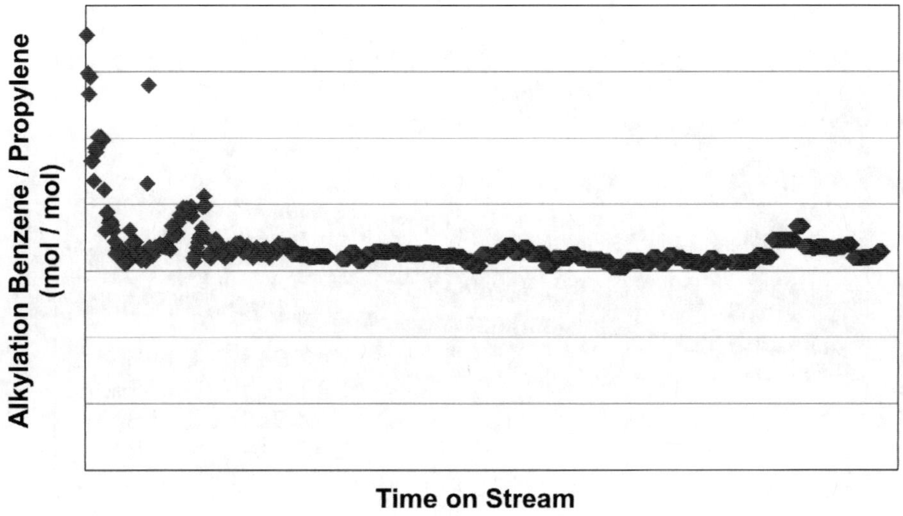

Steady operation at low B/P with no issues

13

Unit C Data – Cumene Purity

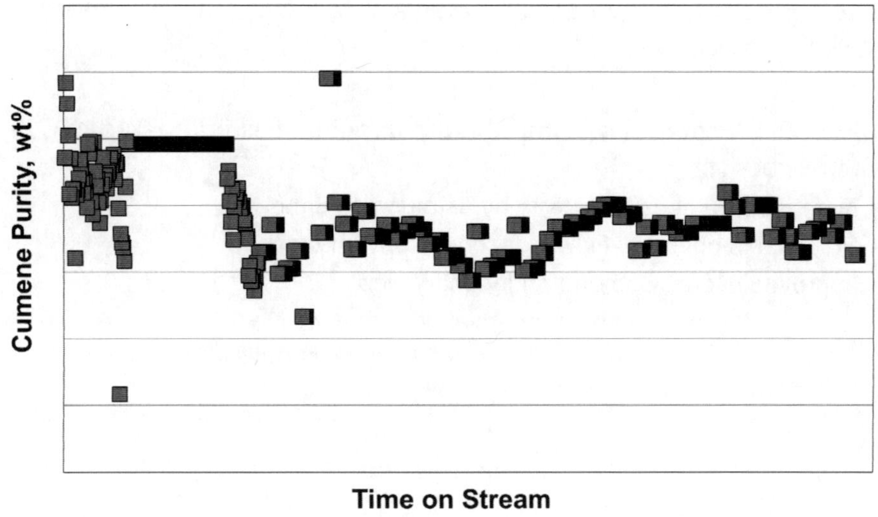

High Cumene purity maintained throughout cycle

14

Unit C Data – End of Active Zone

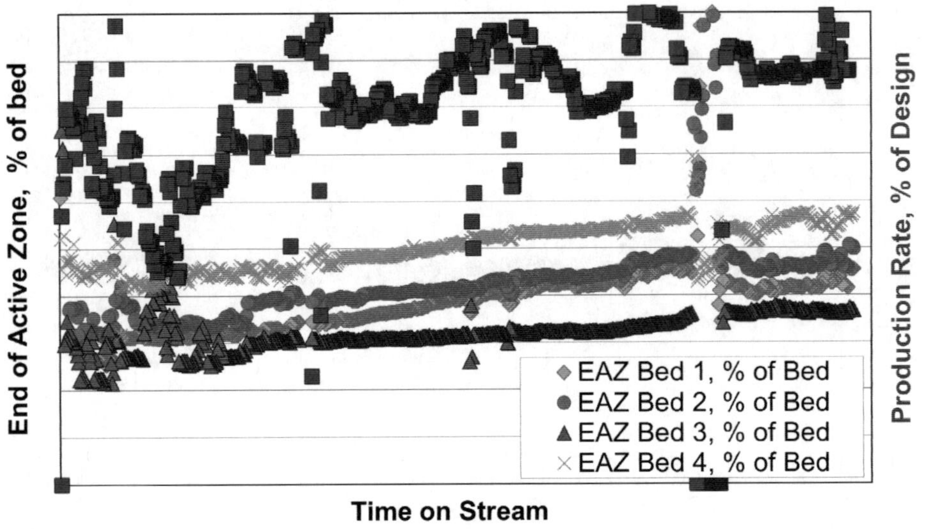

Great stability operating close to maximum design production rate

15

Summary

- UOP has world class R&D capabilities that have enabled the development of state-of-the-art catalysts for the Q-Max™ Cumene Alkylation process

- QZ-2500H™ catalyst has improved characteristics compared to QZ-2500™ catalyst:
  - Increased Cumene selectivity by as much as 6 mol.%
  - Potential to operate at low Benzene/Propylene ratios
  - Improved overall Cumene purity and yields
  - Outstanding on-stream commercial performance
    - Loaded in 4 commercial units with a 5th start-up planned in 2018

Honeywell UOP

Q&A

UOP 7116-16

480265 Production of Renewable Styrene Using Catalytic Microwave Depolymerization

Abstract:

According to the EPA [1], less than 10 % of the plastic produced in the USA is recycled. The principal reason is the incapacity to reprocess the plastic waste in such way that the output can be reused into virgin plastic applications, such as production of food and beverage containers and other consumer goods. Although technologies exist for polyethylene terephthalate (PET) where it can be reintroduced into some FDA compliant applications [2], it is not the case for most of other post-consumer plastics such as polystyrene, especially when they are contaminated, e.g., organic materials and paints. The present paper shows how catalytic microwave depolymerization (CMD) can be applied at commercial scale for recycling post-consumer contaminated polystyrene and reuse them in food packaging applications. The mechanism of action is simple: the use of CMD breaks down the polystyrene into monomers that can then be reintroduced in the existing ecosystem of refining and polymerization industries. One of the differences of CMD from other conventional pyrolysis processes is the heating rate which is faster because energy is transferred on a volumetric basis, whereas other conventional approaches are slower because the heating occurs on a conduction/convection basis. The decomposition step also allows handling of a broader amount of contaminants [3] compared to mechanical/physical technologies.

This paper discusses the change in the heating mechanism inferred by the use of microwaves and how it affects the heating profile and the kinetic of decomposition. Results show that the Arrhenius ratio E_A

the same ratio obtained by conventional conductive/convective heating. Where there are no explanations that can support a change in activation energy E_A, which is the normal way to approach kinetic analysis, this paper investigates the possibility that the local temperature might be significantly higher to explain a change in this ratio.

The phenomenon is approached by considering resonance between the microwave electrical field and the molecular structure of the polymers being decomposed. This resonance is believed to increase the molecular kinetic energy of the polymer molecules and to increase its molecular energy. This increased level of energy is believed to make it more susceptible to break chemical bonds and generate smaller sets of molecules. These newly formed smaller molecules would also absorb some of the energy through resonance with the microwave field and break apart into even smaller molecules. The production of rather high levels of monomers, less amounts of dimers and trimers, and almost no chains of monomers longer than three molecules also suggest that the absorption of the microwave energy might have something to do with the molecular weight. The long polymer chains would absorb microwave energy to break into smaller monomer chains up to a point that the small chains no longer resonate with the electrical field and remain only subject to conduction and convection heating.

To support this assumption, the absorption of microwave energy was investigated through the analysis of dielectric properties, which is a measure of how much of the absorbed microwave energy is converted into heat. The dielectric properties of the material were investigated as a function of temperature and molecular weight. The results show that the dielectric properties of the polymer material change as a function of both the temperature and the molecular weight and that the dielectric loss tangent can vary up to a factor 20. This supports the claim that at higher temperature and depending on the polymer chains length, the material absorbs more energy over a wider range of modes which increases its kinetic energy (local temperature) and initiates a rapid decomposition. This is considered as a theory to support the variation in product yields obtained with CMD over polystyrene feedstock versus the yields obtained using conductive/convective heating. CMD produces small amounts of trimers, more dimers than trimers and predominantly styrene monomer, whereas conventional heating produces predominantly ethylbenzene and other aromatics.

1. U.S. Environmental Protection Agency (EPA). Advancing Sustainable Materials Management: Facts and Figures Report. 2015

2. U.S. Food and Drug Administration (FDA). Guidance for Industry: Use of Recycled Plastics in Food Packaging: Chemistry Considerations. 2006

3. J. Doucet, J.-P. Laviolette, S. Farag and J. Chaouki. Distributed Microwave Pyrolysis of Domestic Waste. *Waste and Biomass Valorization.* 2013;5(1):1-10

480177 Achieving Targeted Product Mix Using Optimization Tool for Chemical Plants with Multiple Products

Abstract:

Many chemical manufacturing plants are designed to produce multiple products which may be differentiated by factors such as chain length or degree of substitution. Market demand will determine the actual production requirement for each product, and may vary widely over the course of a month, year, or longer. Multiple controllable conditions can be used to achieve the desired production mix and suppress unwanted by-products. The sheer number of variables and process constraints make meeting multiple production requirements a challenging task. Combined with market data, near-real time optimization control can maximize the overall profit of the plant. Moreover, coordination with other plants is also very important to meet the strategy of the business. Currently, this multi-dimensional control process highly relies on operators' experience with forward numerical calculation using tools like Microsoft Excel®. Thus, to find the optimal operation conditions for each plant in real time is inefficient and may not yield the optimal result. The goal of this research work is to develop a backward optimization tool, which uses desired production mix to calculate the corresponding control conditions, to help operators make the decision more efficiently and reliably. This optimization tool will combine existing numerical models and safety constraints. Potentially, this tool can help plant to increase production rate and achieve additional margin.

Refining/Petrochemical Integration – FCC Gasoline and Aromatics are Key and Related Areas

2017 AIChE Spring Meeting – San Antonio
Abstract ID #481803

Rick Kolodziej PE, **Joseph C. Gentry, PE and Sachin Joshi**
GTC Technology US, LLC, 900 Threadneedle, Suite 800, Houston, TX. 77079 USA

Presenting author email: rkolodziej@gtctech.com

Abstract

This presentation covers the important topic of upgrading FCC gasoline and by doing so, side benefits can exist in terms of recovery aromatics. Many refiners recover propylene from the light ends of the FCC, but are ignorant of the commercial process to recover aromatics or other products from the naphtha stream. Coincident with product upgrading opportunities is the need to reduce sulfur to very low levels, which this same technology accomplishes nicely. Traditional technologies use selective hydrodesulfurization to upgrade FCC gasoline, which results in an objectionable octane loss in the gasoline.

Refining/Petrochemical Integration – FCC Gasoline and Aromatics are Key and Related Areas covers the commercial application of the GT-BTX Plu**S**® technology which is used to meet the stringent gasoline specifications in the most economical manner. Additional case studies show how to gain value from FCC gasoline processing, rather than to downgrade products or suffer losses by conventional processing schemes.

Introduction

Sulfur reduction is a problem for many refiners. It is an onerous requirement to stay in business, and yet there is no economic benefit gained. However, projects to accomplish sulfur reduction can be an opportunity to take advantage of innovative process technology to produce valuable petrochemicals that are otherwise missed.

Sulfur

Regulations around the world are requiring the reduction of the sulfur content in finished gasoline to < 10 ppm. FCC gasoline is the dominant source of the sulfur, so methods are developed to pre-treat the FCC feed or post-treat the FCC gasoline. The techniques for sulfur removal depend on the type of sulfur molecule. Mercaptans are easily removed by conversion to disulfides and extracted, or by mild hydrotreating. Sulfides are relatively easy to remove by hydrotreating; and thiophenes are removed by more severe hydrotreating. As the sulfur requirement decreases, nearly all of the incremental sulfur removed is of the thiophenic type, which results in unintended olefin saturation with commensurate octane loss. Each olefinic molecule converted to the corresponding paraffin will lose approximately 20 octane numbers.

Here are some details about the distribution of sulfur species in FCC gasoline, and the respective processes to remove them.

Mercaptans are the dominant species in the lighter fraction, and thiophenes are dominant in the heavier cuts. Olefins are highest in the light fraction, and lowest in the heavy fraction. Thiophene sulfur and some olefins are in the mid-cut. Sulfur removal from the light cut is the easiest, and the heaviest fraction has hardly any olefins to matter in hydrodesulfurization. Hence, the mid-cut is most susceptible to olefin saturation and octane loss via hydrodesulfurization.

The olefinic raffinate can go to simple aromatization to produce more aromatics, recycle cracked in the FCC to produce propylene, or blend into gasoline with no change in octane value.

Figure 1: Sulfur and Olefin distribution by boiling points

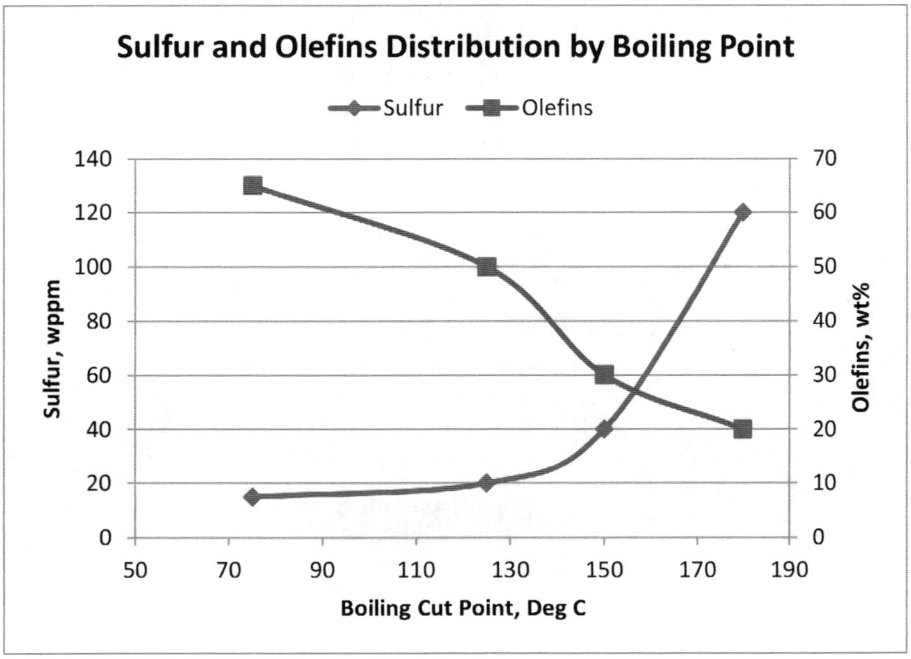

Reference: Shorey, S.W., D. A. Lomas, and W. H. Keesom (1999). *Hydrocarbon Processing*, v. 78, p. 11, 1999.

Conventional sulfur removal processes use hydrodesulfurization, in a scheme shown as follows:

Figure 2: Conventional FCC gasoline desulfurization

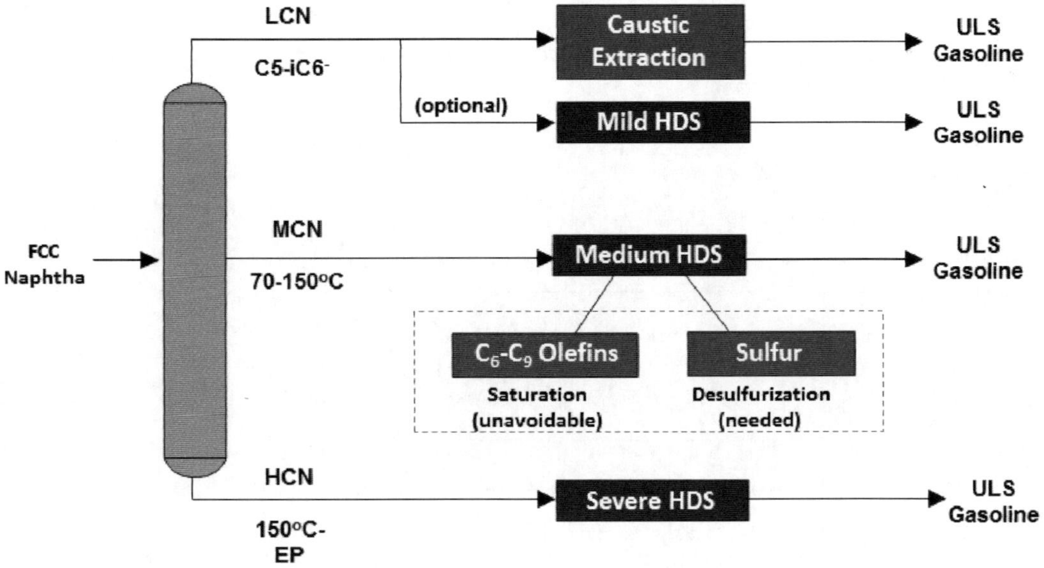

The thiophene sulfurs are ring-shaped components, similar to aromatics. The novel method presented here is extraction of the sulfur along with the aromatics in the patented GT-BTX PluS® process using extractive distillation. With the olefins rejected from the aromatics plus sulfur, the extract stream can be desulfurized even to 1 ppm with insignificant saturation of the hydrocarbons.

Figure 3: GT-BTX PluS® scheme for FCC gasoline desulfurization

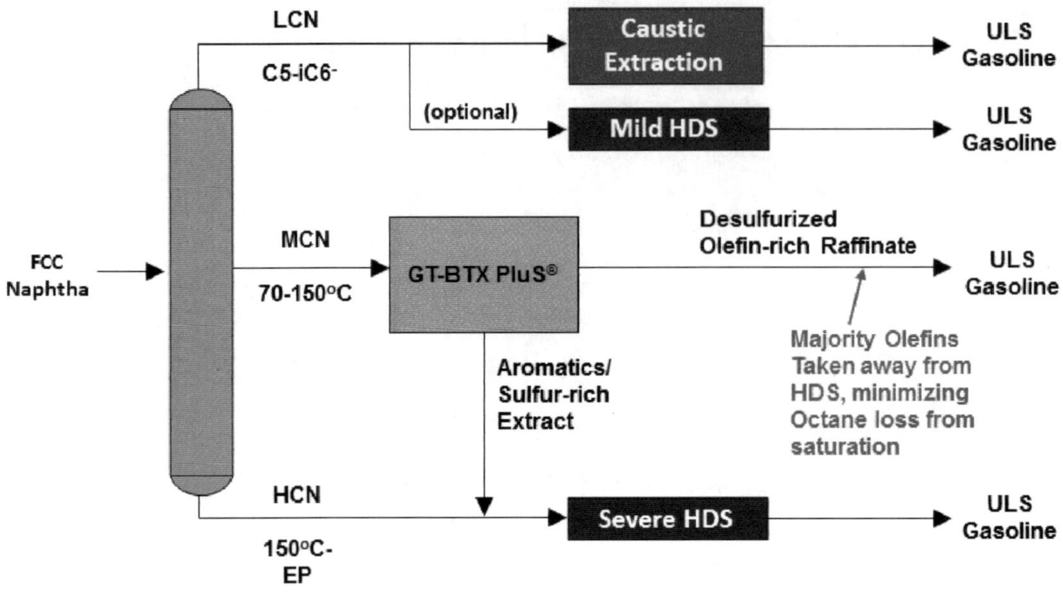

Octane

The obvious value proposition to the GT-BTX Plu**S**® process is preservation of octane. Octane-bbl costs are increasing. According to a recent report "Understanding octane value in America" by Adi Analytics' Uday Turaga and Tyler Wilson, the value of an octane barrel has risen significantly in recent years, going from $2.30 in 2010 to $3.58 in 2015. Some of the reasons for the increase in octane-barrel are higher use of paraffinic crude oils and higher compression automobile motors. There is also a phase down on allowable aromatics in the gasoline pool.

In order to hydrotreat the thiophenic sulfur to < 10 ppm, there will be olefin saturation resulting in an octane loss of 2-4 or more numbers.

Figure 4: Dependence of the loss of octane number on the level of hydrosulfurization in HDS section of Prime G based on data by Axens

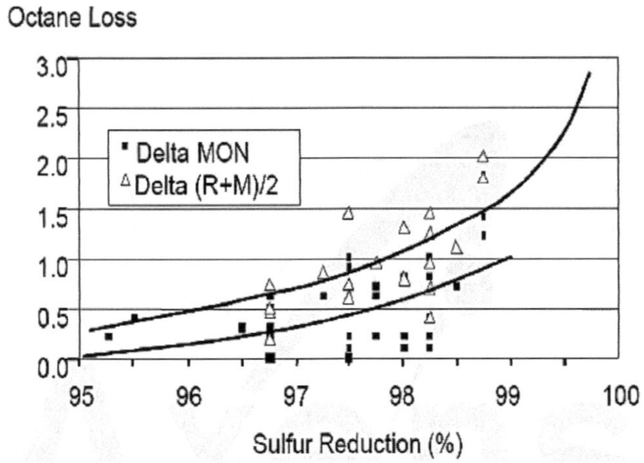

At $3.00 per octane-bbl, a typical desulfurization unit would lose $20 million per year, for each drop in octane number.

Petrochemicals

While aromatics are being phased down in motor gasoline there is a continued demand for BTX in the petrochemical sector. The GT-BTX Plu**S**® process automatically produces high quality aromatics after the HDS section, rich in toluene, xylenes, and C_9 aromatics, which can all be converted into paraxylene.

Additionally, this process creates a new stream, rich in olefins but essentially void of sulfur and aromatics. The olefinic-rich raffinate from the GT-BTX Plu**S**® process can go to any of three dispositions, each one having higher value than with the traditional processing scheme:

1) Process in a fixed-bed aromatization unit to generate additional aromatics. The C_4 and C_5 fractions from the FCC can also go to this unit.
2) Recycle to the FCC unit to re-crack into mainly propylene; or use a segregated unit such as Gasolfin^{sm} to re-crack into propylene and butylenes.
3) Blend into gasoline as a low-sulfur, low-benzene fraction having higher octane value than from traditional systems.

The cracked products generated in the FCC have higher value than the corresponding paraffins. The refiner doesn't have to destroy these in order to reduce the sulfur.

Application

Removal of aromatics and sulfur from cracked gasoline cannot be easily done with liquid-liquid extraction, because of objectionable internal recycles of the olefins, and problems with phase separation. Instead, the GT-BTX Plu**S**® process uses extractive distillation with a proprietary Techtiv®-DS solvent. Techtiv®-DS is stable and non-reactive with the unsaturated components, capable to operate with feed impurities, environmentally acceptable, and performs well across a wide-boiling range of feed. This gives the highest level of sulfur removal, with simultaneous rejection of olefins.

Figure 5: Liquid-Liquid Extraction vs Extractive System for extraction of sulfur with rejection of olefinic species. Objectionable recycles and phase separation problems occur with LLE.

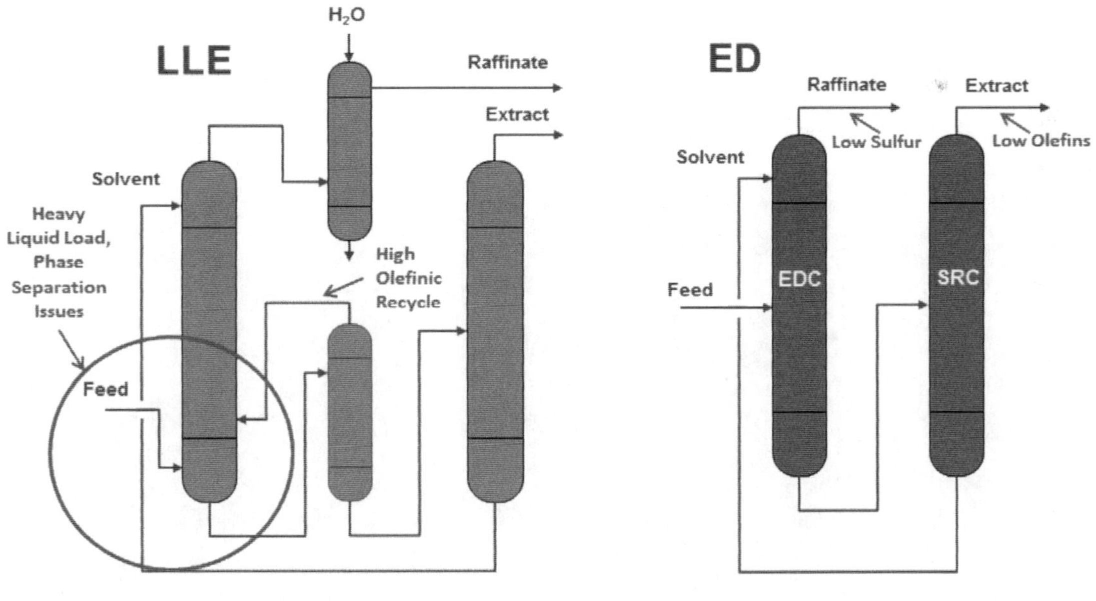

Figure 6: An expanded rendition of the flow scheme

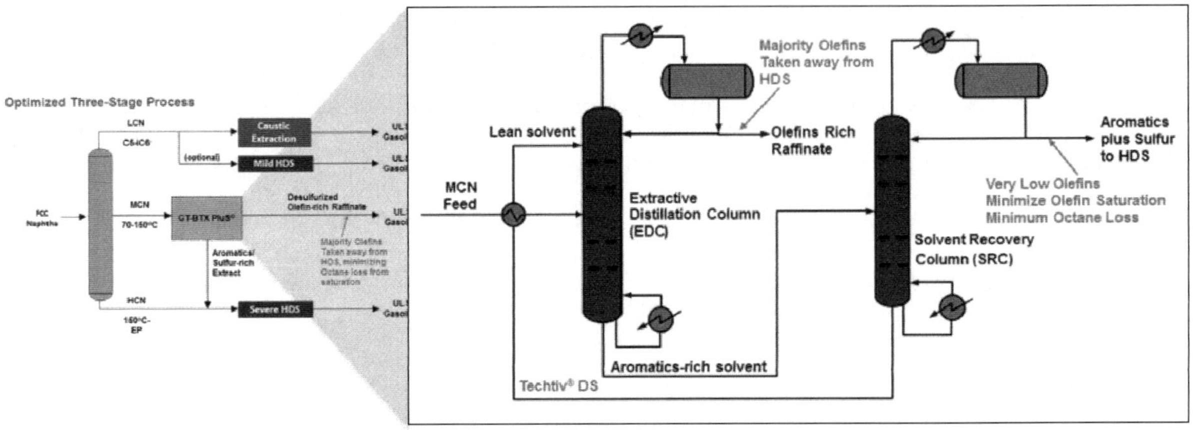

CAPEX and OPEX for a 1 MM ton per annum unit is:

GT-BTX PluS®		
1 MMTPA Unit, USGC 2015 Prices		
CAPEX	17.0	MMUS$
OPEX	3.0	MMUS$

Case Study

Recently a 400 KTA GT-BTX Plu**S**® Unit started up in Dongying, China. This unit was added as a drop-in to the scheme with an existing FCC gasoline HDS unit, With the addition of the GT-BTX Plu**S**® unit, octane loss was reduced from 4, to 0.6; final gasoline sulfur content was reduced from the level of 600 ppm to 2-7 ppm, and H_2 consumption was reduced by 60%.

Figure 7: GT-BTX Plu**S**® unit in Dongying, China

The GT-BTX PluS® configuration is quite simple. In this case, there were only three fractionating columns added, with re-use of the existing hydrodesulfurization reactor. Other schemes are also possible to conveniently retrofit. In the grassroots case, one may avoid the selective HDS section altogether and use a standalone GT-BTX PluS®.

Figure 8: GT-BTX PluS® with three fractionating columns

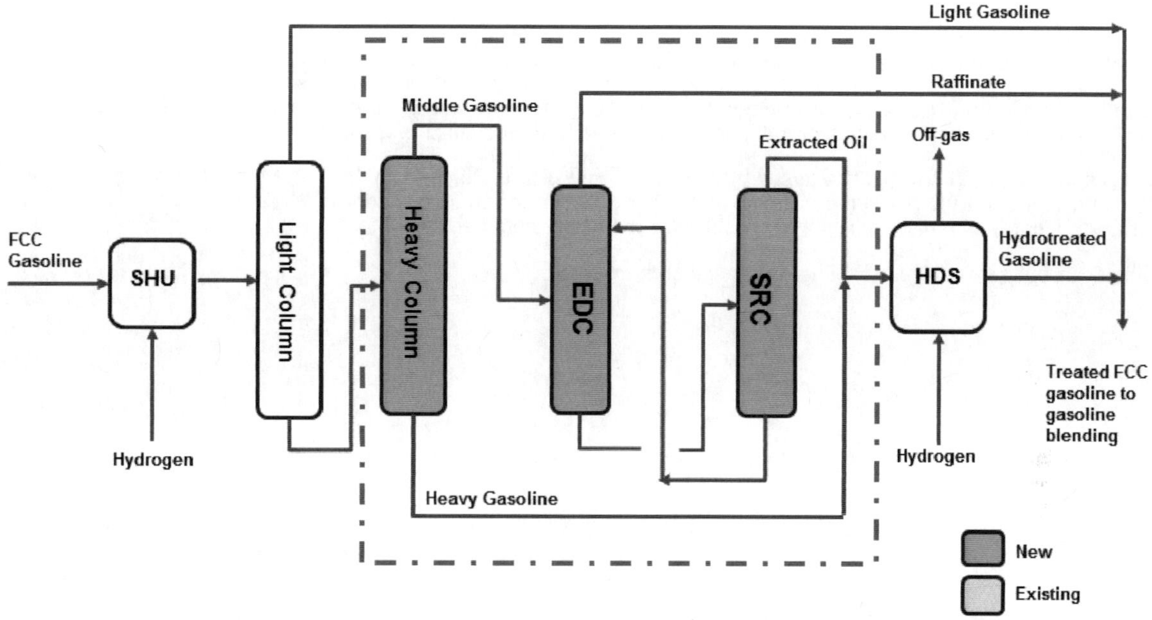

Summary

Removing sulfur from gasoline does not have to be a drain on refinery profitability. The key is to extract the sulfur species first, then hydrotreat the extract. This way, the olefinic fraction does not go to hydrotreatment which would cause substantial octane loss. The hydrotreated extract is an excellent feedstock for paraxylene production. The olefinic raffinate can go to simple aromatization to produce more aromatics; recycle cracked in the FCC to produce propylene; or blend into gasoline with no change in octane value.

Reduce sulfur. Preserve octane value. Produce petrochemicals.

481063 Research Challenges in the Light Component Separation for the Methane Future Feedstock Projects

Abstract:

Methane as an alternative feedstock has been studied in the research division for many years. Most of the projects are diminishing as the challenges of reducing high process cost has been increased with the low ethane cost on the market in these years. Methane reformer is one of the successful technology that has been implemented for a few commercial routes in basic chemical industries. Other technologies, like OCM, MTM, MCTO, and combustion pyrolysis, are all under the research and development stages.

In this communication, the authors would like to share the overviews and to discuss the main challenges of these technologies. In particular, for the MCTO technology, the authors will introduce an innovative process design for the light hydrocarbon separation and an optimal integrated process to identify the major milestones required for the future commercial success.

INHA UNIVERSITY

Modelling and Optimization of Environment-Friendly Process at Oil Offshore Platform

2017.03.27

Graduate School of Chemistry and Chemical Engineering, Inha University

Yongheon Cho, Soojin Kwon, Junghoon Kim and Sungwon Hwang

Contents

1. Backgrounds
2. Modeling and Simulation
3. Optimization procedure
4. Optimization results
5. Summary

1. Backgrounds

1. Background

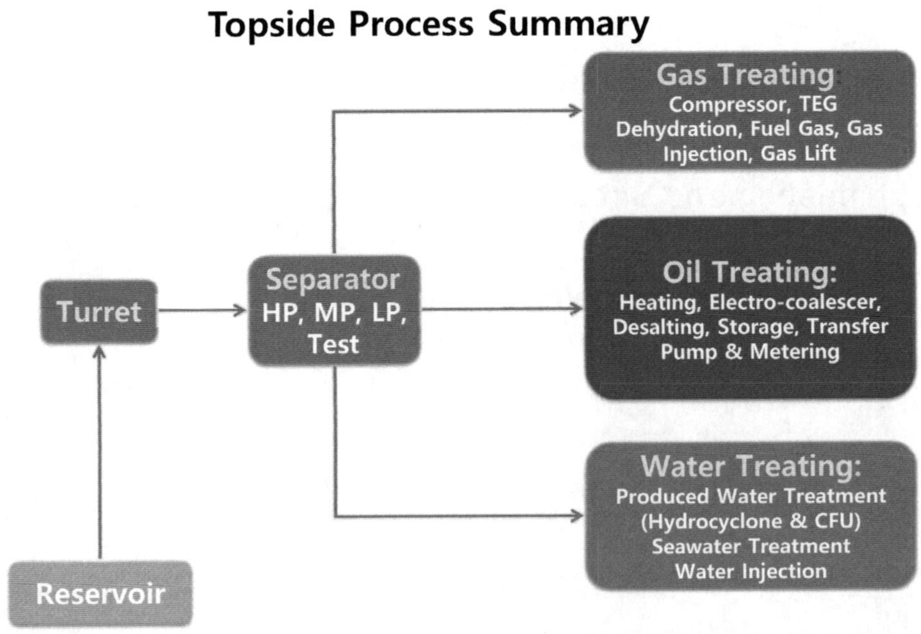

INHA UNIVERSITY

2. Modeling and Simulation

2. Modeling and Simulation

Simulation of Case 1

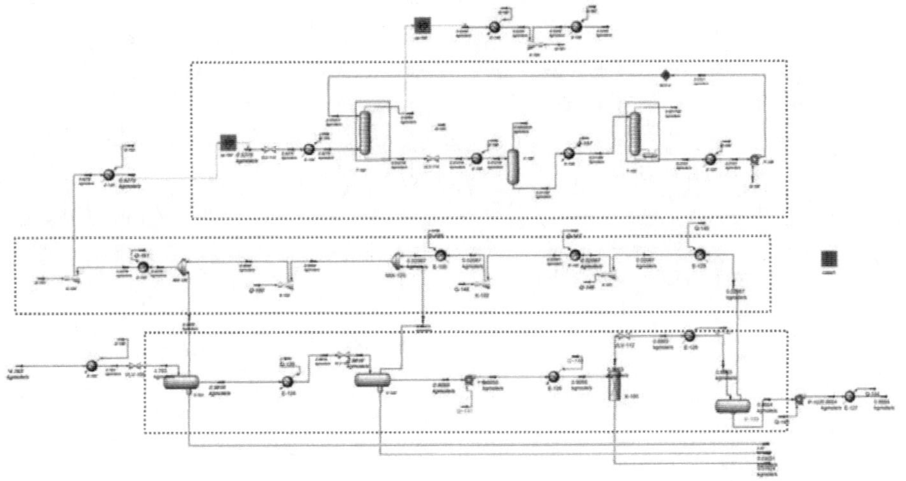

2. Modeling and Simulation

Case 1 : 3 separators without scrubber

- Use 3 separators
- Gas separated from separator is compressed without condensate recycling
- Variables : Pressures of V-102, V-103 and discharge pressure of K-101, K-102 and K-103

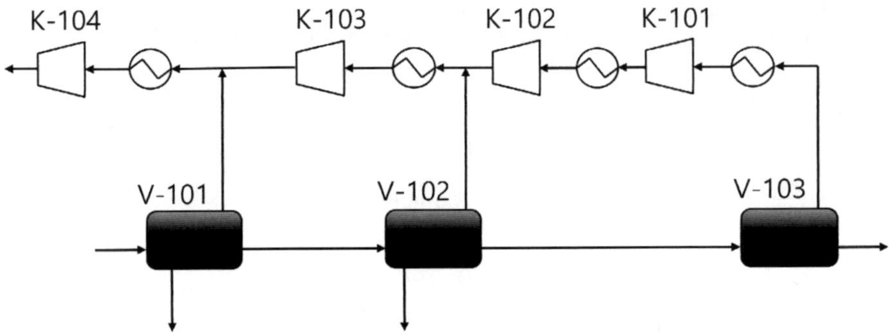

2. Modeling and Simulation

Simulation of Case 2

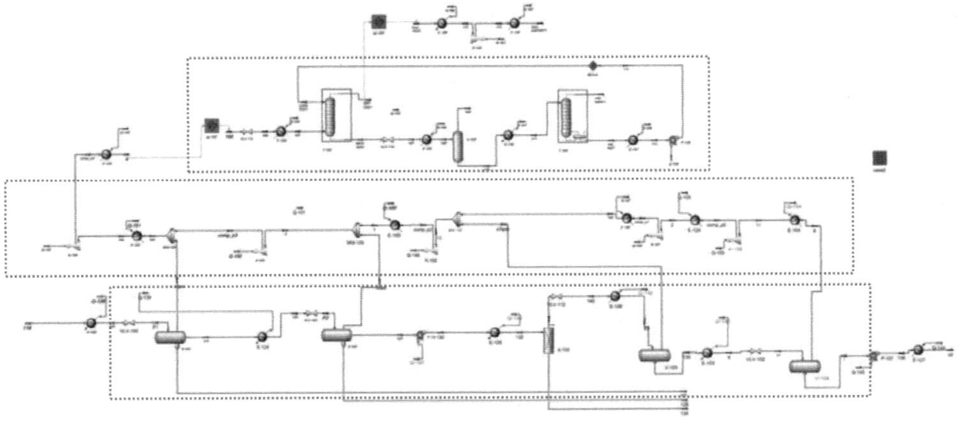

2. Modeling and Simulation

Case 2 : 4 separators without scrubber

- Use 4 separators
- Gas separated from separator is compressed without condensate recycling
- Variables : Pressures of V-102, V-103 , V-104 and discharge pressures of K-101, K-102, K-103 and K-104

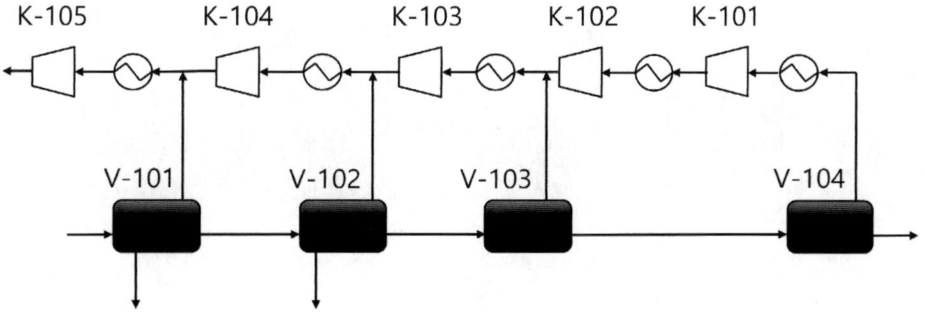

2. Modeling and Simulation

Simulation of Case 3

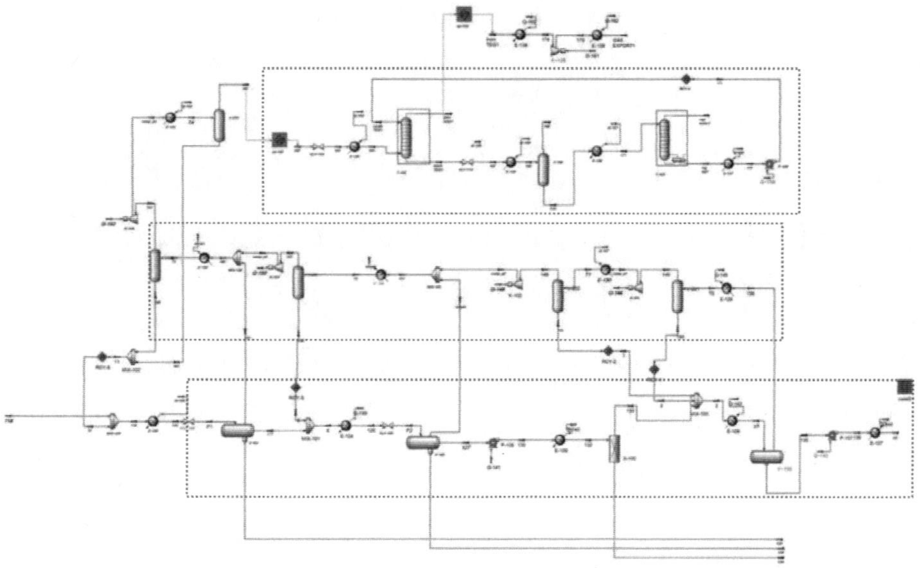

2. Modeling and Simulation

Case 3 : 3 separators with recycling

- Use condensate scrubbers
- Gas separated from separator is compressed with condensate recycling
- Operating variables : Pressures (V-102 and V-103) discharge pressures (K-101, K-102, K-103, K-104), Temperatures (V-201, V-202, V-203 and V-204)

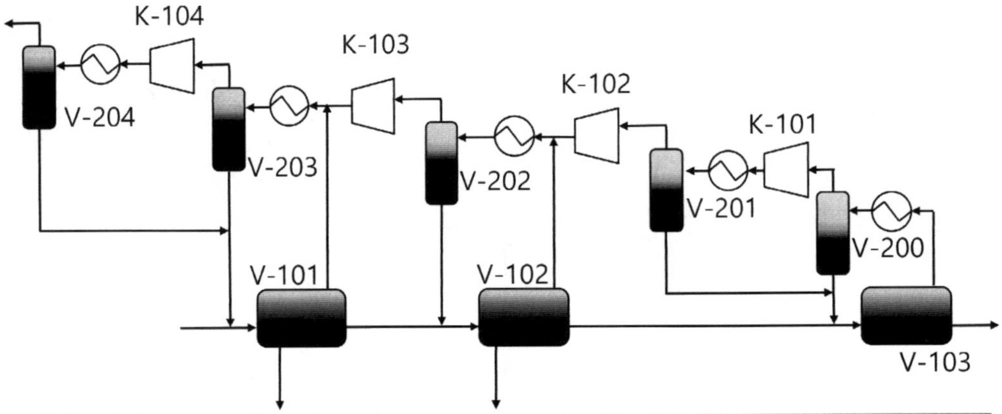

2. Modeling and Simulation

Simulation of Case 4

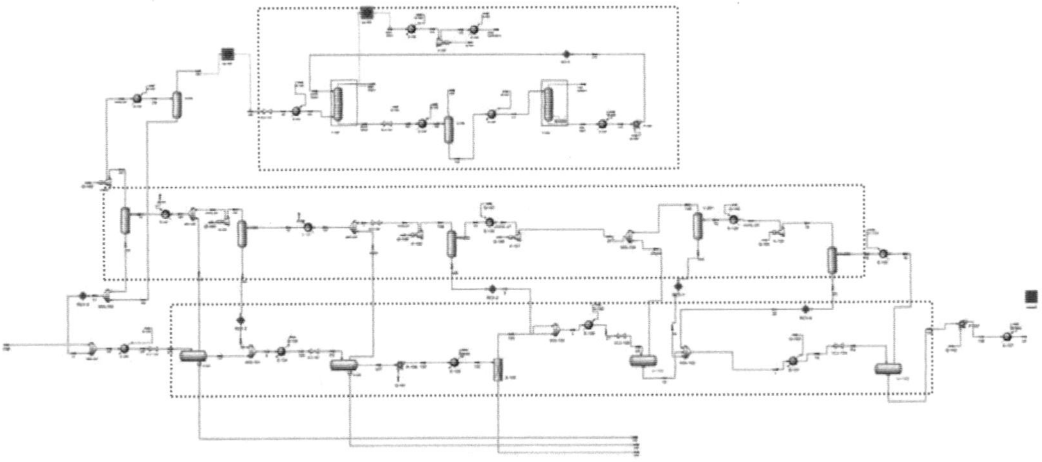

2. Modeling and Simulation

Case 4 : 4 separators with recycling

- Use condensate scrubbers
- Gas separated from separator is compressed with condensate recycling
- Variables : Pressures (V-102, V-103, V-104), discharge pressure (K-101, K-102, K-103 and K-104), Temperatures (V-201, V-202, V-203, V-204 and V-205)

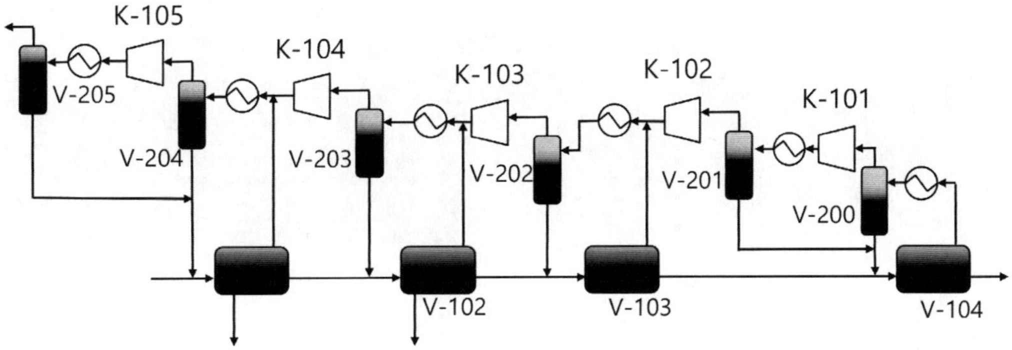

INHA UNIVERSITY

3. Optimization procedure

3. Optimization procedure

1. Simulation Based Optimization Method

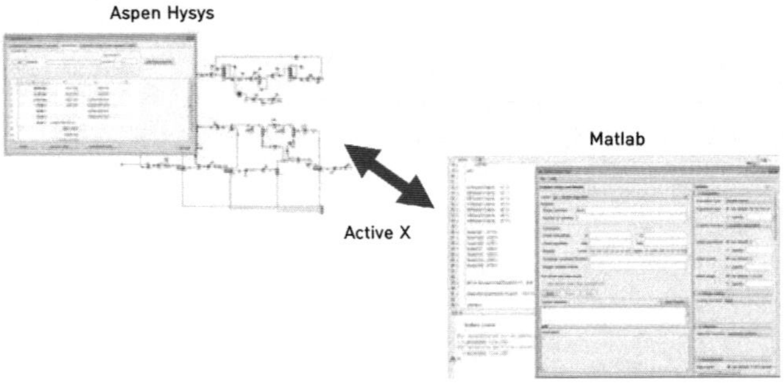

GA : Based on the mechanics of biological evolution

A simulated population of randomly selected individuals is generated and then allowed to evolve.

Express the problem in terms of a bit string
X = (1001010101011100)

Assign a fitness to each individual and Repeat until n children have been produced

INHA UNIVERSITY 15 Clean Energy Process Integration Lab

3. Optimization procedure

2. Optimization Algorithm

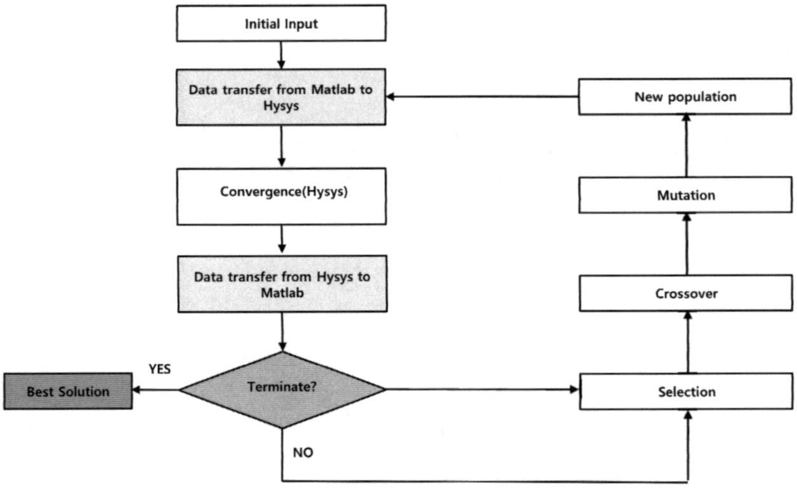

INHA UNIVERSITY 16 Clean Energy Process Integration Lab

3. Optimization procedure

3. Operating Variables and Bounds

		case1			case2			case3			case4		
	notation	peak oil	peak gas	peak water	peak oil	peak gas	peak water	peak oil	peak gas	peak water	peak oil	peak gas	peak water
1st separator pressure	P1 (bar)	30	30	30	30	30	30	30	30	30	30	30	30
2nd separator pressure	P2 (bar)	12 - 30	12 - 30	12 - 30	12 - 30	12 - 30	12 - 30	12 - 30	12 - 30	12 - 30	12 - 30	12 - 30	12 - 30
3rd separator pressure	P3 (bar)	1 – 6	1 – 11	1 – 6	6 – 20	10 – 20	6 - 20	1 – 6	1 – 11	1 – 11	6 – 20	10 – 20	10 – 20
4th separator pressure	P4 (bar)	-	-	-	1 – 6	1 – 11	1 – 6	-	-	-	1 – 6	1 – 11	1 – 11
1st separator temperature	T1	29.8	59.2	29.9	29.8	59.2	29.9	29.8	59.2	29.8	29.8	59.2	29.9
2nd separator temperature	T2	60	60	60	60	60	60	60	60	60	60	60	60
3rd separator temperature	T3	72	72	72	72	72	72	72	72	72	72	72	72
4th separator temperature	T4	-	-	-	85	85	85	-	-	-	85	85	85
compressor P1	comP1(bar)	1 – 7	1 – 11	1 – 7	1 – 9	1 – 11	1 – 9	1 – 7	1 – 11	1 – 11	1 – 9	1 – 12	1 – 11
compressor P2	comP2(bar)	7 - 20	11 - 20	7 - 20	9 - 20	11 - 20	9 - 20	7 - 20	11 - 20	11 - 20	7 - 20	12 - 20	11 - 20
compressor P3	comP3(bar)	20 - 30	20 - 30	20 - 30	9 - 20	11 - 20	9 - 20	20 - 30	20 - 30	20 - 30	7 - 20	12 - 20	11 - 20
compressor P4	comP4(bar)	75	75	75	20 - 30	20 - 30	20 - 30	75	75	75	20 - 30	20 - 30	20 - 30
compressor P5	comP5(bar)	-	-	-	75	75	75	-	-	-	75	75	75
scrubber temperature 1	scr_T1 (℃)	-	-	-	-	-	-	38 – 60	38 – 60	38 – 60	38 – 60	38 – 60	38 – 60
scrubber temperature 2	scr_T2 (℃)	-	-	-	-	-	-	38 – 60	38 – 60	38 – 60	38 – 60	38 – 60	38 – 60
scrubber temperature 3	scr_T3 (℃)		-	-	-	-	-	38 – 60	38 – 60	38 – 60	38 – 60	38 – 60	38 – 60
scrubber temperature 4	scr_T4 (℃)	-	-	-	-	-	-	38 – 60	38 – 60	38 – 60	38 – 60	38 – 60	38 – 60
scrubber temperature 5	scr_T5 (℃)	-	-	-	-	-	-	-	-	-	38 – 60	38 – 60	38 – 60

3. Optimization procedure

4. Objective function : Maximize profit

$$\text{Max } f_{profit}(x) = Sales_{oil}(x) + Sales_{gas}(x) - Cost_{utility}(x)$$

Prices of Oil, Gas and utilities

item	price
gas product	\$3.047/MMBtu
oil product	\$43/bbl
steam(100pisg)	\$7.42/1000lb
electricity	\$0.1/kWh

* From U.S Energy Information Administration.
* Cooling water expense was ignored.

4. Optimization results

Maximum Profit and Optimum Operating Conditions

Peak oil result

	notation	case1	case2	case3	case4
2nd separator pressure	P2 (kPa)	1835.4	1864.9	1769.2	1868.8
3rd separator pressure	P3 (kPa)	515.4	695.4	513.3	671.2
4th separator pressure	P4 (kPa)	-	548.5	-	538.2
compressor P1	comP1(kPa)	532.4	681.5	560.8	567.1
compressor P2	comP2(kPa)	782.2	1130.5	1675.9	1343.3
compressor P3	comP3(kPa)	2252.3	1863.9	2999.8	1864.9
compressor P4	comP4(kPa)	-	2934.5	-	2988.0
scrubber temperature 1	T1 (℃)	-	-	38.3	41.9
scrubber temperature 2	T2 (℃)	-	-	38.3	38.0
scrubber temperature 3	T3 (℃)	-	-	38.3	38.4
scrubber temperature 4	T4 (℃)	-	-	58.1	38.2
scrubber temperature 5	T5 (℃)	-	-	-	56.5
steam cost	$/day	42422.0	62438.6	42476.3	62518.1
electric cost	$/day	52319.5	51513.6	49482.8	49371.2
oil product sales	$/day	27847786.8	27847987.6	27897460.2	27897297.5
gas product sales	$/day	532768.6	532762.2	533410.7	533666.7
overall profit	$/day	28285813.9	28266797.5	28338911.7	28319075.0
oil RVP	kPa	68.0	67.8	67.8	66.6

4. Optimization results

Maximum Profit and Optimum Operating Conditions

Peak gas result

	notation	case1	case2	case3	case4
2nd separator pressure	P2 (kPa)	2444.0	2410.1	1977.0	2475.6
3rd separator pressure	P3 (kPa)	950.9	1367.6	937.5	1324.1
4th separator pressure	P4 (kPa)	-	990.4	-	987.2
compressor P1	comP1(kPa)	984.9	990.9	960.4	1198.4
compressor P2	comP2(kPa)	1995.5	1424.5	1635.2	1474.0
compressor P3	comP3(kPa)	2967.2	1955.9	2999.7	1761.7
compressor P4	comP4(kPa)	-	2995.0	-	2976.7
scrubber temperature 1	T1 (℃)	-	-	38.6	42.9
scrubber temperature 2	T2 (℃)	-	-	38.0	41.7
scrubber temperature 3	T3 (℃)	-	-	38.4	41.3
scrubber temperature 4	T4 (℃)	-	-	60.0	38.4
scrubber temperature 5	T5 (℃)	-	-	-	58.6
steam cost	$/day	10783.2	35443.3	11216.3	35947.2
electric cost	$/day	49572.8	47549.1	45843.5	45987.5
oil product sales	$/day	33453293.1	33452471.8	33575109.0	33575180.4
gas product sales	$/day	510671.0	510802.8	512648.0	512659.3
overall profit	$/day	33903608.0	33880282.0	34030697.0	34005905.0
oil RVP	kPa	68.0	67.6	68.0	68.0

4. Optimization results

Maximum Profit and Optimum Operating Conditions

Peak water result

	notation	case1	case2	case3	case4
2nd separator pressure	P2 (kPa)	1362.6	2233.4	2425.1	2492.0
3rd separator pressure	P3 (kPa)	464.3	909.1	819.0	1061.3
4th separator pressure	P4 (kPa)	-	500.5	-	872.7
compressor P1	comP1(kPa)	569.0	701.3	1074.7	1045.4
compressor P2	comP2(kPa)	1329.6	1997.4	1799.7	1377.5
compressor P3	comP3(kPa)	2977.9	1252.9	2999.4	1703.7
compressor P4	comP4(kPa)	-	2999.8	-	2995.3
scrubber temperature 1	T1 (℃)	-	-	40.1	54.1
scrubber temperature 2	T2 (℃)	-	-	39.3	50.6
scrubber temperature 3	T3 (℃)	-	-	38.2	44.0
scrubber temperature 4	T4 (℃)	-	-	56.0	38.6
scrubber temperature 5	T5 (℃)	-	-	-	55.0
steam cost	$/day	7088.9	9803.2	7197.0	9934.8
electric cost	$/day	33923.1	33811.2	32181.1	32194.6
oil product sales	$/day	3699292.4	3699305.4	3748972.6	3747785.0
gas product sales	$/day	369463.5	369455.2	369233.2	369161.0
overall profit	$/day	4027744.0	4025146.0	4078828.0	4074817.0
oil RVP	kPa	67.9	67.9	66.5	67.1

4. Optimization results

Summary of Overall Profit - peak oil

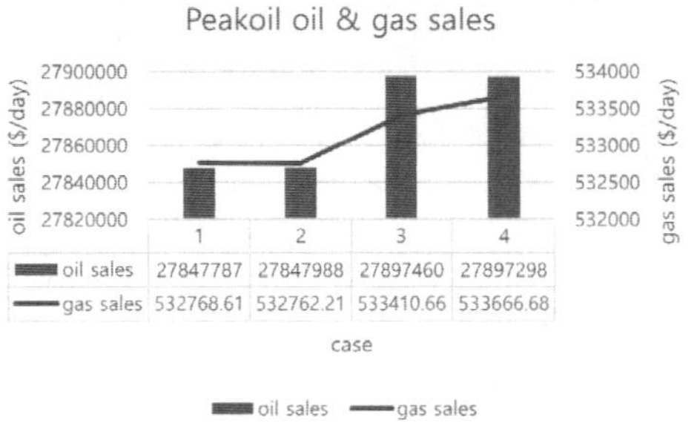

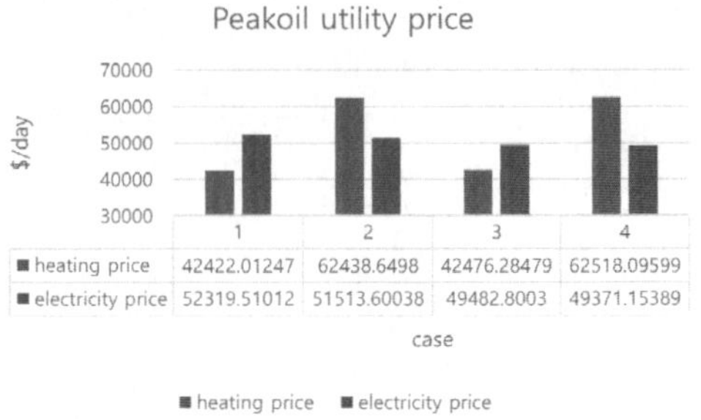

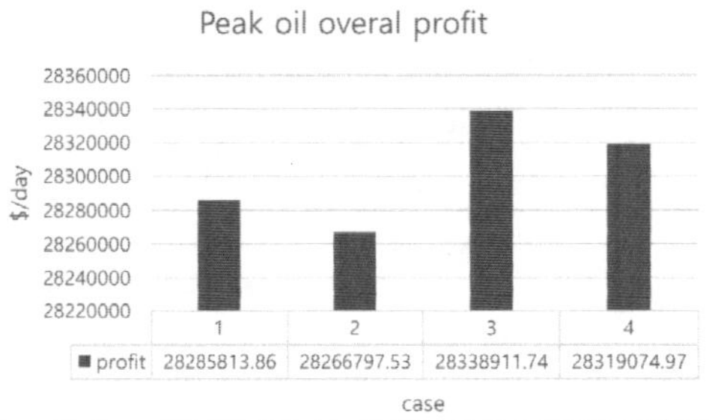

4. Optimization results

Summary of Overall Profit - peak gas

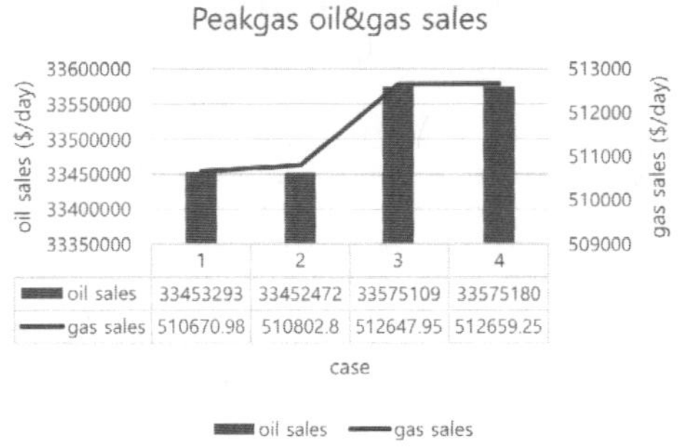

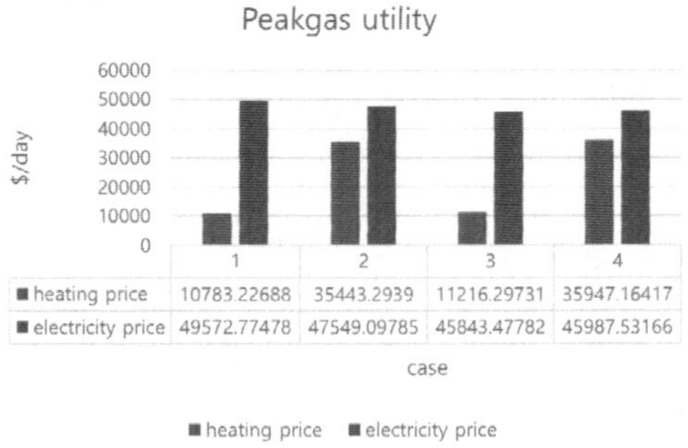

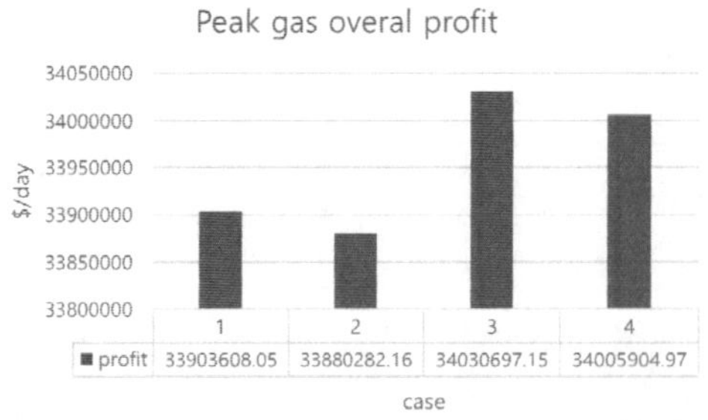

4. Optimization results

Summary of Overall Profit - peak oil

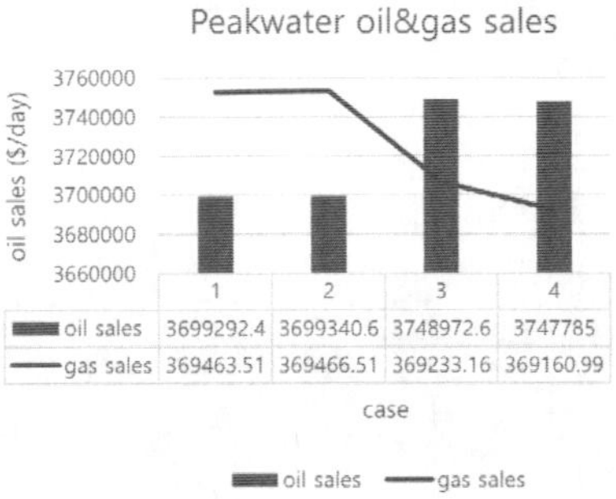

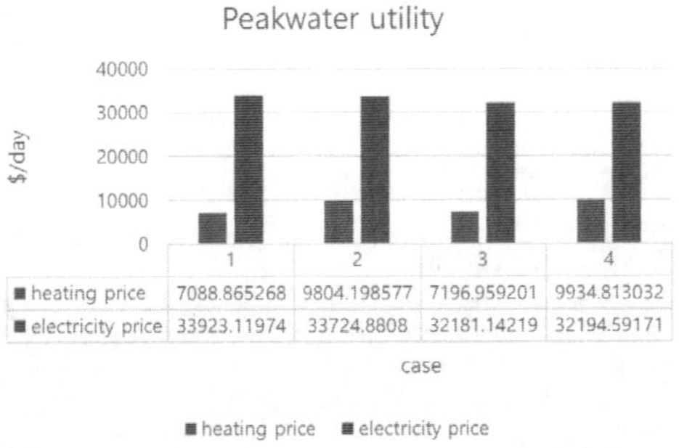

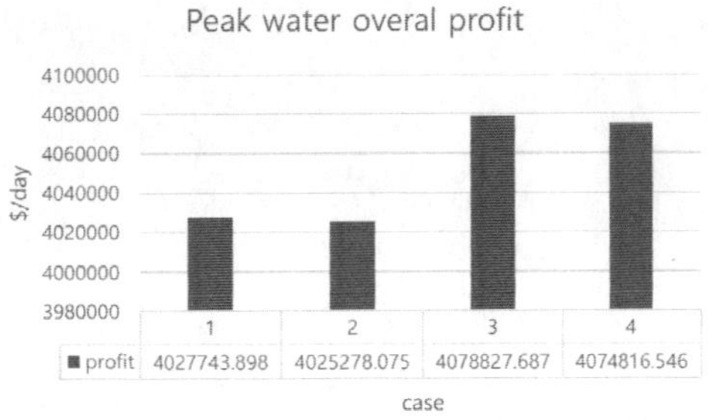

4. Optimization results

Summary of Capital cost

	peakoil	peakgas	peakwater
case1	34635100	42132100	25384600
case2	37942900	45539900	25775800
case3	35000000	36362000	25781100
case4	39394500	42268900	28623100

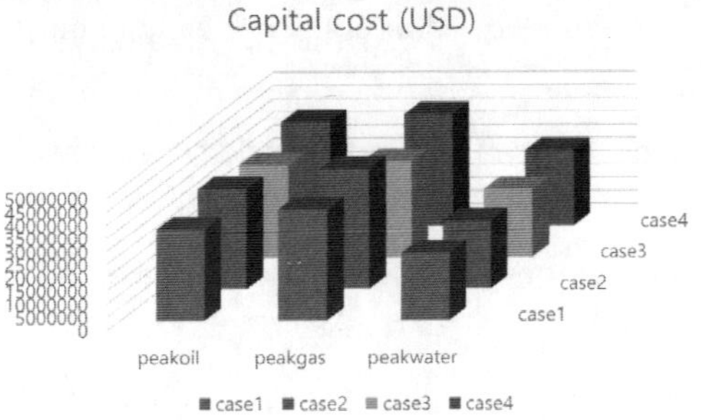

* Capital cost and installed cost were calculated by Aspen Process Economic Analyzer

5. Summary & Future Plans

4. Summary

Summary

- Case 3 and 4 have condensate recycle scheme to make the process produce more oil product. Heavy hydrocarbon components carried over to gas flow are recycled to oil flow and only light hydrocarbon components flow toward gas process.
  It means that separation efficiency of whole process was improved and hydrocarbon recovery was increased.

- As more oil flow enter the separator through recycling, duty of pre-heaters increased while power of compressors decreased. Total utility price is almost constant whether it has recycle.

- As separator is added, heat duty increases and electricity decreased. But Total utility price increases when the process has more separator.

- Comparison including capital cost, operating cost, product sales shows that Case 3 is the most economically feasible.

- Best operating condition was drawn through Simulation based optimization using GA .

MEETING ADDITIONAL PLANT CAPACITY REQUIREMENTS WITHOUT ADDING ADDITIONAL VESSELS

Gene Eberhardt
Woven Metal Products - Alvin, TX USA

Gene Eberhardt is the Vice President of the Reactor Division for Woven Metal Products in Alvin, Texas. He holds a Bachelor of Science Degree in Business from Saint Mary's University of Minnesota, a Master's Degree in Business Administration (MBA) from Purdue University and a Master's Degree in International Business Management from TiasNimbas Business School - The Netherlands. He holds a patent for Catalyst Bed Support systems and has several patents pending. He continues to develop industry changing technologies and methods in collaboration with Purdue University.

Given today's demanding throughput needs to maximize plant profits, plant de-bottlenecking quickly becomes an area ideal for identification and immediate return on investments. Today's gas plants require systems optimization to safely produce and deliver predictable, profitable products, longer predictable drying cycles and increased molecular sieve bed performance. These are not off-the-shelf solutions and require robust engineered solutions that are performance driven, industry proven and readily available for demanding plant outage requirements.

Molecular sieve bed support is one of, if not the most critical mechanical components within the dehydration vessel. Once the bed support requirements are understood, all options can be evaluated and a competent molecular sieve bed plan can be developed.

As with many natural gas plants commissioned over the years, molecular sieve bed support and proper molecular sieve bed utilization have been a secondary concern in these plant designs. In many cases, operational conditions have not allowed for these units to be properly evaluated or engineered to meet the operational needs of the plant. It has become an industry standard to install an outlet collector (elephant stool) in the outlet nozzle of the vessel with very little engineering emphasis given to the outlet collector, the effects that the outlet collector has on the molecular sieve bed, the performance of the molecular sieve bed or the effects the outlet collector has on the life of the molecular sieve bed. In many applications, the outlet collector serves only as a containment device and has not been engineered to aid in the bed performance or in the bed longevity. Many of the outlet collectors utilized today include an assortment of materials and configurations that include (but not limited to) slotted pipe, rolled plate, engineered support frames and a variety of material coverings including – wire mesh in a variety of material grades, wire mesh sizes and open area. In addition to the many wire mesh offerings many licensors and fabricators select profile wire in a wide range of stainless steel material grades and slot openings all of which do not observe any industry standards or a standard engineering design for these applications.

In some applications, a dehydration unit will have a horizontal bed support installed near the tangent line or the head seam of the vessel with a variety of materials used for bed containment including layers of wire mesh wire tied over grating and profile wire grid arrangements. In these bed configurations, the outlet collector is often installed as a secondary containment device as most vessels are not rolled perfectly round creating challenges to accomplish a full perimeter seal. The horizontal bed support configuration is a significant upgrade to the traditional outlet collector only configuration, however the molecular sieve bed design and perimeter sealing must be fully understood with emphasis on a secondary containment device to prevent total bed migration if the perimeter seal becomes compromised. The horizontal bed support option does offer much better bed utilization; however, this does sacrifice the total bed volume as the head of the vessel (at a minimum) becomes a void space.

This paper will focus on a recent installation of the Woven Metal Products / Haldor Topsoe co-developed (U.S. Patent Number 9409137) Outlet Support Grid (OSG) bed support system for the Blue Racer Midstream Company – Natrium West Virginia Complex, which consists of two cryogenic natural gas processing plants, each of which has 200 MMcf/d of processing capacity. This plant design includes 3 industry standard dehydration vessels that have an inner diameter of 84 inches (7 feet) and a total seam length (not including vessel heads) of 164 inches (13 feet 8 inches). When operational, this plant has been designed for parallel adsorption (Down Flow) and one vessel in regeneration (Up Flow) / standby.

Woven Metal Products Outlet Support Grid U.S. Patent 09409137

The Outlet Support Grid now offers an engineered total bed solution utilizing the performance of a horizontal bed support with the containment reliability that an outlet collector provides. Often the additional bed volumes realized with the Outlet Support Grid system installed will provide 15% or more additional bed volumes depending on the elevation of the horizontal bed configuration it replaces. As shown in Figure 1 the Computational Fluid Dynamics (CFD) modeling of the Outlet Support Grid shows a very uniform flow pattern which is critical for optimized bed performance, combined with full bed containment the Outlet Support Grid system is a reliable operational solution.

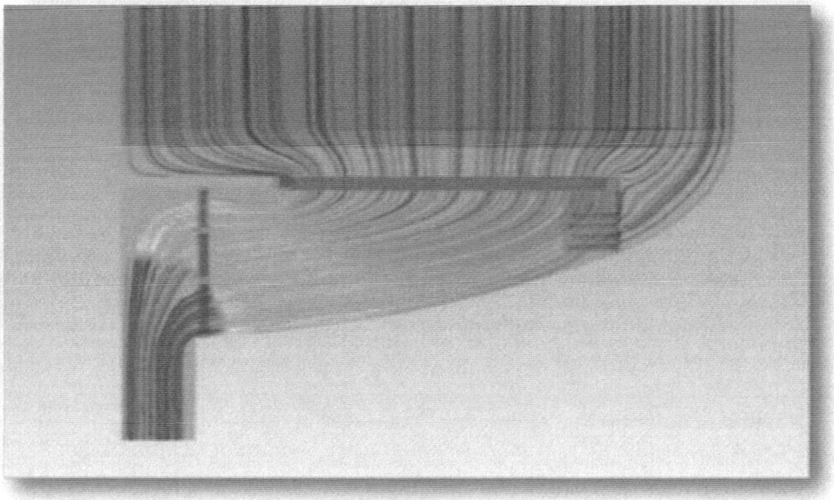

Figure 1.

History

This CB&I natural gas plant design was commissioned in April 2015, at which time operations had encountered / suspected a bed lifting (bed "fluffing") that resulted in an unusually high pressure reading across all three molecular sieve beds of 130 psi ΔP indicated. This unusually high ΔP was also accompanied with heavy dusting and plugging in the units down-stream filters, which prompted the operations team to suspect molecular sieve crushing to have incurred during the startup of the plant.

New molecular sieve was promptly ordered for the three dehydration vessels and the unit outage was taken for the replacement of the suspected crushed molecular sieve. During this outage, the unit beds were not fully emptied or cleaned only the molecular sieve was replaced and then the beds were put back in service with the existing bed support media undisturbed and no inspections or cleaning of the outlet baskets or inert media was performed. These three units were restarted and the high ΔP remained at 130 psi and continued to climb.

The Blue Racer operations team was now faced with the task of discovering any and all options to correct the high ΔP situation and restore the plant to a safe operational rate, quickly. The Blue Racer operations and engineering teams contacted their local UNIVAR field representative to source not only the new molecular sieve and new inert media support balls for each dehydration bed but they also started looking for replacement options for the potential of a semi crushed or fully crushed outlet basket in each of the three dehydration units. Woven Metal Products was contacted and was informed of the current operational situation and the challenges facing the Blue Racer team.

Collectively the Blue Racer, Woven Metal Products, UNIVAR, Zeochem and Catalyst Force teams addressed the following items; safety risks of continued operations, critical project timeline, operational limitations, current bed configuration, and equipment history of the installed equipment.

Each of the three dehydration vessels were originally equipped with outlet baskets sized at 15 inches outer diameter, 10 inches overall height with a profile wire covering, then welded to the outlet nozzle. These outlet baskets were constructed of 304 stainless steel, designed for 50 PSIG and 30% - 40% open area. Each dehydration vessel was loaded as follows:

Pre – Installation Bed Loading			
Media	**Pounds**	**Cubic Feet^3**	**Feet / Inches**
Ceramic Inert Balls (size unknown)	820 (estimate)	19	0.5 Feet
Guard Bed Media	1,653	37.57	0.98 Feet
4 x 8 mesh 1/8" Molecular Sieve	12,600	273.91	7.12 Feet
8 x 12 mesh 1/16" Molecular Sieve	7,200	153.19	3.98 Feet
Ceramic Inert Balls (size unknown)	820 (estimate)	19	0.5 Feet
Totals	23,093 (estimate)	502.67	13.08 Feet

The Blue Racer operational records reveal an average ΔP of 134 PSIG with a maximum ΔP of 180 PSIG indicated in table 1 (below) given the very high ΔP the actual condition of the outlet basket and molecular sieve bed was unknown. The Blue Racer operations reported a high frequency of heavy filter dusting / plugging, exchangers plugging, high system pressure requirements, dew point control issues, overall output product controls and longer bed regeneration times.

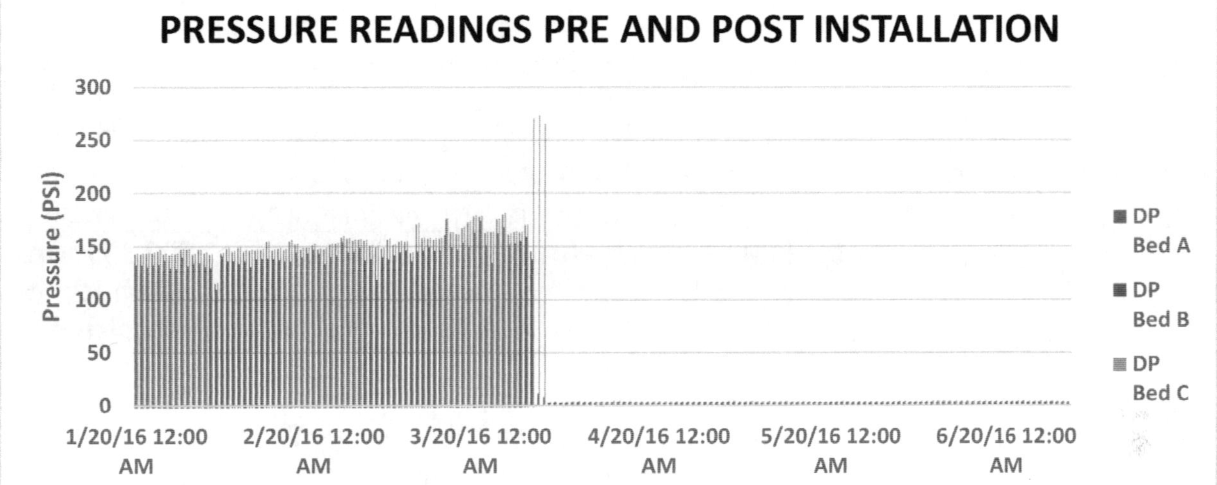

Table 1.

The following options were considered given the current vessel configuration and the critical path for both fabrication and the unit outage requirements;

- Replace the current outlet baskets with a design that would allow more open area and sized to the vessel.
- A horizontal support grid option was not considered due to the dehydration vessels not originally equipped with a support ring or beams and the time required to modify the vessels.
- Outlet Support Grid (OSG) system.

Given Blue Racer's operational experience with the outlet baskets the evaluation and decision was made to install the Outlet Support Grid solution. Woven Metal Products began engineering the OSG solution with fabrication drawings provide to the client for approvals in two weeks and full fabrication / delivery promised in just four weeks. The team confirmed that the Zeochem molecular sieve and bed support media would be onsite for the outage date and the Catalyst Force team would be able to service the outage with the required staff to execute a safe and timely outage. Schedules were drafted for the expedited fabrication and for the unit outage, both with very tight time lines. The following table (Table 2) shows the outage plan with just 26 hours allocated for a blind to blind per bed installation plan.

Event	Scheduled Time	Actual Time	Difference
Install Blinds	2.0 Hours	2.0 Hours	
N2 Blowdown	2.0 Hours	2.0 Hours	
Open Top Manway	1.0 Hours	0.75 Hours	0.25 Hours
Obtain Entry Permit / Safety review Meeting	1.0 Hours	1.0 Hours	
Equipment Rigging	2.0 Hours	2.0 Hours	
Unloading Bed	3.0 Hours	2.0 Hours	1.0 Hours
Vessel Cleaning	1.0 Hours	0.5 Hours	0.5 Hours
Remove Outlet Basket	2.0 Hours	0.5 Hours	1.5 Hours
Install OSG	4.0 Hours	1.5 Hours	2.5 Hours
Loading Molecular Sieve Bed	4.0 Hours	2.5 Hours	1.5 Hours
Close Manways / Un-install Blinds	4.0 Hours	3.5 Hours	0.5 Hours
Total Hours **Table 2.**	**26 Hours**	**18.25 Hours**	**7.75 Hours (Schedule Positive)**

The Outlet Support Grid is designed and installed without the need to weld to the vessel, no rope packing, no sealing materials and no thermal blankets are required to form a reliable sealing solution to the vessel head. The Outlet Support Grid mechanical sealing system preforms extremely well over the vessel and process thermal cycles which is critical in the full bed containment solution preventing media migration.

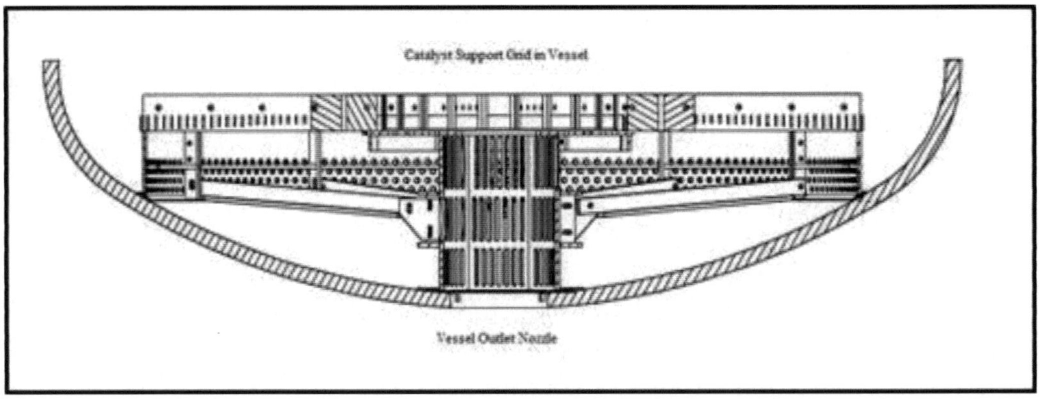

Installed View

Given the safety concerns of the high ΔP of 130 PSIG and rising, Woven Metal Products Outlet Support Grid was engineered to address these concerns and meet the bed design pressure drop of 3.08 PSIG at startup. With properly designed profile wire applied on all grid surfaces, the Outlet Support Grid has shown and will continue to perform at a very low and predictable ΔP as shown in table one since start-up. The Outlet Support Grid is designed to be in direct contact with the molecular sieve or any other media as the profile wire slot spacing is designed for the catalyst size loaded into the bed. Any support media utilized further assists in the bed hydraulics and should be designed and loaded for this purpose. As loaded the following inert balls were utilized 1/2" (bottom), ¼" (mid layer) and 1/8" (top layer) in contact with the molecular sieve, each application should be carefully examined and designed by catalyst providers.

Post – Installation Bed Loading			
Media	**Pounds**	**Cubic Feet^3**	**Feet / Inches**
Ceramic Inert Balls ½"	828 (estimate)	19	0.5 Feet
Guard Bed Media	2,200	50	1.8 Feet
4 x 8 mesh 1/8" Molecular Sieve	10,500	228.26	6.1 Feet
8 x 12 mesh 1/16" Molecular Sieve	6,000	127.66	3.98 Feet
Ceramic Inert Balls 1/8" (Bottom – Top)	414 (estimate)	10	0.25 Feet
Ceramic Inert Balls ¼" (Bottom – Middle)	414 (estimate)	10	0.25 Feet
Ceramic Inert Balls ½" (Bottom – Bottom)	414 (estimate)	15	0.25 Feet
Totals	20,770 (estimate)	459.92	13.13 Feet

Blue Racer operations utilize a timer method for bed cycling; previous timer cycles were set at adsorption for 12 hours and regeneration for 6 hours. Post installation adsorption cycle is now set at 16 hours and regeneration is set at 6 hours. With the molecular sieve bed hydraulics optimized, the bed timers are being increased steadily over time while watching for potential bed break through. The initial bed performance indications and the plant records for these dehydration beds show minimal filter dusting after one year of continuous processing. These are all positive indications that these molecular sieve beds will meet or exceed the estimated 3 years of bed life giving the Blue Racer operations a much safer and very predicable operation.

This project and the installation were considered very successful with the dehydration vessel and the bed support concerns addressed and corrected including;

- Maximum volume of molecular sieve for the unit
- Longer molecular sieve bed adsorption cycles which enhance throughput / plant de-bottlenecking
- Fewer bed regeneration cycles reduces the amount of molecular sieve breakdown over time
- Full bed containment
- Full bed utilization
- Consistent dew point control
- Predictable bed performance
- Reduce fuel gas consumption

Outside the vessel benefits can be examined and appreciated with full bed utilization, fewer bed regeneration cycles will be required. The operational impact of fewer bed regeneration cycles is easily calculated in the reduction of valve cycles – given each bed includes 6 inlet /

outlet valves, incoming wet gas feed valves, steam and heater control valves essentially doubling the operational life of these valves greatly reducing the required maintenance costs for each valve. Each plant should consult with their valve manufacture / supplier to fully understand these operational valve benefits and the potential maintenance cost savings. Full bed containment aids not only in the valve life with less contamination present, full bed containment greatly reduces downstream filter plugging issues, cryogenic expander unit plugging and the need to bring the operational plant down, open and clean these units at un-scheduled times and loosing valuable production profits.

Conclusion

As with all plants, processes and capital projects many factors must be considered and evaluated. Return on investment (ROI) calculations can include many of the areas evaluated and discussed in this field installation and within this paper. Many factors will need to be evaluated including personnel and plant safety, natural gas prices, operational needs, operational costs, opportunity costs associated with lost production, lost efficiencies in molecular sieve beds and the installed equipment that are not engineered to maximize production capabilities and capacities. More information on the Outlet Support Grid and many other equipment solutions can be found at our website www.wovenmetal.com

Development of Novel Approach to Pretreat Natural Gas at Off-Shore Plants

2017. 03. 27

Soojin Kwon

Index

1. Introduction of natural gas pretreatment
2. Research contents
   2.1. Modeling of Membrane
   2.2. Case study
   2.3. Results analysis
3. Summary
4. Future works

1. Introduction of natural gas pretreatment

1. Introduction - natural gas process configuration

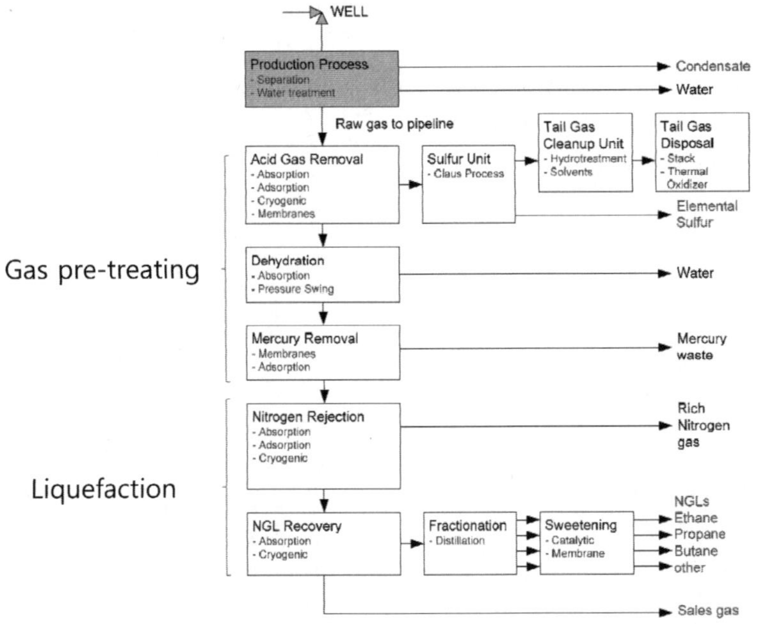

1. Introduction – Acid gas removal

- Absorption

Undesired compounds in the gas stream must be slightly soluble in the liquid, which is categorized as an absorbent.

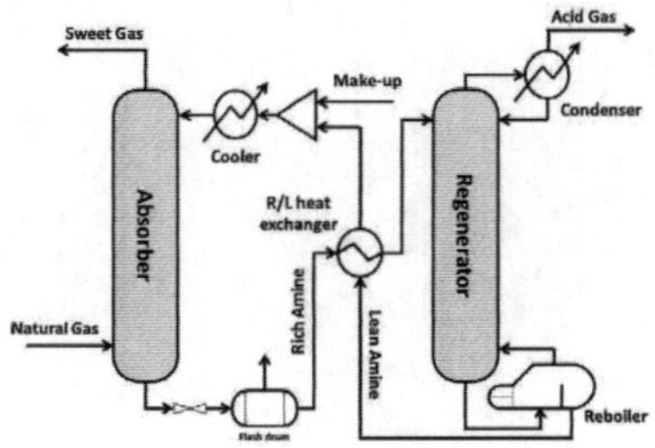

A typical amine-based acid gas removal unit.

- Membrane

- Energy saving
- Quite simple
  no mooving parts, compact, easy to scale-up, relatively easy to operate and control

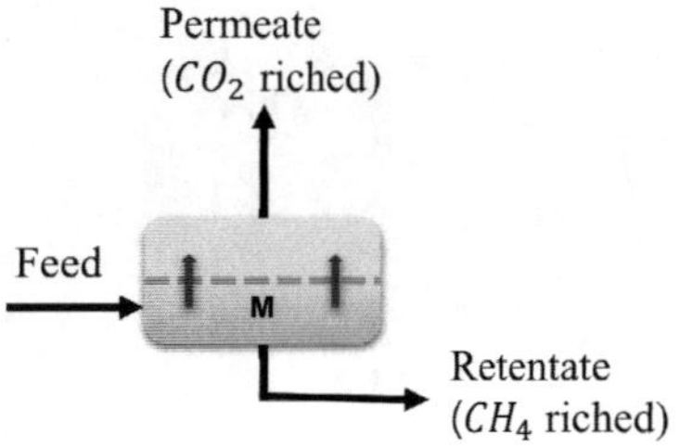

1. Introduction – various configuration for acid gas removal

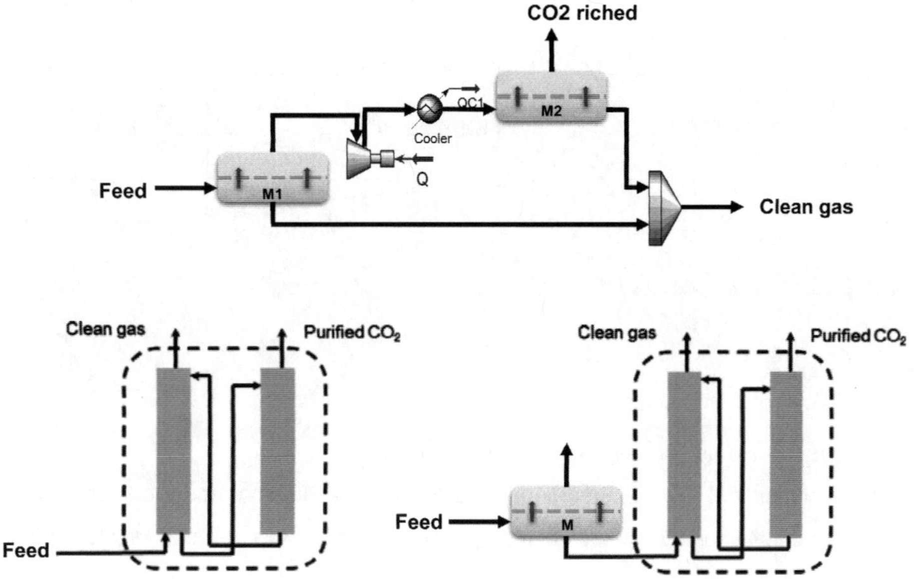

2. Research contents

2.1. Modeling of membrane system

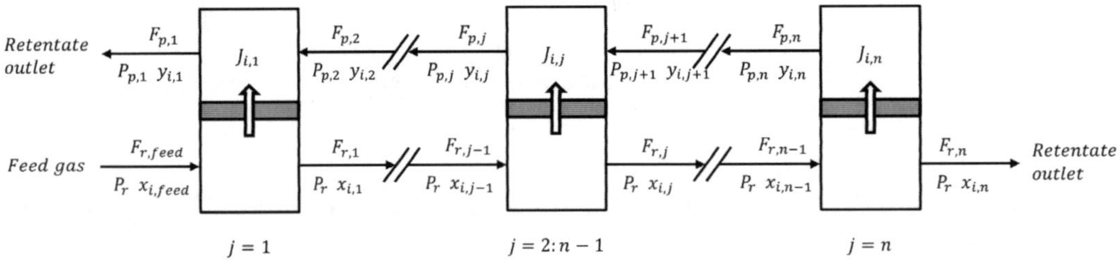

Schematic diagram of hollow fiber membrane separation.

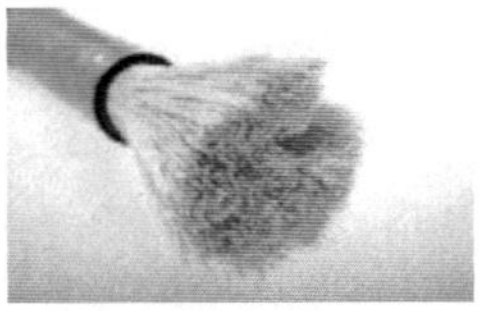

- Asymmetric cellulose acetate hollow fiber was modeled.
- This study considers counter-current configuration
- For the longitudinal flow configuration (considering variation of compositions and flow rates of both shell and tube sides along the axial direction), this model is divided into cells in the axial direction.

2.2. Case Study

Operating and feed conditions used in the membrane modeling

Component	Feed composition (mole fraction)	Operating conditions	
CO_2	0.172	Feed flow	$32.8Nm^3/s$
CH_4	0.7759	Feed pressure	80 bar
C_2H_6	0.0417	Permeate pressure	4 bar
C_3H_8	0.0104	Flow pattern	counter-current

Membrane properties	
Inner diameter	200μm
Outer diameter	500μm
Effective length	3m
Product purity target	$x_{CO_2} \leq 2mol\%$

Permeance and selectivity (Pan, 1986)

Permeance [10^{-8}mol/sm^{2}Pa]		Selectivity	
		CO_2/CH_4	36
CO_2	1.34	C_2H_6/CH_4	0.27
		C_3H_8/CH_4	0.054

Note : Feed condition data is extracted from gas reservoir that is located at west coast of Australia.

2.2. Case Study

Four configurations used to compare each methane recovery of product stream and economic evaluation.

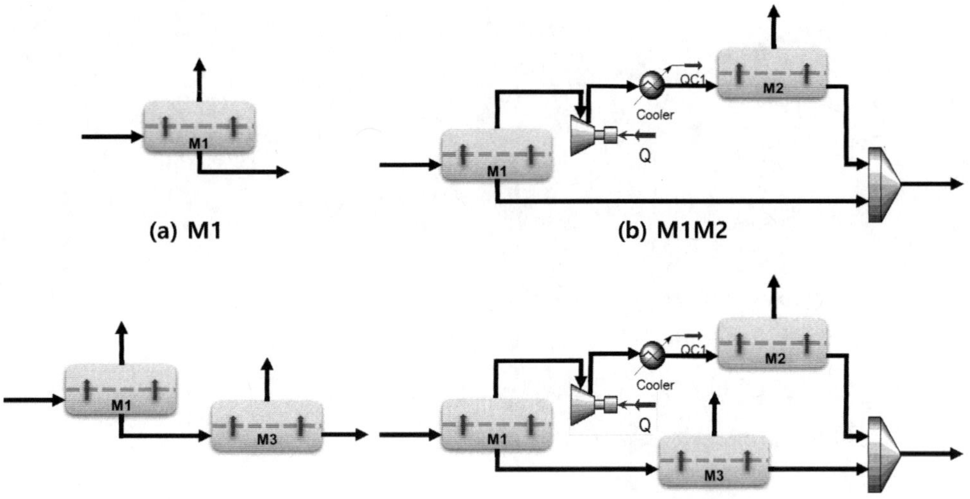

(a) M1 (b) M1M2

(c) M1M3 (d) M1M2M3

2.2. Case Study

Economic parameters and assumptions for computation of GPC (Hao et al., 2002)

Total Plant Investment	**TPI**	**TPI = TFI + SC**
Membrane module cost	MC	\$ 10/ft2 (includes cost of membrane element)
Fixed cost	FC	MC + CC
Base Plant Cost	BPC	1.12 × FC (includes home office cost : 0.12FC)
Project Contingency	PC	0.2 × BPC
Total facilities investment	TFI	BPC + PC
Start up cost	SC	0.10 × VOM
Annual Variable Operating and Maintenance Cost	**VOM**	**VOM = CMC + LTI + DL + LOC + MRC + UC**
Contract and Material Maintenance Cost	CMC	0.05 × TFI
Local Taxes and insurance	LTI	0.015 × TFI
Direct Labor cost	DL	\$ 15/h
Labor Overhead Cost	LOC	1.15 ×z DL
Membrane Replacement costs	MRC	\$ 3/ft2 of membrane
Annual Cost of CH4 Lost in Permeate	**CH_4LS**	**CH_4LS = NGLS * NHV * NWP**
heating value of natural gas	NHV	1000 Btu/ft3
wellhead price of crude natural gas	NWP	\$ 3.026 /MMBTU
Annual Natural Gas Lost	NGLS	NGLS = 365 × OSF × Qf × yp(CH4) × xf(CH4)
Gas Processing Cost	**GPC**	**GPC = (CRC + CH_4LS + VOM)/[365 × OSF × Qf × (1-SCE) × 100]**
Annual Capital Related Cost	CRC	0.2 × TPI
Membrane Life	t	4 years
Heating Value of Natural Gas	1000	
On stream factor	OSF	96%
stage-cut equivalent (P_N/F_N)	SCE	
net feed flow rate	FN	
net permeate flow rate	PN	

The costs of compressor and cooler with utilities refer to Aspen Process Economic Analyzer (APEA).

2.3 Results Analysis - (a) M1

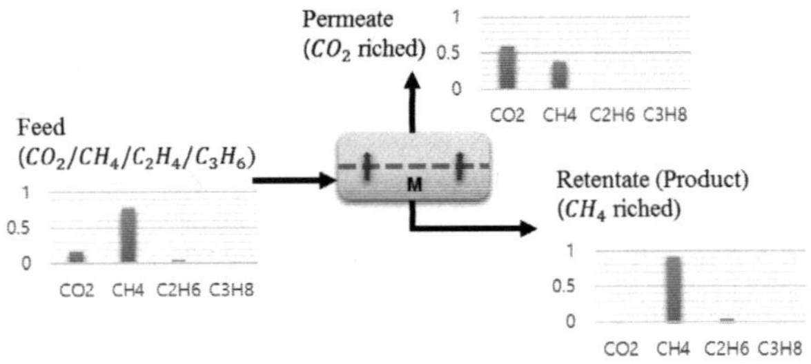

- CH_4 recovery : 87.1%
- ROI : 87.3%
- The number of fibers : 12,380,000
- Membrane area : 58,186 m^2

2.3 Results Analysis - (b) M1M2

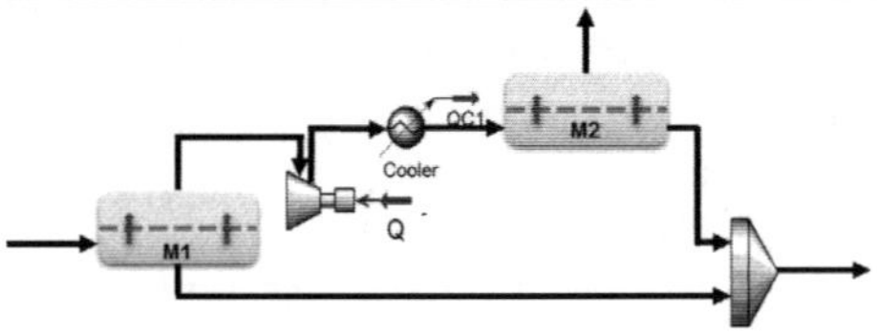

Methane recovery and ROI variation due to changes of the ratio of the membrane size of M2 membrane to that of M1 membrane (M2/M1).

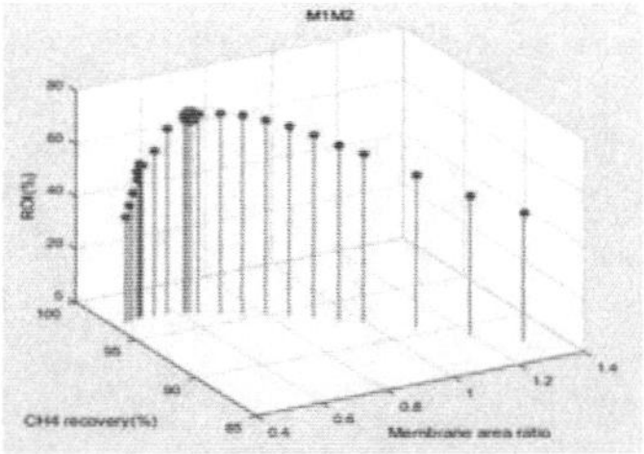

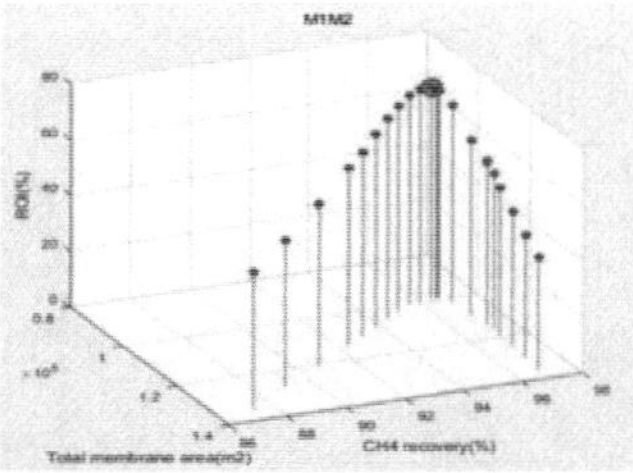

- A large amount of CO_2 is mostly removed in M1, and the size of M2 membrane directly affects the methane recovery of product stream.
- If M2/M1 value is more than 1.4, the amount of retentate flow of M2 (Fr2) becomes negative (no flow).
- The optimum point that shows the highest value of methane recovery with the highest profit is confirmed from M1M2 configuration.

2.3 Results Analysis - (c) M1M3

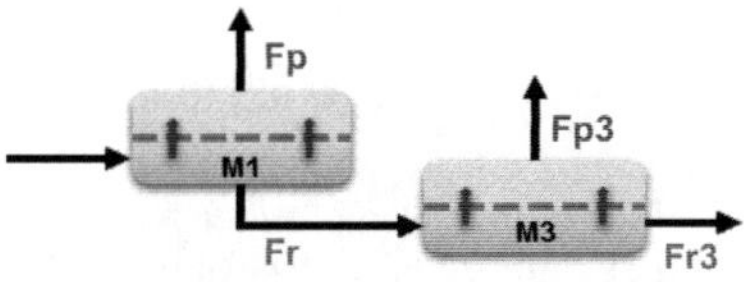

Methane recovery and ROI variation due to changes of the ratio of the membrane size of M3 membrane to that of M1 membrane (M3/M1).

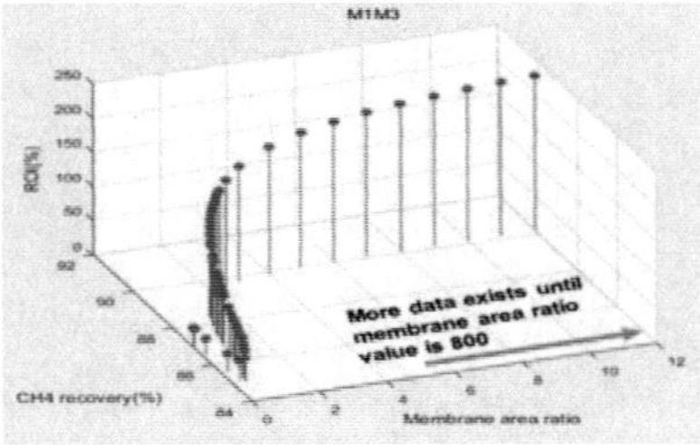

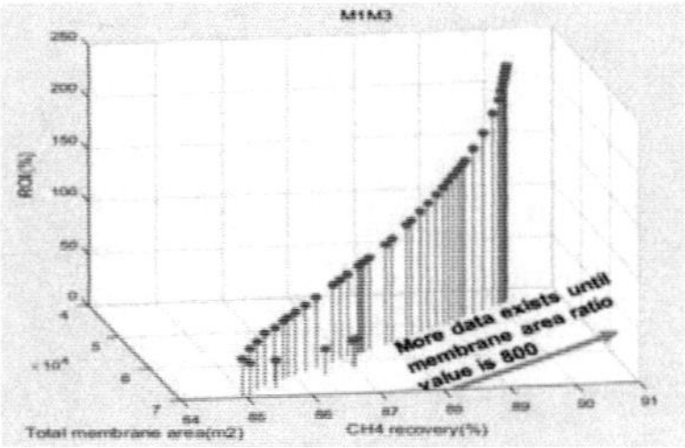

- If M3/M1 value is less than 0.05, the amount of permeate flow of M3 (Fp3l) becomes negative (no flow).
- If M3/M1 value is more than 10, the variation of ROI and methane recovery values can be considered negligible. That means, if M3 have a pretty big memabrane area in comparison with the area of M1, M3 acts like all alone.
- The total membrane area (M3+M1) become almost same especially over the point of M3/M1=10 until the amount of permeate flow (Fp1) reaches to the value of 0 (no flow) at about M3/M1=1000.

2.3 Results Analysis - (d) M1M2M3 (fixed M3/M1)

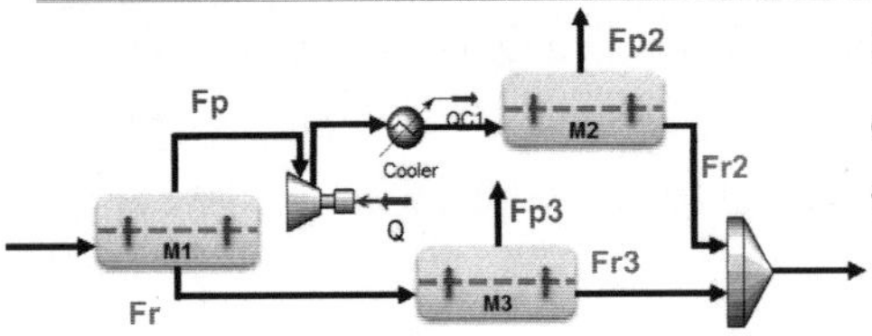

Methane recovery and ROI variation due to changes of the ratio of the membrane size of M2 membrane to that of M1 membrane (M2/M1) with fixed M3/M1 at 1.3 (the most reasonable value by trial and error).

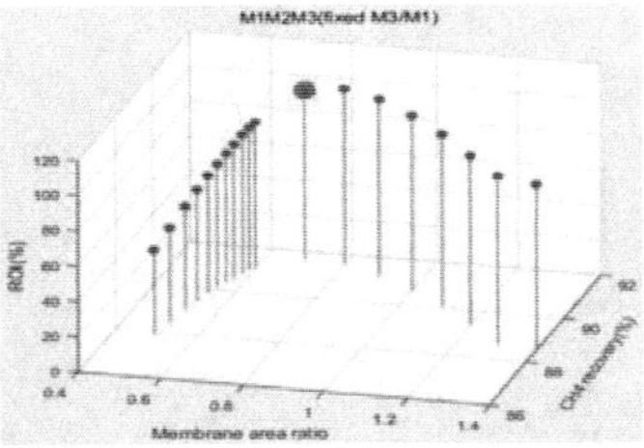

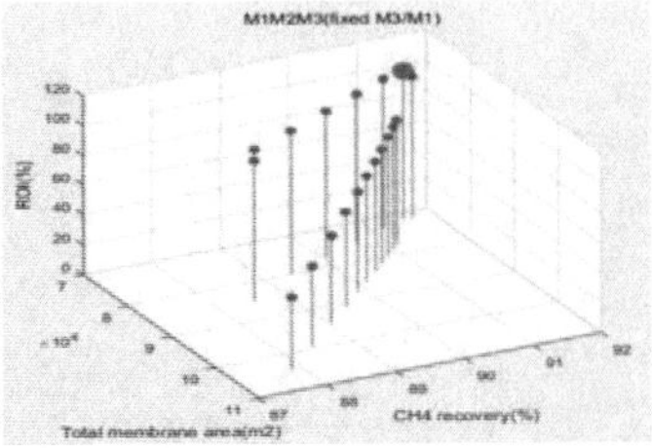

- If M2/M1 value is more than 1.5, the amount of retentate flow of M2 (Fr2) becomes negative (no flow).
- If M2/M1 value is less than 0.5, the membrane size of in M1 increases considerably in order to meet the pipeline specification of CO2 concentration (<2mol%). It causes the total profit to become negative (no profit) because of the capital cost for membrane area.
- The optimum point that shows the highest value of methane recovery with the highest profit is confirmed in M1M2M3 (fixed M3/M1) configuration.

2.3 Results Analysis - (d) M1M2M3 (fixed M2/M1)

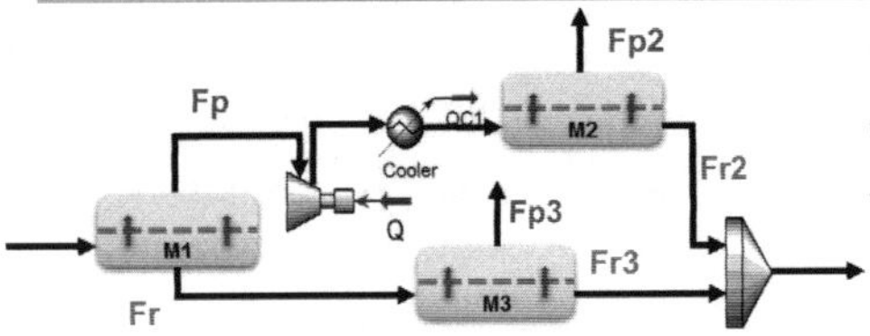

Methane recovery and ROI variation due to changes of the ratio of the membrane size of M3 membrane to that of M1 membrane (M2/M1) with fixed M3/M1 at 0.5 (the most reasonable value by trial and error).

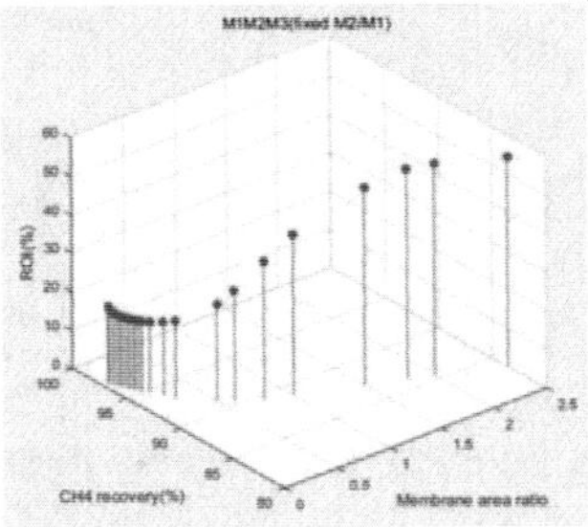

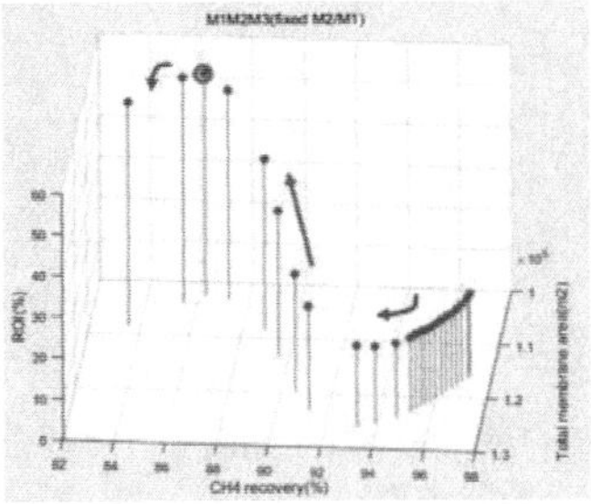

- If M3/M1 value is less than 0.05, the amount of permeate flow of M3 (Fp3l) becomes negative (no flow).
- M1M2M3 (fixed M2/M1) configuration has S-shape (middle section of graph shows inverse relationship between CH4 recovery and ROI) that shows higher profit with lower methane recovery.

2.3 Results Analysis - (d) M1M2M3 (fixed M2/M1)

Membrane area in M1 decreases (**M1↓**) ⟹ Less permeate flow in M1 (**Fp↓**), More retentate flow in M1 (**Fr↑**)

Membrane area in M3 increases (**M3↑**) ⟹ More permeate flow in M3 (**Fp3↑**), Less retentate flow in M3 (**Fr3↓**)

} Less product flow = Less methane recovery = **Less product profit**

<

Decreased total membrane area in both M1 and M2, **-Δ(M1+M2)**

V

Increased membrane area in M3, **Δ(M3)**

} Less membrane area in total membranes **(M1+M2+M3 ↓)** = **Less capital cost**

Total profit increases (ROI ↑)

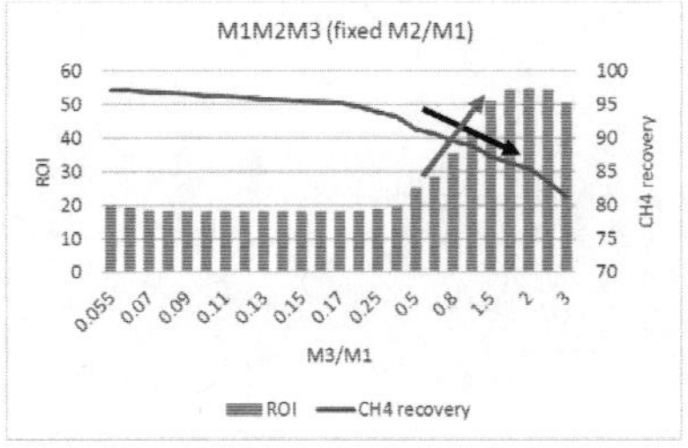

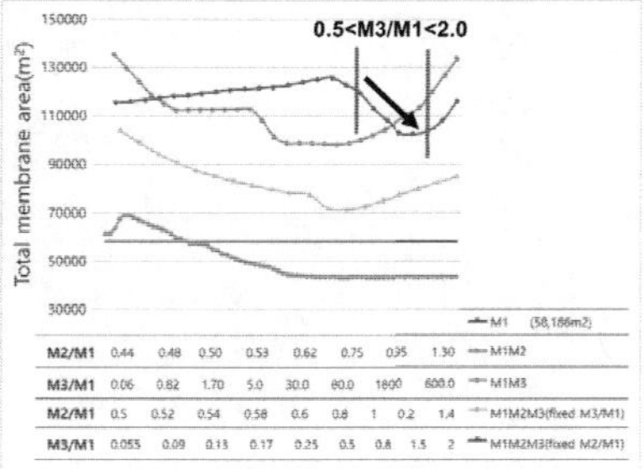

3. Summary

INHA UNIVERSITY

3. Summary

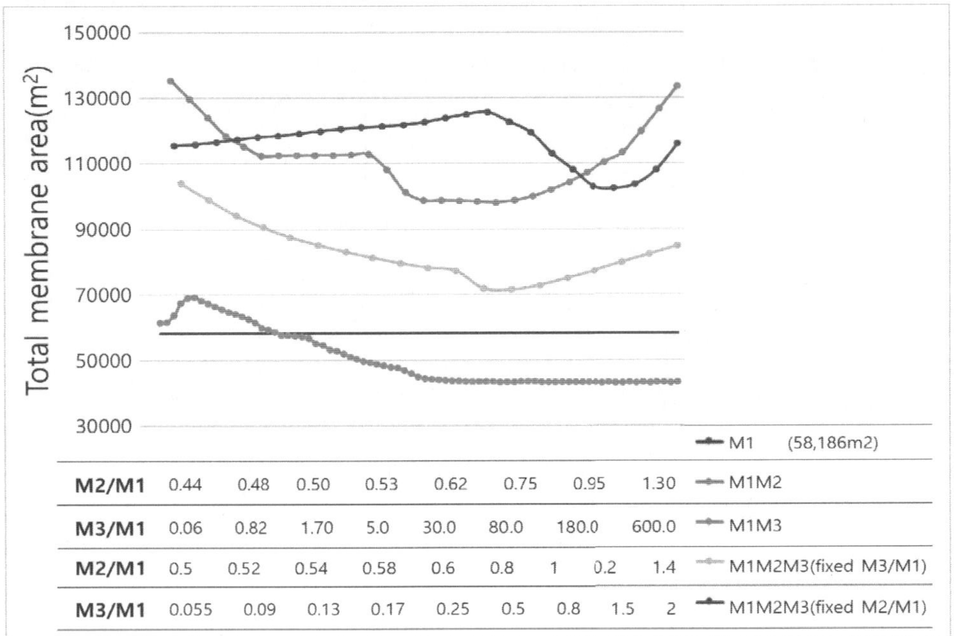

3. Summary

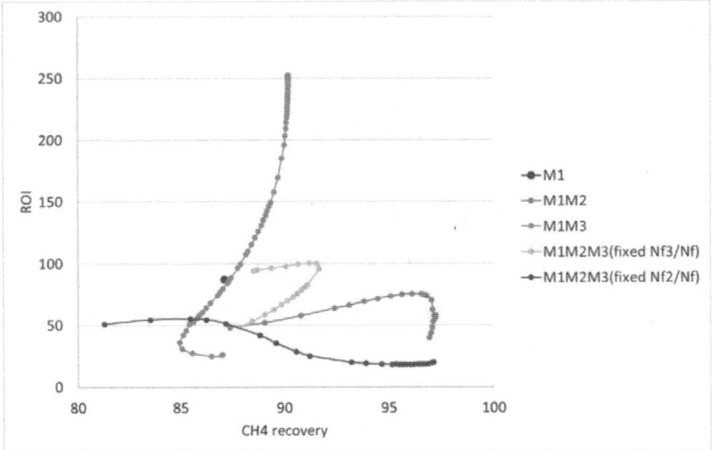

It provides the solution for the optimization of membrane configurations to maximize economic profit depending on required methane recovery meeting pipeline specification such as minimum CO2 concentration.

4. Future works

4. Future works

- Comparison economic analysis of membrane configurations with amine based absorption process including hybrid configuration.
- Continuous development of various configurations for natural gas treatment.

479398 Depressurization of Process Equipment during Emergency or Planned Shutdown on a High Potential CO2-Solid-Forming Stream Utilizing the Latest Dynamic Simulation and Blowdown Models

Abstract:

Natural gas produced from many major reservoirs can contain significant amounts of carbon dioxide (CO2) and must be treated to meet typical specifications for pipelines or liquefaction plant feed. These treatment processes often require high-pressure operations forming highly concentrated CO2-rich streams. Pressure protection for these systems has been challenging to date because of the potential for CO2 solids generation upon pressure let down and the consequent potential for plugging.

Blowdown, the emergency or planned depressuring of process equipment, is a critical process safety operation. It may be necessary, in the event of a fire, leak, pipe rupture or other hazardous situation, as well as for a planned shutdown. Devices such as control valves, relief valves, restriction orifices, rupture disks, and safety valves transfer the potentially dangerous contents of process equipment to a safe lower-pressure location, or to the flare system for controlled combustion.

To ensure blowdown can be executed safety and effectively, a number of design concerns must be addressed, such as solid CO2 identification, low temperature (for both process and equipment material). Rapid depressuring and gas expansion can potentially putting equipment at risk of brittle fracture and if the construction material goes below its ductile-brittle transition temperature as well as potential plugging, due to CO2 solid formation. In addition, the entire pressure relief system, including safety valves, relief orifices, flare piping and knockout drums, must be sufficiently sized to handle the flowrates that occur during blowdown, in addition to the piping and capacity of the flare system.

For new installations, accurately predicting the minimum vessel wall temperature during blowdown is important for selecting the appropriate construction material, for helping reduce overdesign and consequently for lowering project cost. Similarly, having an accurate prediction of the maximum flow rate during blowdown reduces overdesign associated with the relief valve/network, without compromising on safety.

The paper will address the potential of solid CO2 formation based on commercially available software for blowdown and propose some mitigation plan with respect to solid CO2 formation.

Inherently Safer Piping – Best Practices

Steve Streblow
ACS Engineering
10914 Milano Ct.
Richmond, TX 77406
streblowsr@yahoo.com

Prepared for Presentation at
American Institute of Chemical Engineers
2017 Spring Meeting and 13th Global Congress on Process Safety
San Antonio, TX
March 26 – 29, 2017

Inherently Safer Piping – Best Practices

Steve Streblow
ACS Engineering
10914 Milano Ct.
Richmond, TX 77406
streblowsr@yahoo.com

Keywords: 49 CFR 192, 49 CFR 195, API 1102, ASME B31, buried, midstream, pipe, piping, pressure, relief, safety, upstream

Abstract

This paper presents lesser known items in the upstream and midstream industries important to onshore metal piping stress and overpressure protection.

Well known RAGAGEP presented are 49 CFR 195, 49 CFR 192 and the ASME piping codes B31.1/3/4/8. Specific items include corrosion allowance, equations, manufacturing tolerance, maximum allowable pressures, overpressure protection devices, stress values and thread depth. Inspection, testing and maintenance of over pressure protection devices is also discussed. The implications of B31.8 references to B31.1 for water and steam and B31.3 for air, lube oil and hydraulic power are discussed. Many times, design pressures are set below the code minimum design pressures, so those are presented for steam, feedwater, blow piping and drains.

Items detailed outside the CFRs and ASME piping code that should be taken into account, such as surface dead loads, surface live loads, and cyclic loads on buried pipe fatigue are also presented. Failure mechanisms, equations, and acceptance criteria for deflection, ovality, through wall bending, crushing, ring buckling, fatigue, longitudinal, thermal expansion and fluid transients are described. Sources of this information are API Recommended Practice 1102, Steel Pipelines Crossing Railroads; and Highways and American Lifelines Alliance, Guidelines for the Design of Buried Steel Pipe.

Using these presented methods results in inherently safer piping and safer pressure relief system design.

1 Introduction

In having worked with and for many upstream and midstream engineering and operating companies, I have seen that the onshore piping design (for both gases and liquids) is often not done in enough detail. I have always seen that the pipe wall thickness is checked for maximum internal operating pressure, but even then, there are some factors that are often times overlooked.

For gas piping, the category (production, gathering, transmission or distribution) is first determined, and then a determination if the piping is regulated or not is made. The main regulations for gas piping, 49 CFR 192, the minimum requirement, is discussed.

The main regulation for liquid piping, 49 CFR 195, the minimum requirement, is discussed.

Then, some best practices are presented, starting with ASME B31.8 for Gas Piping and Stations such as Compression. The best practice for liquid piping, ASME B31.4 is presented. The sections of ASME B31.1 and B31.3 which are best practices for Compressor station piping are presented.

Some best practices for piping, regardless of fluid and location are presented, starting with API 1102 and then finishing with Guidelines for the Design of Buried Steel Pipe.

Using these documents allows for many more factors such as corrosion allowance, manufacturing tolerance, stress values and thread depth to be taken into account.

Surface dead loads, surface live loads, and cyclic loads on buried pipe fatigue are also presented. Failure mechanisms, equations, and acceptance criteria for deflection, ovality, through wall bending, crushing, ring buckling, fatigue, longitudinal, thermal expansion and fluid transients are described.

Using these presented methods results in inherently safer piping and safer pressure relief system design.

2 What is the Gas Piping Category?

For natural gas piping, one of the first steps is to determine if the pipeline is production, gathering, transmission or distribution. Reference 1, section 2.2, is used to determine the split between production and gathering.

The gathering classification can’t extend beyond the first natural gas processing plant, when it switches to transmission. Transmission then eventually switches to distribution.

3 Regulated Gas Gathering or Not?

In reference 2, there is a parameter called the class location, which accounts for risk due to buildings and occupancy near the pipeline. The class location ranges from 1, for lowest risk to 4 for highest risk. Details in reference 2, paragraph 192.5.

If the gas piping is gathering, one must then determine if it is regulated, per reference 2, paragraph 192.8. Regulated gathering depends on the hoop stress, class location and safety buffer. Regulated gathering lines have to follow certain aspects of reference 2.

4 CFR 192 for Gas Piping – Part 1

So, one needs to follow reference 2 for regulated gathering and transmission piping.

Per reference 2, paragraph "192.103 General. Pipe must be designed with sufficient wall thickness, or must be installed with adequate protection, to withstand anticipated external pressures and loads that will be imposed on the pipe after installation." However, reference 2 does not provide any formulas or methods to satisfy this requirement. This is one area that I have commonly seen overlooked.

Per reference 2, paragraph "192.105 Design formula for steel pipe. (a) The design pressure for steel pipe is determined in accordance with the following formula:"

P = (2 S x t/D) x F x E x T, where

P =Design pressure in pounds per square inch gage.
S =Yield strength in pounds per square inch determined in accordance with §192.107.
t = Nominal wall thickness of the pipe in inches. Additional wall thickness required for concurrent external loads in accordance with §192.103 may not be included in computing design pressure.
D =Nominal outside diameter of the pipe in inches.
F =Design factor determined in accordance with §192.111.
E =Longitudinal joint factor determined in accordance with §192.113.
T =Temperature derating factor determined in accordance with §192.115.

Per reference 2, paragraph "192.107 Yield strength (S) for steel pipe."…"The yield strength to be used in the design formula in §192.105 is the SMYS stated in the listed specification."

Per reference 2, paragraph "192.111 Design factor (F) for steel pipe.

Table 1. Design factor (F) for steel pipe.

Class location	Design factor (F)
1	0.72
2	0.60
3	0.50
4	0.40

A design factor of 0.60 or less must be used in the design formula in paragraph 192.105 for steel pipe in Class 1 locations that: (1) Crosses the right-of-way of an unimproved public road, without a casing: (2) Crosses without a casing, or makes a parallel encroachment on, the right-of-way of either a hard surfaced road, a highway, a public street, or a railroad: (3) Is supported by a vehicular, pedestrian, railroad, or pipeline bridge: or (4) Is used in a fabricated assembly, (including separators, mainline valve assemblies, cross-connections, and river crossing headers) or is used within five pipe diameters in any direction from the last fitting of a fabricated assembly, other than a transition piece or an elbow used in place of a pipe bend which is not associated with a fabricated assembly. (c) For Class 2 locations, a design factor of 0.50, or less, must be used in the design formula in §192.105 for uncased steel pipe that crosses the right-of-way of a hard surfaced road, a highway, a public street, or a railroad. (d) For Class 1 and Class 2 locations, a de-sign factor of 0.50, or less, must be used in the design formula in §192.105 for–(1) Steel pipe in a compressor station, regulating station, or measuring station." These more specific design factors for specific locations and types of facilities is one area that I have commonly seen overlooked.

Per reference 2, "§192.113 Longitudinal joint factor (E) for steel pipe. The longitudinal joint factor to be used in the design formula in §192.105 is determined in accordance with the following table:"

5 E factor comparison

The E factor table from 49 CFR 192 was combined with the E factor table from 49 CFR 195 and the E factor from ASME B31.8 and is presented immediately below. Comparison of the three tables shows that there are a few differences.

Table 2. Factor (E) for steel pipe.

<table>
<tr><th>Specification</th><th>Pipe Class</th><th>49 CFR 192 Longitudinal Joint Factor (E)</th><th>49 CFR 195 Seam Joint Factor (E)</th><th>ASME B31.8</th></tr>
<tr><td rowspan="4">ASTM A53/A53M</td><td>Seamless</td><td colspan="3">1.00</td></tr>
<tr><td>Electric resistance welded</td><td colspan="3">1.00</td></tr>
<tr><td>Furnace lap welded</td><td>NA</td><td>0.80</td><td>NA</td></tr>
<tr><td>Furnace butt welded</td><td colspan="3">0.60</td></tr>
<tr><td>ASTM A106</td><td>Seamless</td><td colspan="3">1.00</td></tr>
<tr><td>ASTM A134</td><td>Electric-fusion arc-welded</td><td colspan="2">NA</td><td>0.80</td></tr>
<tr><td>ASTM A135</td><td>Electric resistance-welded</td><td colspan="2">NA</td><td>1.00</td></tr>
<tr><td>ASTMA139</td><td>Electric-fusion arc-welded</td><td colspan="2">NA</td><td>0.80</td></tr>
<tr><td rowspan="3">ASTM A333/A333M</td><td>Seamless</td><td>NA</td><td>1.00</td><td>1.00</td></tr>
<tr><td>Electric resistance welded</td><td>1.00</td><td>NA</td><td>1.00</td></tr>
<tr><td>Welded</td><td>NA</td><td>1.00</td><td>NA</td></tr>
<tr><td>ASTM A381</td><td>Double submerged arc welded</td><td colspan="3">1.00</td></tr>
<tr><td rowspan="2">ASTM A671</td><td>Electric-fusion welded
Classes 13, 23, 33, 43, 53</td><td colspan="2">1.00</td><td>0.80</td></tr>
<tr><td>Electric-fusion welded
Classes 12, 22, 32, 42, 52</td><td colspan="3">1.00</td></tr>
<tr><td rowspan="2">ASTM A672</td><td>Electric-fusion welded
Classes 13, 23, 33, 43, 53</td><td colspan="2">1.00</td><td>0.80</td></tr>
<tr><td>Electric-fusion welded
Classes 12, 22, 32, 42, 52</td><td colspan="3">1.00</td></tr>
</table>

Table 2. Factor (E) for steel pipe. (continued)

Specification	Pipe Class	49 CFR 192 Longitudinal Joint Factor (E)	49 CFR 195 Seam Joint Factor (E)	ASME B31.8
ASTM A691	Electric-fusion welded Classes 13, 23, 33, 43, 53	1.00		0.80
	Electric-fusion welded Classes 12, 22, 32, 42, 52	1.00		
ASTM A984	Electric-resistance-welded	NA		1.00
ASTM A1005	Double submerged-arc-welded	NA		1.00
ASTM A1006	Laser beam welded	NA		1.00
API 5L	Seamless	1.00		
	Electric resistance welded	1.00		
	Electric flash welded	1.00		
	Submerged arc welded	1.00		
	Furnace lap welded	NA	0.80	NA
	Furnace butt welded	0.60		
Other	Pipe over 4 inches (102 millimeters)	0.80	NA	
Other	Pipe 4 inches (102 millimeters) or less	0.60	NA	

6 CFR 192 for Gas Piping – Part 2

Per reference 2, "192.115 Temperature derating factor (T) for steel pipe. The temperature derating factor to be used in the design formula in §192.105 is determined as follows:

Table 3. Temperature derating factor (T) for steel pipe.

Gas Temperature in degrees Fahrenheit	Temperature derating factor (T)
250 or less	1.000
300	0.967
350	0.933
400	0.900
450	0.867

For intermediate gas temperatures, the derating factor is determined by interpolation."

Per reference 2, "§192.149 Standard fittings. (a) The minimum metal thickness of threaded fittings may not be less than specified for the pressures and temperatures in the applicable standards referenced in this part, or their equivalent." However, reference 2 does not provide any formulas or methods to satisfy this requirement. This is one area that I have commonly seen overlooked.

Per reference 2, "§192.159 Flexibility. Each pipeline must be designed with enough flexibility to prevent thermal expansion or contraction from causing excessive stresses in the pipe or components, excessive bending or unusual loads at joints, or undesirable forces or moments at points of connection to equipment, or at anchorage or guide points." However, reference 2 does not provide any formulas or methods to satisfy this requirement. This is one area that I have commonly seen overlooked.

Per reference 2, "§192.161 Supports and anchors.

(a) Each pipeline and its associated equipment must have enough anchors or supports to: (1) Prevent undue strain on connected equipment; (2) Resist longitudinal forces caused by a bend or offset in the pipe; and, (3) Prevent or damp out excessive vibration.

(b) Each exposed pipeline must have enough supports or anchors to protect the exposed pipe joints from the maximum end force caused by internal pressure and any additional forces caused by temperature expansion or contraction or by the weight of the pipe and its contents.

(c) Each support or anchor on an exposed pipeline must be made of durable, noncombustible material and must be designed and installed as follows: (1) Free expansion and contraction of the pipeline between supports or anchors may not be restricted. (2) Provision must be made for the service conditions involved." However, reference 2 does not provide any formulas or methods to satisfy this requirement. This is one area that I have commonly seen overlooked.

It is important to note that reference 2 describes minimum standards, not best practices. Read on to learn about best practices.

7 CFR 195 for Liquid Piping – Part 1

Hazardous liquids transportation pipelines are often required to follow reference 3 for several aspects, including design.

From reference 3, "§195.1 Which pipelines are covered by this part?" "(2) Transportation through any pipeline, other than a gathering line, that has a maximum operating pressure (MOP) greater than 20-percent of the specified minimum yield strength;" "(4) Transportation of petroleum in any of the following onshore gathering lines: (i) A pipeline located in a non-rural area; (ii) To the extent provided in §195.11, a regulated rural gathering line defined in §195.11."

From reference 3, "(b) Excepted. This part does not apply to any of the following:" "(8) Transportation of a hazardous liquid or carbon dioxide through onshore production (including flow lines), refining, or manufacturing facilities or storage or inplant piping systems associated with such facilities;"

From reference 3, "§195.11 What is a regulated rural gathering line and what requirements apply?...(a) Definition. As used in this section, a regulated rural gathering line means an onshore gathering line in a rural area that meets all of the following criteria—(1) Has a nominal diameter from 6⅝ inches (168 mm) to 8⅝ inches (219.1 mm);...(3) Operates at a maximum pressure established under §195.406 corresponding to—(i) A stress level greater than 20-percent of the specified minimum yield strength of the line pipe"

From reference 3, "§195.106 Internal design pressure. (a) Internal design pressure for the pipe in a pipeline is determined in accordance with the following formula:
$P = (2\ S \times t/D) \times E \times F$", where

- P = Internal design pressure in p.s.i. gage.
- S = Yield strength in pounds per square inch determined in accordance with paragraph (b) of this section.
- t = Nominal wall thickness of the pipe in inches.
- D = Nominal outside diameter of the pipe in inches.
- E = Seam joint factor determined in accordance with paragraph (e) of this section.
- F = A design factor of 0.72

8 Comparing Wall Thickness for Gas and Liquid Piping

Comparing the internal design pressure formulas of 49 CFR 192 and 195 shows that the formula are slightly different. 49 CFR 192 design factors range from 0.72 to 0.4, while 49 CFR 195 design factor is set at 0.72. Therefore it can be seen that for a given internal design pressure, with the same piping, the gas piping oftentimes needs to have a thicker wall. 49 CFR 192 temperature derating factor ranges from 1.000 to 0.867. while 49 CFR 195 does not have a temperature derating factor, therefore, for a given operating temperature with the same piping, the gas piping needs to have a thicker wall.

9 CFR 195 for Liquid Piping – Part 2

From reference 3, "(b) The yield strength to be used in determining the internal design pressure under paragraph (a) of this section is the specified minimum yield strength."

From reference 3, "(d) The minimum wall thickness of the pipe may not be less than 87.5 percent of the value used for nominal wall thickness in determining the internal design pressure under paragraph (a) of this section. In addition, the anticipated external loads and external pressures that are concurrent with internal pressure must be considered in accordance with §§195.108 and 195.110 and, after determining the internal design pressure, the nominal wall thickness must be increased as necessary to compensate for these concurrent loads and pressures." However, reference 3 does not provide any formulas or methods to satisfy this requirement. This is one area that I have commonly seen overlooked.

From reference 3, "(e) The seam joint factor used in paragraph (a) of this section is determined in accordance with the following table:" See Table 2 above.

Per reference 3, "§195.108 External pressure. Any external pressure that will be exerted on the pipe must be provided for in designing a pipeline system." However, reference 3 does not provide any formulas or methods to satisfy this requirement. This is one area that I have commonly seen overlooked.

Per reference 3, "§195.110 External loads. (a) Anticipated external loads (e.g.), earthquakes, vibration, thermal expansion, and contraction must be provided for in designing a pipeline system. In providing for expansion and flexibility, §419 of ASME/ANSI B31.4 must be followed. (b) The pipe and other components must be supported in such a way that the support does not cause excess localized stresses. In designing attachments to pipe, the added stress to the wall of the pipe must be computed and compensated for." However, reference 3 does not provide any formulas or methods to satisfy this requirement. I This is one area that I have commonly seen overlooked.

It is important to note that reference 3 describes minimum standards, not best practices. Read on to learn about best practices.

10 Best Practice For Gas Piping and Stations such as Compression – ASME B31.8

What if the gas pipeline or facility, such as production or non-regulated gathering, is not subject to reference 2? How does one account for factors and loads not detailed in CFR 192? One can then follow reference 4, which has the following applicability "piping transporting products that are predominately gas between sources and terminals, including compressor, regulating, and metering stations; gas gathering pipelines."

From reference 4, paragraph "802.1 Scope…(b) This Code does not apply to…(4) piping in oil refineries or natural gasoline extraction plants, gas treating plant piping other than the main gas stream piping in dehydration, and all other processing plants installed as part of a gas transmission system, gas manufacturing plants, industrial plants, or mines (See other applicable sections of the ASME Code for Pressure Piping, B31.) (5) vent piping to operate at substantially atmospheric pressures for waste gases of any kind."

From reference 4, paragraph "833.2 Calculation of Longitudinal Stress Components
(a) The longitudinal stress due to internal pressure in restrained pipelines is
$S_P = 0.3S_H$, where S_H is the hoop stress, psi
(b) The longitudinal stress due to internal pressure in unrestrained pipeline is
$S_P = 0.5S_H$, where S_H is the hoop stress, psi
(c) The longitudinal stress due to thermal expansion in restrained pipe is $S_T = E\ a\ (T_1 - T_2)$, where
E = the elastic modulus, psi, at the ambient temperature, which can be found in reference 4, Table 832.5-1.
T_1 = the pipe temperature at the time of installation, tie-in, or burial, °F
T_2 = the warmest or coldest pipe operating temperature, °F
a = the coefficient of thermal expansion, 1/°F, which can be found in reference 4, Table 832.1-1
If a section of pipe can operate either warmer or colder than the installed temperature, both conditions for T_2 may need to be examined." This is one area that I have commonly seen overlooked.

From reference 4, paragraph "833.3 Summation of Longitudinal Stress in Restrained Pipe
(a) The net longitudinal stresses in restrained pipe are $S_L = S_P + S_T + S_X + S_B$
Note that S_B, S_L, S_T, or S_X can have negative values.
(b) The maximum permitted value of $|S_L|$ is 0.9ST, where S is the specified minimum yield strength, psi, per para. 841.1.1(a), and T is the temperature derating factor per para. 841.1.8." This is one area that I have commonly seen overlooked.

From reference 4, paragraph "833.4 Combined Stress for Restrained Pipe
(a) The combined biaxial stress state of the pipeline in the operating mode is evaluated using the calculation in either (1) or (2) below. (1) max. $\{|S_H - S_L|, |S_H|, |S_L|\}$ or (2) $[S_L^2 - S_LS_H + S_H^2]^{1/2}$
The maximum permitted value for the combined biaxial stress is kST where S is the specified minimum yield strength, psi, per para. 841.1.1(a), T is the temperature derating factor per para. 841.1.8, and k is defined in (b) and (c) below. (b) For loads of long duration, the value of k shall not exceed 0.90. (c) For occasional nonperiodic loads of short duration, the value of k shall not exceed 1.0. (d) S_L in (a) above is calculated considering both the tensile and compressive values of S_B. (e) Stresses induced by loads that do not occur simultaneously need not be considered to be additive. (f) The biaxial stress evaluation described above applies only to straight sections of pipe." This is one area that I have commonly seen overlooked.

From reference 4, paragraph "833.6 Summation of Longitudinal Stresses in Unrestrained Pipe
(a) The net longitudinal stress in unrestrained pipe is $S_L = S_P + S_X + S_B$, psi (b) The maximum permitted longitudinal stress in unrestrained pipe is $S_L \leq 0.75ST$, where S is the specified minimum yield strength, psi, per para. 841.1.1(a), and T is the temperature derating factor per para. 841.1.8." This is one area that I have commonly seen overlooked.

From reference 4, paragraph "833.8 Flexibility Stresses and Stresses Due to Periodic or Cyclic Fatigue Loading (b) The cyclic stress range $S_E <= S_A$, where
$S_A = f\,[1.25\,(S_c + S_h) - S_L]$, where
$f = 6N^{-0.2} \leq 1.0$
N = equivalent number of cycles during the expected service life of the piping system
$S_c = 0.33\ S_uT$ at the minimum installed or operating temperature
$S_h = 0.33\ S_uT$ at the maximum installed or operating temperature
S_L = longitudinal stress calculated according to para. 833.6(a), psi
S_u = specified minimum ultimate tensile strength, psi
T = temperature derating factor per para. 841.1.8"
This is one area that I have commonly seen overlooked.

Per reference 4, paragraph "833.9 Local Stresses (b) The maximum allowable sum of circumferential stress due to internal pressure and circumferential through-wall bending stress caused by surface vehicle loads or other local loads is 0.9ST, where S is the specified minimum yield strength, psi, per para. 841.1.1(a), and T is the temperature derating factor per para. 841.1.8." This is one area that I have commonly seen overlooked.
Per reference 4, paragraph "833.10 Buckling and Lateral Instability. In order to prevent buckling in the form of wrinkling of the pipe wall or lateral instability, the maximum allowable net compressive stress is 2⁄3 of the critical buckling stress estimated using a suitable stability criterion." This is one area that I have commonly seen overlooked.

Per reference 4, see "Table 841.1.6-2 Design Factors for Steel Pipe Construction". Consider these in addition to those from reference 2.

From reference 4, paragraph "841.1.9 Additional Design Information or Instructions…
(g) Pressure/Flow Control Facilities…
(1) Overpressure protection shall be provided by the use of (-a) a monitor regulator in series with a controlling regulator (each regulator run). (-b) adequately sized relief valve(s) downstream of the controlling regulator(s). (-c) overpressure shutoff valve(s) upstream or downstream of the controlling regulator(s). Installation of alarm devices that indicate primary (controlling) regulator failure is useful and should be considered for monitoring regulator systems…
(k) Other (Nonmandatory) Considerations for Pressure Cycle Fatigue of Longitudinal Pipe Seams. Line pipe longitudinal seams are not generally regarded as susceptible to fatigue due to operational pressure cycles in most natural gas service. Evaluation of the potential for fatigue crack growth due to pressure cycles is recommended for pipe containing longitudinal seams where the expected lifetime accumulation of full MAOP cycles may exceed the following number of occurrences: $N = 6.0 \times 10^{17} / ((F \times S)^3 \times t^{0.5})$ where
F = design factor from Table 841.1.6-1
N = equivalent number of cycles during the expected service life of the piping system
S = specified minimum yield strength, psi
t = nominal wall thickness, in."
This is one area that I have commonly seen overlooked.

From reference 4, paragraph "843.3.4 Pressure-Limiting Requirements in Compressor Stations

(a) Pressure Relief. Pressure relief or other suitable protective devices of sufficient capacity and sensitivity shall be installed and maintained to ensure that the maximum allowable operating pressure of the station piping and equipment is not exceeded by more than 10%.
(b) Pressure Relief Valves. A pressure relief valve or pressure-limiting device, such as a pressure switch or unloading device, shall be installed in the discharge line of each positive displacement transmission compressor between the gas compressor and the first discharge block valve. If a pressure relief valve is the primary overprotection device, then the relieving capacity shall be equal to or greater than the capacity of the compressor.
(c) Venting. Vent lines provided to exhaust the gas from the pressure relief valves to atmosphere shall be extended to a location where the gas may be discharged without undue hazard. Vent lines shall have sufficient capacity so that they will not inhibit the performance of the relief valve. For additional design considerations, see para. 841.1.9(d)." This is one area that I have commonly seen overlooked.

From reference 4, paragraph "843.4.3 Air Piping System. (a) All air piping within gas compressing stations shall be constructed in accordance with ASME B31.3". This is one area that I have commonly seen overlooked.

From reference 4, paragraph "843.4.4 Lubricating Oil Piping. All lubricating oil piping within gas compressing stations shall be constructed in accordance with ASME B31.3." This is one area that I have commonly seen overlooked.

From reference 4, paragraph "843.4.5 Water Piping. All water piping within gas compressing stations shall be constructed in accordance with ASME B31.1." This is one area that I have commonly seen overlooked.

From reference 4, paragraph "843.4.6 Steam Piping. All steam piping within gas compressing stations shall be constructed in accordance with ASME B31.1." This is one area that I have commonly seen overlooked.

From reference 4, paragraph "843.4.7 Hydraulic Piping. All hydraulic power piping within gas compressing stations shall be constructed in accordance with ASME B31.3." This is one area that I have commonly seen overlooked.

From reference 4, paragraph "844.5 General Provisions Applicable to Both Pipe-Type and Bottle-Type Holders…Relief valves shall be installed in accordance with provisions of this Code that will have relieving capacity adequate to limit the pressure imposed on the filling line and thereby on the storage holder to 100% of the design pressure of the holder or to the pressure that produces a hoop stress of 75% of the specified minimum yield strength of the steel, whichever is the lesser." This is one area that I have commonly seen overlooked.

Paragraph "845.3 Requirements for Design of Pressure Relief and Pressure-Limiting Installations. (c) The discharge stacks, vents, or outlet ports of all pressure relief devices shall be located where gas can be discharged into the atmosphere without undue hazard." This is one area that I have commonly seen overlooked.

From reference 4, paragraph "845.4.1 Required Capacity of Pressure-Relieving and Pressure-Limiting Stations...(a) Each pressure relief station, pressure-limiting station, or group of such stations installed to protect a piping system or pressure vessel shall have sufficient capacity and shall be set to operate to prevent the pressure from exceeding the following levels: (1) Systems With Pipe or Pipeline Components Operating at Hoop Stress Levels Over 72% of the SMYS. The required capacity is the maximum allowable operating pressure plus 4%. (2) Systems With Pipe or Pipeline Components Operating at Hoop Stress Levels at or Below 72% of the SMYS Other Than in Low-Pressure Distribution Systems. The required capacity is the lesser of the following two items: (-a) the maximum allowable operating pressure plus 10% (-b) the pressure that produces a hoop stress of 75% of the specified minimum yield strength." This is one area that I have commonly seen overlooked.

From reference 4, paragraph "845.4.2 Proof of Adequate Capacity and Satisfactory Performance of Pressure-Limiting and Pressure Relief Devices. Where the safety device consists of an additional regulator that is associated with or functions in combination with one or more regulators in a series arrangement to control or limit the pressure in a piping system, suitable checks shall be made. These checks shall be conducted to determine that the equipment will operate in a satisfactory manner to prevent any pressure in excess of the established maximum allowable operating pressure of the system, should any one of the associated regulators malfunction or remain in the wide open position." This is one area that I have commonly seen overlooked.

From reference 4, paragraph "849.2 Steel Service Lines. (b) Steel service pipe shall be designed in accordance with the requirements of paras. 841.1.1 and 841.1.9(a). Where the pressure is less than 100 psig, the steel service pipe shall be designed for at least 100 psig pressure." This is one area that I have commonly seen overlooked.

From reference 4, paragraph "853.1 Compressor Station Maintenance...853.1.2 Inspection and Testing of Relief Valves. All pressure-relieving devices in compressor stations shall be inspected and/or tested in accordance with para. 853.3, and all devices except rupture disks shall be operated periodically to determine that they open at the correct set pressure. Any defective or inadequate equipment found shall be promptly repaired or replaced. All remote control shutdown devices shall be inspected and tested at least annually to determine that they function properly." This is one area that I have commonly seen overlooked.

From reference 4, paragraph "853.3 Maintenance of Pressure-Limiting and Pressure-Regulating Stations...853.3.1 Condition and Adequacy. All pressure limiting stations, relief devices, and other pressure regulating stations and equipment shall be subject to systematic, periodic inspections and suitable tests, or reviewed to determine that they are...(2) At least once each calendar year, a review shall be made to ensure that the combined capacity of the relief devices on a piping system or facility is adequate to limit the pressure at all times to values prescribed by this Code. This review should be based on the operating conditions that create the maximum probable requirement for relief capacity in each case, even though such operating conditions actually occur infrequently and/or for only short periods of time. If it is determined that the relieving equipment is of insufficient capacity, steps shall be taken to install new or additional equipment to provide adequate capacity." This is one area that I have commonly seen overlooked.

From reference 4, paragraph "853.3.3 Stop Valves (a) An inspection and/or test of stop valves shall be made to determine that the valves will operate and are correctly positioned. (Caution shall be used to avoid any undesirable effect on pressure during operational checks.) The following shall be included in the inspection and/or test: (1) station inlet, outlet, and bypass valves (2) relief device isolating valves (3) control, sensing, and supply line valves (b) The final inspection procedure shall include the following: (1) a check for proper position of all valves. Special attention shall be given to regulator station bypass valves, relief device isolating valves, and valves in control, sensing, and supply lines. (2) restoration of all locking and security devices to proper position." This is one area that I have commonly seen overlooked.

From Reference 4, see "MANDATORY APPENDIX D, SPECIFIED MINIMUM YIELD STRENGTH FOR STEEL PIPE, COMMONLY USED IN PIPING SYSTEMS"

11 Best Practice for Liquid Piping – ASME B31.4

What if the liquid pipeline or facility, such as production or non-regulated gathering, is not subject to reference 3? How does one account for factors and loads not detailed in CFR 195? One can then follow reference 5, which has the following applicability "piping transporting products that are predominately liquid between plants and terminals and within terminals, pumping, regulating, and metering stations."

Reference 5 shows that various categories of loads need to be considered, such as paragraph "401.1.2 Sustained Loads…401.1.3 Occasional Loads…401.1.4 Construction Loads and 401.1.5 Transient Loads." However, reference 5 does not provide any formulas or methods to satisfy this requirement. This is one area that I have commonly seen overlooked.

Reference 5 shows some subcategories of loads needing consideration, such as paragraph "401.2.2.2 Internal Design Pressure…401.2.2.3 External Hydrostatic Pressure…401.2.2.4 Weight Effects…401.2.2.5 Residual Loads…401.2.2.6 Subsidence…401.2.3.1 Earthquakes, 401.2.3.2 Wind Loads…401.2.3.3 Ice Loads…401.2.3.4 Road and Rail Traffic…401.2.3.5 Vibration…401.2.3.6 Waves and Currents…401.2.3.7 Temperature Effects…401.2.4.1 Installation Loads and 401.2.4.2 Hydrostatic Testing." However, reference 5 does not provide any formulas or methods to satisfy this requirement. This is one area that I have commonly seen overlooked.

From reference 5, paragraph "401.3 Combining of Loads" needs to be considered. This is one area that I have commonly seen overlooked.

From reference 5, paragraph "402 CALCULATION OF STRESSES...402.1 General...Circumferential, longitudinal, shear, and equivalent stresses shall be considered, taking into account stresses from all relevant sustained, occasional, construction, and transient loads, including vibration, resonance, and subsidence." This is one area that I have commonly seen overlooked.

From reference 5, paragraph "403.2 Criteria for Pipe Wall Thickness and Allowances...403.2.1 Criteria. The nominal wall thickness of straight sections of steel pipe shall be equal to or greater than t_n determined in accordance with the following equation: $t_n \geq t + A$, where
A = sum of allowances for threading, grooving, corrosion, and erosion as required in paras. 403.2.2 through 403.2.4, and increase in wall thickness if used as protective measure in para. 403.1
t_n = nominal wall thickness satisfying requirements for pressure and allowances
t = pressure design wall thickness"

This is one area that I have commonly seen overlooked.

In reference 5, see "Table 403.3.1-1 Allowable Values for Pipeline System Stresses." This is one area that I have commonly seen overlooked.

From reference 5, Paragraph "452.2 Controls and Protective Equipment (a) Controls and protective equipment, including pressure-limiting devices, regulators, controllers, relief valves, and other safety devices, shall be subjected to systematic periodic inspections and tests, at least annually, except as provided in para. 452.2(b), to determine that they are (1) in good mechanical condition (2) adequate from the standpoint of capacity and reliability of operation for the service in which they are employed (3) set to function at the correct pressure (4) properly installed and protected from foreign materials or other conditions that might prevent proper operation (b) Relief valves on pressure storage vessels containing LPG or liquid anhydrous ammonia shall be subjected to tests at least every 5 yr." This is one area that I have commonly seen overlooked.

12 Best Practice for Compressor Station Piping- ASME B31.1

From reference 4, paragraph "843.4.5 Water Piping. All water piping within gas compressing stations shall be constructed in accordance with ASME B31.1." This is one area that I have commonly seen overlooked.

From reference 4, paragraph "843.4.6 Steam Piping. All steam piping within gas compressing stations shall be constructed in accordance with ASME B31.1." This is one area that I have commonly seen overlooked.

So, to follow the best practice for compressor station water liquid and vapor piping, refer to reference 7, which has the following applicability "piping typically found in electric power generating stations, in industrial and institutional plants, geothermal heating systems, and central and district heating and cooling systems."

From reference 7, paragraph, "102.4.2 Threading and Grooving. The calculated minimum thickness of piping (or tubing) which is to be threaded shall be increased by an allowance equal to thread depth; dimension h of ASME B1.20.1 or equivalent shall apply. For machined surfaces or grooves, where the tolerance is not specified, the tolerance shall be assumed to be 1⁄64 in. (0.40 mm) in addition to the specified depth of cut. The requirements of para. 104.1.2(C) shall also apply." This is one area that I have commonly seen overlooked.

From reference 7, see "Table 102.4.3 Weld Joint Efficiency Factors"

From reference 7, paragraph "104.1.2 Straight Pipe Under Internal Pressure, Design pressure shall not exceed $P = 2SE(t_m - A) / (D_o - 2y(t_m - A))$, where
P = internal design pressure, psig
SE = maximum allowable stress in material due to internal pressure and joint efficiency at the design temperature, psi. The value of SE shall not exceed that given in Appendix A, for the respective material and design temperature. These values include the weld joint efficiency E.
tm = minimum required wall thickness, in. If pipe is ordered by its nominal wall thickness, the manufacturing tolerance on wall thickness must be taken into account. After the minimum pipe wall thickness tm is determined, this minimum thickness shall be increased by an amount sufficient to provide the manufacturing tolerance allowed in the applicable pipe specification or required by the process. The next heavier commercial wall thickness shall then be selected from thickness schedules such as contained in ASME B36.10M or from manufacturers' schedules for other than standard thickness. (A.1.4) Where ends are subject to forming or machining for jointing, the wall thickness of the pipe, tube, or component after such forming or machining shall not be less than tm minus the amount provided for removal by para. 104.1.2 (A.6.1).
A = additional thickness, in. to compensate for material removed in threading, grooving, etc., required to make a mechanical joint.
Do = outside diameter of pipe, in. For design calculations, the outside diameter of pipe as given in tables of standards and specifications shall be used in obtaining the value of tm. When calculating the allowable working pressure of pipe on hand or in stock, the actual measured outside diameter and actual measured minimum wall thickness at the thinner end of the pipe may be used to calculate this pressure.
y = coefficient having values as given in Table 104.1.2(A)". This is a temperature derating factor.

From reference 7, "(C.1) Where steel pipe is threaded and used for steam service at pressure above 250 psi or for water service above 100 psi with water temperature above 220°F, the pipe shall be seamless having the minimum ultimate tensile strength of 48,000 psi and a weight at least equal to Schedule 80 of ASME B36.10M." This is one area that I have commonly seen overlooked.

From reference 7, paragraph "104.1.3 Straight Pipe Under External Pressure. For determining wall thickness and stiffening requirements for straight pipe under external pressure, the procedures outlined in UG-28, UG-29, and UG-30 of Section VIII, Division 1 of the ASME Boiler and Pressure Vessel Code shall be followed." This is one area that I have commonly seen overlooked.

From reference 7, see paragraph "104.8.1 Stress Due to Sustained Loads formula." This is one area that I have commonly seen overlooked.

From reference 7, see paragraph "104.8.2 Stress Due to Occasional Loads formula." This is one area that I have commonly seen overlooked.

From reference 7, see paragraph "104.8.3 Thermal Expansion Stress Range." This is one area that I have commonly seen overlooked.

From reference 7, paragraph "122.1.2 Steam Piping. The design pressure shall not be taken at less than 100 psig for any condition of service or material." This is one area that I have commonly seen overlooked.

From reference 7, paragraph "122.1.3 Feedwater Piping. For piping from the boiler to and including the required stop valve and the check valve, the minimum value of P except as permitted in para. 122.1.3(A.4) shall exceed the maximum allowable working pressure of the boiler by either 25% or 225 psi, whichever is the lesser. (A.3) The value of P in the formula shall not be taken at less than 100 psig for any condition of service or material, and shall never be less than the pressure required to feed the boiler." This is one area that I have commonly seen overlooked.

From reference 7, paragraph "122.1.4 Blowoff and Blowdown Piping. (A.1) The value of P to be used in the formulas in para. 104 shall exceed the maximum allowable working pressure of the boiler by either 25% or 225 psi whichever is less, but shall be not less than 100 psig." This is one area that I have commonly seen overlooked.

From reference 7, paragraph "122.1.5 Boiler Drains. (D) Drain piping from the drain connection, including the required valve(s) or the blanked flange connection, shall be designed for the temperature and pressure of the drain connection. The remaining piping shall be designed for the expected maximum temperature and pressure. Static head and possible choked flow conditions shall be considered. In no case shall the design pressure and temperature be less than 100 psig and 220°F, respectively." This is one area that I have commonly seen overlooked.

From reference 7, see "NONMANDATORY APPENDIX II - RULES FOR THE DESIGN OF SAFETY VALVE INSTALLATIONS....II-2.3 Reaction Forces From Valve Discharge" and "II-3.0 BENDING MOMENT COMPUTATIONS". These are two areas that I have commonly seen overlooked.

From reference 7, see "NONMANDATORY APPENDIX VII - PROCEDURES FOR THE DESIGN OF RESTRAINED UNDERGROUND PIPING." This is one area that I have commonly seen overlooked.

13 Best Practice for Compressor Station Piping- ASME B31.3

From reference 4, paragraph "843.4.3 Air Piping System. (a) All air piping within gas compressing stations shall be constructed in accordance with ASME B31.3". This is one area that I have commonly seen overlooked.

From reference 4, paragraph "843.4.4 Lubricating Oil Piping. All lubricating oil piping within gas compressing stations shall be constructed in accordance with ASME B31.3." This is one area that I have commonly seen overlooked.

From reference 4, paragraph "843.4.7 Hydraulic Piping. All hydraulic power piping within gas compressing stations shall be constructed in accordance with ASME B31.3." This is one area that I have commonly seen overlooked.

So, to follow the best practice for compressor station air, lube oil and hydraulic piping, refer to reference 6, which has the following applicability "piping typically found in petroleum refineries; chemical, pharmaceutical, textile, paper, semiconductor, and cryogenic plants; and related processing plants and terminals."

From reference 6, see "NONMANDATORY APPENDIX II" for piping material information.

From reference 6, see "Table 302.3.4 Weld Quality Factor."

From reference 6, paragraph "304.1.1 General (a) The required thickness of straight sections of pipe shall be determined in accordance with $tm = t + c$, where

tm = minimum required thickness, including mechanical, corrosion, and erosion allowances
t = pressure design thickness, as calculated in accordance with para. 304.1.2 for internal pressure or as determined in accordance with para. 304.1.3 for external pressure
c = sum of the mechanical allowances (thread or groove depth) plus corrosion and erosion allowances. For threaded components, the nominal thread depth (dimension h of ASME B1.20.1, or equivalent) shall apply. For machined surfaces or grooves where the tolerance is not specified, the tolerance shall be assumed to be 0.5 mm (0.02 in.) in addition to the specified depth of the cut."

From reference 6, paragraph "304.1.2 Straight Pipe Under Internal Pressure
(a) For $t < D/6$, the internal pressure design thickness for straight pipe shall be not less than that calculated in accordance with eq. $t = PD / (2(SEW + PY))$ where
P = internal design gage pressure
D = outside diameter of pipe
S = stress value for material from Table A-1
E = quality factor from Table A-1A or A-1B
W = weld joint strength reduction factor in accordance with para. 302.3.5(e)
Y = coefficient from Table 304.1.1, valid for $t < D/6$ and for materials shown. The value of Y may be interpolated for intermediate temperatures. For $t \geq D/6$, $Y = (d + 2c) / (D + d + 2c)$ where
d = inside diameter of pipe. For pressure design calculation, the inside diameter of the pipe is

the maximum value allowable under the purchase specification.
(b) For $t \geq D/6$ or for $P/SE > 0.385$, calculation of pressure design thickness for straight pipe requires special consideration of factors such as theory of failure, effects of fatigue, and thermal stress."

From reference 6, See "Table 314.2.1 Minimum Thickness of External Threaded Components." This is one area that I have commonly seen overlooked.

14 Best Practice for Piping – API 1102

API 1102, reference 9, is RAGAGEP that I have commonly seen overlooked.

From reference 9, "Table 2" is a good flowchart of the design process considering loads, stresses and fatigue.

From reference 9, paragraph "4.7.2.1 Stresses Due to Earth Load" has a calculation formula (1). Use Figures 3, 4 and 5 to get values to use in formula (1).

From reference 9, paragraph "4.7.2.2.1 Surface Live Loads", has guidance on rail road and highway road loads. Use equation (2) to determine the applied surface design pressure.

From reference 9, paragraph "4.7.2.2.2 Impact Factor" has guidance on impact factor to be used with live loads.

From reference 9, paragraph "4.7.2.2.3.1" has a has a calculation formula (3) for cyclic circumferential stress due to rail load. Use Figures 7, 8 and 9 to get the values to use in formula (3).

Reference 9, paragraph "4.7.2.2.3.2" has a has a calculation formula (4) for cyclic longitudinal stress for rail load. Use Figures 10, 11, 12 and 13 to get the values to use in formula (4).

Reference 9, paragraph "4.7.2.2.4.1" has a has a calculation formula (5) for cyclic circumferential stress for highway load. Use Table 2, Figures 14 and 15 to get the values to use in formula (5).

Reference 9, paragraph "4.7.2.2.4.2" has a has a calculation formula (6) for longitudinal circumferential stress for highway load. Use Figures 16 and 17 to get the values to use in formula (6).

Reference 9, paragraph "4.7.3" has a has a calculation formula (7) for stress from internal load.

Reference 9, paragraph "4.8.1.1" has a has a calculation formula (8a) to compare the circumferential stress due to internal gas pressure to the factored specified minimum yield strength. It also has a has a calculation formula (8b) to compare the circumferential stress due to internal liquid pressure to the factored specified minimum yield strength.

Reference 9, paragraph "4.8.1.2" has a has a calculation formula (9) to determine the maximum circumferential stress. It also has a calculation formula (10) to determine the maximum longitudinal stress. It also has a calculation formula (11) to determine the maximum radial stress. It also has a calculation formula (12) to determine the total effective stress. It also has a has a calculation formula (13) to compare the total effective stress to the factored specified minimum yield strength.

Reference 9, paragraph "4.8.2.1" has a has a calculation formula (14) to compare design to the girth weld fatigue. Table 3 is used with formula 14.

Reference 9, paragraph "4.8.2.1.1.1" has a has a calculation formula (15) to compare design to the girth weld fatigue for a railroad crossing for distance less than 5 feet.

Reference 9, paragraph "4.8.2.1.1.2" has a has a calculation formula (16) to compare design to the girth weld fatigue for a railroad crossing for distance 5 feet and greater. Figures 18A and B are used with formula 16.

Reference 9, paragraph "4.8.2.1.2" has a has a calculation formula (17) to compare design to the girth weld fatigue for a highway crossing.

Reference 9, paragraph "4.8.2.2.1" has a has a calculation formula (18) to compare design to the longitudinal weld fatigue.

Reference 9, paragraph "4.8.2.2.2" has a has a calculation formula (19) to compare design to the longitudinal weld fatigue for a railroad crossing.

Reference 9, paragraph "4.8.2.2.3" has a has a calculation formula (20) to compare design to the longitudinal weld fatigue for a highway crossing.

Reference 9, "Appendix B" has example design problems.

15 Best Practice for Piping – Guidelines for the Design of Buried Steel Pipe

Guidelines for the Design of Buried Steel Pipe, reference 10, is RAGAGEP that I have commonly seen overlooked.

Reference 10 has some additional useful information on piping design. Section "4.1" covers loads due to airport. Section "4.2" covers ovality, through wall bending, crushing of side walls and ring buckling. Section "5.1" covers maximum impact load. Section "5.2" covers penetration. Section "6" covers buoyancy. Section "11" covers earthquake. Section "12" covers nearby blasting. Section "13" covers fluid transients. "Appendix A" covers suggested acceptance criteria for stress and strain for 13 different loading conditions.

16 Conclusions

This paper presented design portions of CFRs for onshore metal pipelines. It guided the reader through the selection process of production/gathering and regulated/non-regulated in conjunction with API 80. It them went through design aspects of ASME B31.1/3/4/8, some of the equations and factors and the interrelationships between the ASME pipe codes, in particular for compressor stations. There are various loads and stresses that the CFRs and ASME pipe codes require to be considered, but many of the equations and methods are not detailed in those documents. Therefore, API 1102 and Guidelines for the Design of Buried Steel were discussed to fill in some of the gaps in the CFRs and ASME pipe codes, such as surface dead loads, surface live loads, and cyclic loads, failure mechanisms, equations, and acceptance criteria for deflection, ovality, through wall bending, crushing, ring buckling, fatigue, longitudinal, thermal expansion and fluid transients.

Using these presented methods results in inherently safer piping and safer pressure relief system design.

17 References

[1] API, RP 80, Guidance for the Definition of Onshore Gas Gathering Lines, 1st Edition, April 1, 2000, Reaffirmed January 2013.

[2] 49 CFR Part 192, Transportation of Natural and Other Gas By Pipeline: Minimum Federal Safety Standards, Revision 2/11.

[3] 49 CFR Part 195, Transportation of Hazardous Liquids by Pipeline, Revision 2/10.

[4] ASME, B31.8-2014, Gas Transmission and Distribution Piping Systems, ASME Code for Pressure Piping, B31, American Society of Mechanical Engineers, New York, NY, 2014.

[5] ASME, B31.4-2012, Pipeline Transportation Systems for Liquids and Slurries, ASME Code for Pressure Piping, B31, American Society of Mechanical Engineers, New York, NY, 2012.

[6] ASME, B31.3-2014, Process Piping, ASME Code for Pressure Piping, B31, American Society of Mechanical Engineers, New York, NY, 2015.

[7] ASME, B31.1-2004, Power Piping, ASME Code for Pressure Piping, B31, American Society of Mechanical Engineers, New York, NY, 2004.

[8] ASME, B1.20.1-2013, Pipe Threads, General Purpose (Inch), American Society of Mechanical Engineers, New York, NY, 2013.

[9] API, Recommended Practice 1102, Steel Pipelines Crossing Railroads and Highways, 6th Edition, American Petroleum Institute, Washington, D.C., April 1993, Reaffirmed July 2002.

[10] American Lifelines Alliance, Guidelines for the Design of Buried Steel Pipe, American Lifelines Alliance, July 2001.

Understanding Tolerable Risk Criteria – Considering the Growth of LNG Transportation

Ryan J. Hart, Ph.D., P.E.
Exponent, Inc.
4580 Weaver Parkway, Suite 100
Warrenville, IL
rhart@exponent.com

Delmar "Trey" Morrison, Ph.D., P.E.
Exponent, Inc.
4580 Weaver Parkway, Suite 100
Exponent, Inc.
Warrenville, IL
tmorrison@exponent.com

Prepared for Presentation at
American Institute of Chemical Engineers
2017 Spring Meeting and 13th Global Congress on Process Safety
San Antonio, TX
March 26 – 29, 2017

Understanding Tolerable Risk Criteria – Considering the Growth of LNG Transportation

Ryan J. Hart, Ph.D., P.E.
Exponent, Inc.
4580 Weaver Parkway, Suite 100
rhart@exponent.com

Delmar "Trey" Morrison, Ph.D., P.E.
Exponent, Inc.
tmorrison@exponent.com

Keywords: risk assessment; tolerable risk criteria; transportation risk

Abstract

Cheap natural gas and the aim to lower emissions from combustion systems such as stationary power generation and heavy haul vehicle engines has led to a multi-industry effort to convert to natural gas equipment. Natural gas pipelines can reach most of the U.S., but there are many areas where shipping liquefied natural gas (LNG) via truck or train may be the most economical solution. This article focuses on shipping LNG via rail and benchmarking the risk this poses. Although similar to other hazardous materials, LNG shipping via rail is a new endeavor in the U.S., and the regulations do not currently allow this to happen. The Federal Rail Administration (FRA) is the government entity that regulates rail shipping, and has implemented a requirement for a risk-based analysis in order to get approval to ship high hazard flammable trains (HHFT) after a series of crude oil train accidents over the last decade. Further, the FRA has issued specific guidance related to risk assessment for LNG trains. Quantitative Risk Assessment (QRA) is one tool that can be applied to evaluate the risks posed by shipping LNG. This falls into a special class of QRA called Transportation Risk Analysis.

But how does one evaluate the quantified risk for shipping LNG? How can it be demonstrated that the risk of shipping LNG is acceptable to society? Currently, the U.S. federal government has not explicitly provided risk criteria for evaluating acceptability of shipping LNG via specific routes. In contrast, international regulators and industry organizations have implemented tolerable risk criteria for determining acceptability of new technologies, but their underlying bases may not be apparent. In this paper, we will demonstrate how existing stationary LNG plant tolerable risk criteria can be used to develop criteria for LNG transportation risk. The inappropriateness of directly applying these existing stationary plant risk criteria to transportation will be explained, and another existing HHFT commodity (propane (LPG) for example) will be used to benchmark against the stationary LNG plant criteria and to show how these criteria can be modified to apply to transportation risk. Ultimately, the risks will be quantified for LNG HHFTs and compared to existing HHFTs to show that the newer fuel, LNG, poses reasonable and equivalent risks to many of the other HHFTs currently shipped through the U.S. From this work, a tolerable risk criterion for LNG shipping via rail will be proposed.

1 Introduction

The growth in domestic natural gas production is expected to continue to grow over the next several decades (Figure 1) [1]. The surplus of natural gas is expected to be handled through exports as well as domestic consumption. More and more we will start seeing LNG transport containers being shipped over the road and along the rail. To what extent is yet to be determined, but an increase in transportation of LNG within the U.S. is an almost certainty.

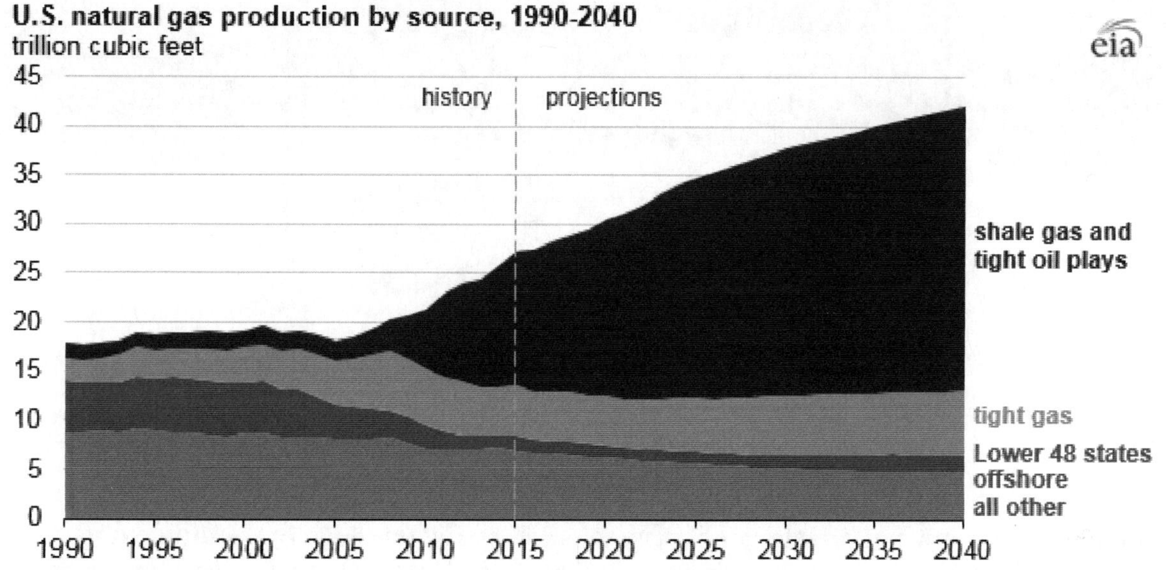

Figure 1. Project domestic natural gas production [1].

Although LNG transportation has existed for decades within the U.S., the anticipated surge in transportation is not without concern. Given the high visibility of transportation incidents involving crude oil, much attention is being paid to LNG transport by rail. Ensuring that the development and implementation of LNG transportation is performed safely is of critical importance to industry, communities, and regulators.

Much of the concern surrounding LNG transport may be attributed to a general lack of understanding regarding the unique and particular hazards of LNG. Quantitative risk assessment (QRA) is a tool that has been used successfully in the past to evaluate the potential risks of emerging technologies and proposed transportation routes. QRA is not a common public-policy tool in the U.S., but it is beginning to grow in popularity with private companies, regulatory agencies, and standards committees alike. Most QRA has been performed for stationary facilities, which are inherently different than transportation projects. In this paper, we discuss tolerable risk criteria for use in evaluating the QRA of transport projects. We anticipate that this approach may be useful for evaluating the blossoming LNG transport in the U.S.

2 Understanding Quantitative Risk Assessments

Risk, simply defined, is the potential to lose something of value. Risk is calculated by taking the product of event likelihood with the event outcome severity. A simplified flow diagram depicting the process for determining risk is provided in Figure 2.

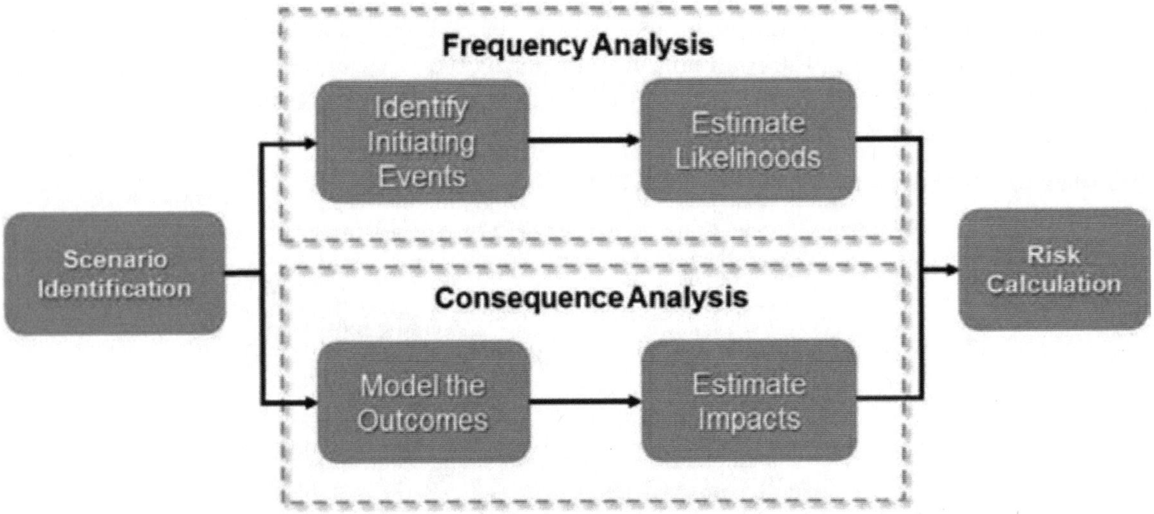

Figure 2. Flow diagram depicting the process for calculating risk.

The likelihood of an event can be estimated using experience relating to given equipment in similar service, industry data, or engineering approximations. A challenge of quantifying risk, or affixing a number to a particular risk level, is determining how to quantify the event outcome portion of the equation. For quantifying risk at industrial facilities and transportation operations, the outcome of an event is typically evaluated as the potential for a fatality or multiple fatalities.

In evaluating the potential for fatality, two metrics are utilized to yield the risk: (1) Individual Risk (IR) and (2) Societal Risk (SR). IR is the frequency (yr^{-1}) where an individual with continuous potential exposure may be expected to sustain a serious or fatal injury. When calculating IR, it is assumed that persons are constantly occupying all space where an incident may express an undesired impact. As such, IR is largely independent of population density except to the extent that the likelihood component of the risk is dependent on population (e.g. likelihood of ignition calculation). IR results are typically presented as frequency contours on aerial maps that illustrate the risk to individuals positioned within those contours. An example of an IR result is presented in Figure 3.

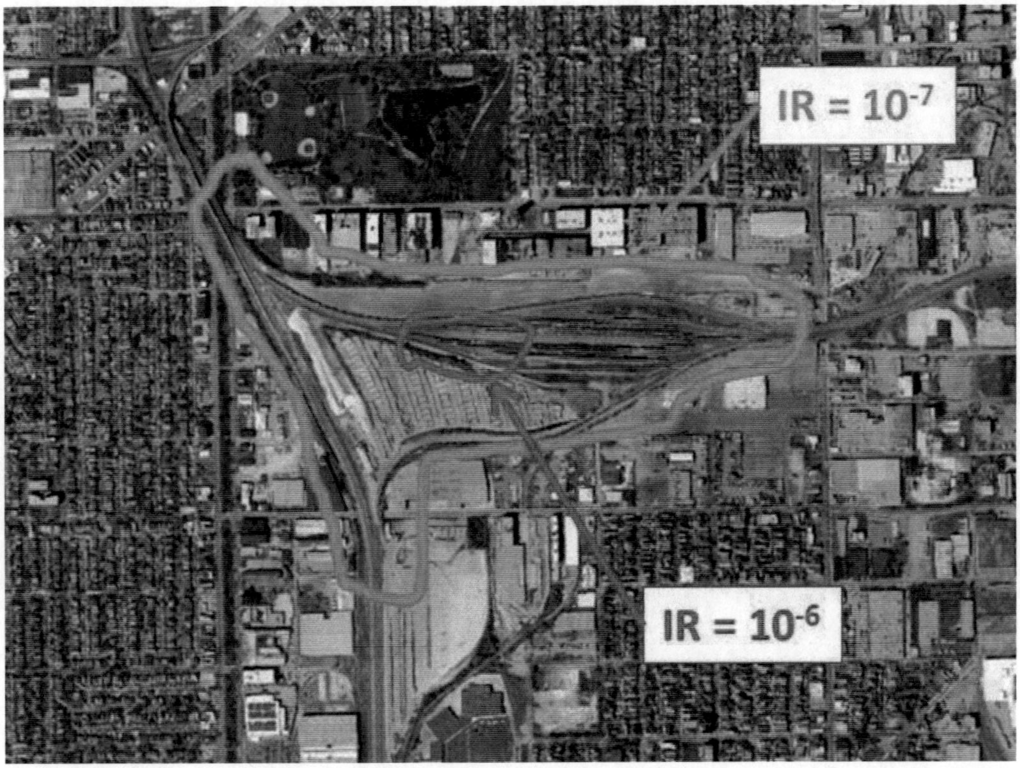

Figure 3. Example of an IR result, presented around a rail yard.

Societal Risk (SR) is another metric for evaluating the risk of a given facility or operation. Unlike IR, the SR calculation considers the relationship between the cumulative number of potential fatalities and the likelihood from a series of potential events. The outcome of a SR analysis is a graph depicting annual frequency on the y-axis and N fatalities on the x-axis, where N is the cumulative number of potential fatalities for all scenarios represented by the corresponding cumulative frequency of events. Unlike IR, the SR calculation takes into account the differences in population density in the vicinity surrounding the incident. Whereas the IR calculation gives insight into the probability of having a fatality at a specific location, the SR calculation gives the likelihood of a number of potential fatalities covering the entire operation. This is especially important for evaluating scenarios with a large potential impact for loss of life, such as train derailments of flammable materials. An example of an SR result for two separate scenarios is provided in Figure 4.

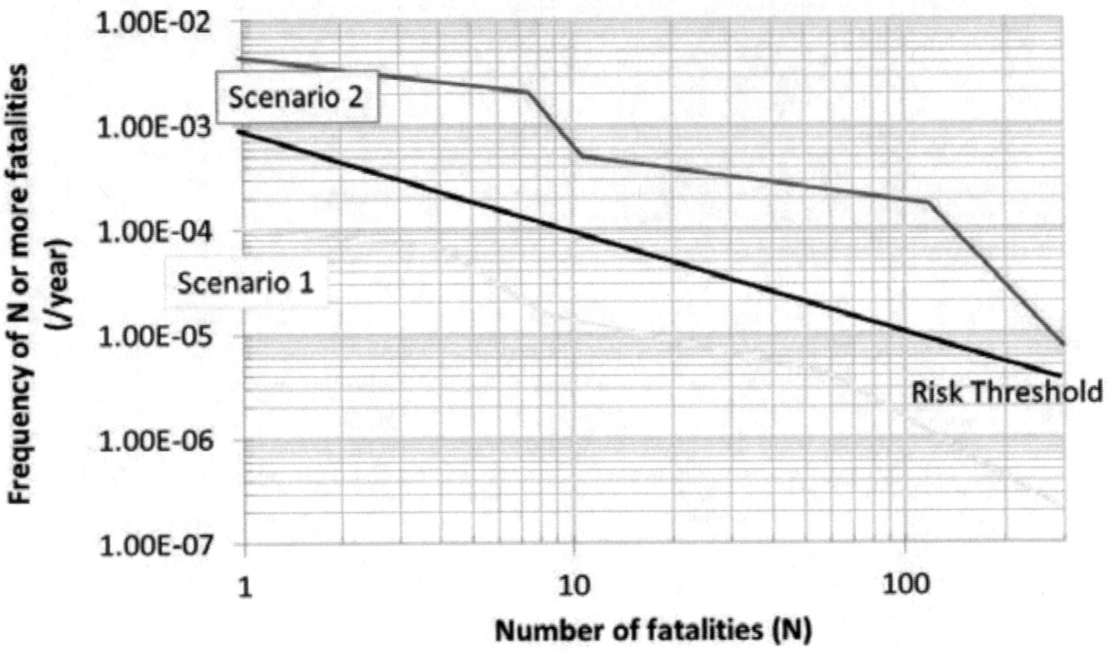

Figure 4. Example of an SR result for two different scenarios.

Quantitative risk assessments may be performed for several reasons:

1. Comparing the relative risk of two or more competing options (e.g. comparing route options for transporting hazardous materials)
2. Identifying high risk aspects of a specific operation
3. Demonstrating that the risks of a certain operation are tolerable or acceptable

When analyzing risks to demonstrate their acceptability or tolerability, the selection of appropriate risk tolerability criteria is paramount. Selection of inappropriate tolerability criteria could potentially result in a false-confidence in the perceived safety (in instances where the risk tolerability may be too high) or unnecessary elimination of a proposed project or operation (in cases where the risk tolerability may be too low).

3 Quantitative Risk Tolerability Criteria

After quantifying risk and presenting the calculations as IR and SR for a given operation or process, the results are evaluated for tolerability (or acceptability). Typically, stakeholders (e.g., government agencies, investors, communities) have a threshold risk level that is deemed acceptable—known as quantitative risk criteria. Currently, the U.S. federal regulatory agencies have not codified quantitative risk criteria for hazardous materials transportation scenarios [2] and QRA analyses are not common regulatory requirements in the U.S. Several foreign regulatory agencies, such as the Dutch government, have been international leaders in utilizing QRA techniques for determining acceptability of fixed facilities and transportation routes.

However, no broadly-accepted risk criteria are employed by domestic communities or industries. There are several foreign and several domestic examples of quantitative risk criteria (see [3]-[5]) and these will be summarized in the next sections for both IR and SR.

Recommendations for QRA of LNG plants were issued in the National Fire Protection Association (NFPA) standard, NFPA 59A *Standard for the Production, Storage, and Handling of Liquefied Natural Gas (LNG)* [6]. In addition to including QRA as a risk assessment tool in NFPA 59A, the standard also includes quantitative risk criteria for fixed LNG facilities. NFPA 59A explicitly applies to LNG plants and stationary facilities; it does not apply to LNG transportation or portable LNG containers. Thus, the tolerable quantitative risk criteria proposed in the standard are not directly applicable to the transportation of LNG, as will be discussed below.

There are significant differences between stationary facilities and transportation projects, for example:

- Size scale of the enterprise. The risk calculation is directly related to the number of vessels or components and the length of piping systems or a route. In a fixed facility, the length scale is fixed and on the order of hundreds to thousands of feet. In transport, the length scale may be hundreds to thousands of miles. Thus, transport is anticipated to create higher risks for similar hazmat.
- Site location. Stationary facilities that are yet to be built typically have the opportunity to survey multiple potential sites, whereas transportation routes options are more restricted (in some cases, only one transportation route may be an option).
- Siting. For stationary facilities, consideration can be given where on the site to place hazardous material operations in order to minimize potential off-site impacts or reduce the likelihood of suffering cascading events. Transportation of hazardous materials has little flexibility in placement along a given route and has no control over adjacent hazards (e.g. a hazmat transport carrier cannot prevent another hazmat transport carrier from overtaking them on the highway).
- Standoff distances. Proposed facilities may have the opportunity to acquire more land for a given project in order to increase the distance from hazmat operations to the property boundary. The separation distances between a transport carrier and the public are already established due to the zoning of the transport route and adjacent properties. Additionally, in some transport modes (like over the road trucking) the transport carrier has no choice but to be a few feet from a public receptor.
- Control of ignition sources. Facilities operating with hazmat have complete control over the abundance and location of ignition sources on their site. Transport routes have no control over ignition sources, and nearly every public transporter can be assumed to be a competent ignition source.
- Training/emergency response. Facilities have the opportunity to train personnel and prepare for an onsite release of a hazmat. Transport carriers may have a highly trained crew for their specific carrier, but the public in the immediate vicinity of the hazard is presumed to have no training with respect to the hazards presented by hazmat. Additionally, hazmat transport relies upon the emergency response of the jurisdiction through which it travels, which may change dozens or hundreds of times on a given

route. Facilities, by comparison, usually have only 1 or 2 emergency responder agencies to coordinate with.

For the reasons stated above, the quantitative risk criteria for stationary facilities should be different than the quantitative risk criteria for transportation routes. The risks to the onsite and offsite receptors associated with transporting hazmat are inherently different than the risks to those populations surrounding a stationary facility. To account for these differences, transportation risks should be evaluated differently than stationary facility risks.

3.1 Individual Risk Criteria

Among the available references for quantitative risk criteria used domestically and internationally, there is a wide disparity in risk criteria for public exposure, with acceptable IR fatality probabilities ranging from 10^{-4} yr^{-1} (or a fatality per 10,000 years) to 10^{-8} yr^{-1} (or a fatality per 100,000,000 years). NFPA 59A identifies three "Zones" representing ranges of quantitative risk criteria for evaluating IR at stationary LNG facilities. Each risk zone reflects general types of public occupancies recommended to be permitted. As the magnitude of the calculated risk increases, the type of occupancy becomes more restrictive. The quantitative risk criteria for IR of LNG plants, according to NFPA 59A, are reproduced in Table 1.

Based on Zone 3 being the most restrictive zone, any IR values less than 3×10^{-7} yr^{-1} are not of concern for the analysis. The IR ranges and associated criteria reported in NFPA 59A appear to be based on guidance provided by the Health and Safety Executive in the U.K. for QRA [7] and do not account for the factors typically considered in a transportation risk analysis. Conversely, a commonly acceptable level of IR for transportation risks on sensitive populations is 10^{-6}, which is the upper threshold for Zone 3 [5].

Table 1. Quantitative risk criteria for IR contours as provided by NFPA 59A (2016).

Criterion Annual Frequency (yr^{-1})	Remarks
Zone 1 $IR > 10^{-5}$	Not permitted: Residential, office, and retail Permitted: Occasionally occupied developments (e.g., pump houses, transformer stations)
Zone 2 $10^{-6} \leq IR < 10^{-5}$	Not permitted: Shopping centers, large-scale retail outlets, restaurants, etc. Permitted: Work places, retail and ancillary services, residences in areas of 7,250 to 23,300 persons/mile^2 density
Zone 3 $3 \times 10^{-7} \leq IR < 10^{-6}$	Not permitted: Churches, schools, hospitals, major public assembly areas, and other sensitive establishments Permitted: All other structures and activities

For flammable material release scenarios, the magnitude of the risk generally increases as the observation point is moved closer to the transport route. The distance to each risk level identified in the table is a result of the compilation of the outcomes calculated from an event tree of many potential fire and explosion events. An example of calculated distances to various IR thresholds for the transportation of LNG and LPG have been reported **Error! Reference source not found.**, reproduced in Table 2. LPG represents an inherently acceptable hazmat commodity, which has been shipped on railroads for a long time. Table 2 shows that the IR of LNG is similar or less than LPG for both road and rail transport.

Table 2. Example calculated IR distances for the road and rail transport of LNG and LPG.

Material	Transport Mode	Pop Density (/mile2)	IR Distance to... 10^{-5} IR (m)	10^{-6} IR (m)	$3x10^{-7}$ IR (m)
LPG	Road	10,000		118	177
		1,000		86	115
		100		84	108
LNG	Road	10,000		119	169
		1,000		62	95
		100		56	84
LPG	Rail	10,000	186	290	334
		1,000	178	290	334
		100	174	290	334
LNG	Rail	10,000	92	193	213
		1,000		145	184
		100		104	132

It should be noted that the analysis in **Error! Reference source not found.** was conservative, especially in the instances of rail risks where it was assumed that a train carrying a hazmat would involve the hazmat rail car in all instances. Despite the conservativeness, the data illustrates that the NFPA 59A Zone 3 criteria for IR (3×10^{-7} yr^{-1}) is much more restrictive than the Zone 2 criteria (1×10^{-6} yr^{-1}). Additionally, the use of the Zone 3 criteria would indicate that the inherently acceptable risks of transporting LPG may not satisfy this criterion in instances where there are sensitive targets within 334 m of railroads and 177 m of roads where LPG is transported.

3.2 Societal Risk Criteria

As with IR, the international and domestic quantitative risk criteria for SR can vary greatly. Most SR criteria present two tolerability lines with three distinct regions for which to evaluate the resulting FN curve. For discussion purposes, the SR criteria from NFPA 59A is provided in Figure 5. There are two lines presented in the figure:

(1) an upper risk criteria line (solid red line), commonly referred to as the maximum risk line, and

(2) a lower risk criteria line (dashed line), commonly referred to as the tolerable risk line.

For calculated FN curves that fall above the upper risk criteria line, the risk is considered "unacceptable." Calculated FN curves that fall below the lower risk criteria line are determined to be acceptable, and FN curves between the two criteria lines fall within the "As Low As Reasonably Practicable (ALARP)" region. Risks falling within the ALARP region are typically evaluated in more detail for opportunities to either mitigate or reduce the perceived risk of the operation, in the form of additional engineering or administrative controls.

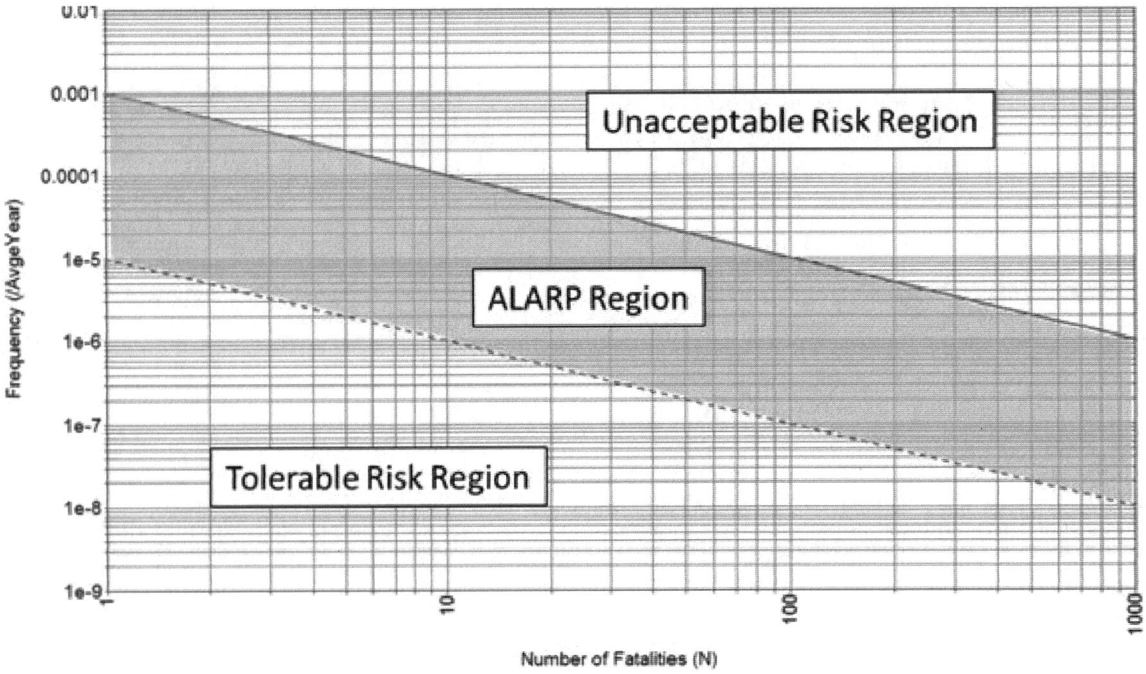

Figure 5. NFPA 59A (2016) SR tolerability criteria.

Most international and domestic examples of SR criteria follow the same general format with both an upper and lower risk bound. Although different regulatory agencies, standards committees, and private companies may have different definitions of upper and lower risk lines, nearly all examples of SR criteria apply to only stationary facilities.

Using a previous reported QRA study for the transport of LNG and LPG [8], we have adopted the data here to demonstrate calculated FN curves. Figure 6 depicts the calculated FN curves for the rail and road transport of LNG and LPG along a 50-mile long route surrounded by a population with a density of 1,000 people/mile^2. The figure also includes the maximum and minimum risk lines from NFPA 59A, and the ALARP region highlighted in grey. The figure illustrates the potential issues when directly applying SR criteria intended for stationary facilities to transportation risks. The risks for all four transport routes fall within the "Unacceptable Risk" region as defined in NFPA 59A. LPG is certainly transported by truck and rail through similar distances and population densities, but the risk to society is acceptable.

Although most SR criteria available both domestically and internationally are applied to stationary facilities, methods have been developed to apply the stationary SR criteria to transportation routes. These methods involve normalizing the scale of the transport project risks and modifying the risk criteria lines. To correct the fixed facility SR criteria for application to a transportation project that may be considerably larger in scale, the SR for transportation routes are calculated for a fixed unit of length. This allows for similar hazard scales to be compared between transport routes and fixed facilities. The common unit for normalizing the FN curves of transportation projects is either 1 mile or 1 km, depending on the common engineering units in the project's location. This can be accomplished by either (1) calculating the FN curve for the entire transportation route and dividing the y-axis (frequency) by the route length or (2) conservatively assuming that the highest-risk section of the transport route bounds the risk along the entire line. Such an approach is sensible because any potential incident can only affect a limited portion of the society along the route for these flammable materials; thus, using a per-mile or per-kilometer basis for SR is reasonable. The same approach may not be acceptable for toxic materials, e.g., chlorine gas, which could project significant hazards over a greater distance. An additional approach is to compare the aggregate SR curve for the route for the new hazmat (LNG) to an existing hazmat (LPG). If the new curve is similar or less, then it poses less risk along the route than the reference material.

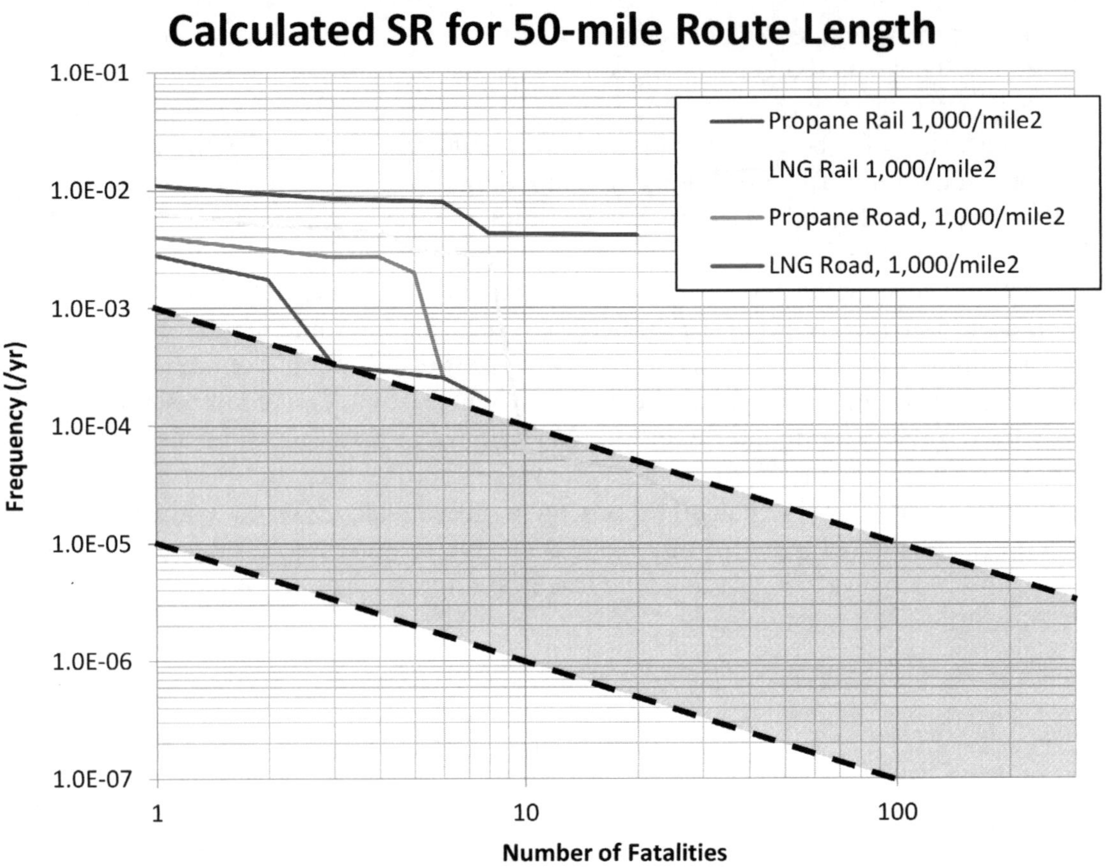

Figure 6. Calculated FN curves for the road and rail transport of LNG and LPG along a 1-mile long route surrounded by a population with a density of 1,000 people/mile^2.

Figure 7 provides the FN curves for the same transport routes above, but normalized on a 1-mile length. All other inputs are identical between the two graphs. When scaling the transportation FN curves to be comparable to the length scale of stationary facilities, the application of the stationary facility SR criteria becomes more reasonable. However, it can be seen that one of the analyzed transportation routes, propane rail transport, still exhibits a portion of the FN curve with the NFPA 59A defined "unacceptable risk" region. As previously mentioned, there are other considerations between project scales that distinguish the risks of stationary facilities from transportation projects. In order to consider for these other differences (such as ignition source control, training, siting), some international organizations increase the maximum and minimum risk lines by an order of magnitude – the implication here is that the tolerability of risks associated with transportation are inherently higher that for exposure to a stationary facility. If this approach were to be taken for the FN curves presented in Figure 7, none of the FN curves would lie within the "unacceptable risk" region (in fact, the LNG Road scenario would lie in the "acceptable risk" region).

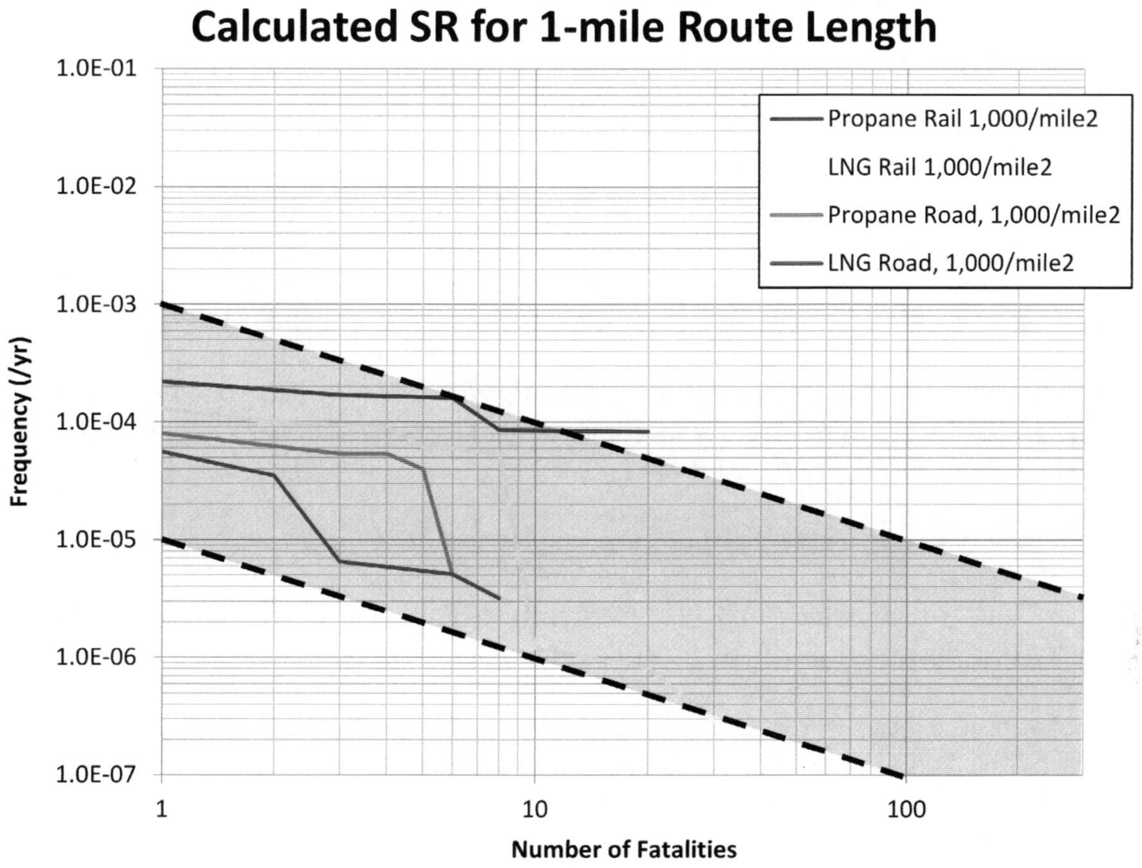

Figure 7. Calculated FN curves for the road and rail transport of LNG and LPG along a 1-mile long route surrounded by a population with a density of 1,000 people/mile^2.

4 Conclusions

Quantitative risk assessments are powerful tools for assessing the risks of emerging and existing hazardous material operations within stationary facilities and along transportation routes. The risks can be assessed through either comparative analyses or by application of tolerable risk criteria. Many international and domestic organizations, both public and private, have identified tolerable risk criteria for evaluating both Individual Risk (IR) and Societal Risk (SR), but considerable variability exists. Additionally, most tolerable risk criteria have been developed for stationary facilities.

As shown in this paper, application of risk criteria developed for stationary facilities to transportation routes may not be appropriate. The risks of transporting hazardous materials is different from fixed facilities, as are the thresholds for risk tolerability that society is inherently willing to accept. The application of improper risk criteria may prevent or delay the proposed transportation of emerging materials, such as LNG.

Based on previously reported IR and SR values for LNG and LPG transport, we have demonstrated that although the risks of LNG transport are less than the risks of transporting LPG, certain risk tolerability criteria may indicate that the risks are unacceptable. This conflicts with the inherently acceptable risks of LPG transport, which were demonstrated to be higher than those for comparable LNG transport.

A survey of domestic and international risk tolerability was performed. Most risk tolerability criteria have been developed for and applied to stationary facilities. The risks of stationary facilities are inherently different that transportation risks for two primary reasons: (1) transportation routes can be significantly larger than stationary facilities and (2) the inherent tolerability of transportation risks is different than for stationary facilities. Based on the analysis of representative LNG and LPG transport risks, compounded with the typical treatment of risk tolerability criteria by international organizations, methods for using stationary facility risk criteria to evaluate transport risks are proposed:

- IR Tolerability Criteria – for transport projects, the use of 10^{-6} yr^{-1} as the IR frequency threshold for impacts to sensitive targets is consistent with an abundance of international sources. This value is approximately one order-of-magnitude higher than the minimum sensitive target tolerability criteria presented in NFPA 59A for stationary LNG facilities (3×10^{-7} yr^{-1}).
- SR Tolerability Criteria – for transport projects, normalizing the FN curves for a representative length (1 mile or 1 km) can allow for re-scaling of transportation risks to a stationary facility. Further, to account for the higher risk tolerability for transportation relative to stationary facilities, the maximum and minimum risk criteria lines may be increased by an order of magnitude.

5 References

[1] U.S. Energy Information Administration, "Annual Energy Outlook 2016," July 7, 2016.

[2] Strang J, "Federal Railroad Administration Risk Reduction Programs," United States Army Corps of Engineers Workshop on Tolerable Risk, March 18-19, 2008, Alexandria, Virginia.

[3] Appendix B: Survey of Worldwide Risk Criteria Applications, *Guidelines for Developing Quantitative Safety Risk Criteria*. Center for Chemical Process Safety, AIChE (2009).

[4] Cornwell JB and MM Meyer, "Risk Acceptance Criteria or 'How Safe is Safe Enough?'" presented at II Risk Control Seminar in Puerto La Cruz, Venezuela, October 13, 1997.

[5] Ham JM, M Struckl, AM Heikkila, E Krausmann, C DiMauro, M Christou, JP Nordvik, "Comparison of Risk Analysis Methods and Development of a Template for Risk Characterisation," Institute for the Protection and Security of the Citizen, European Commission, Directorate-General Joint Research Center (2006).

[6] NFPA 59A, Standard for the Production, Storage, and Handling of Liquefied Natural Gas (LNG), 2016 edition, National Fire Protection Association.

[7] "B.1 Evolution of Land Use Planning Criteria in the UK," in Guidelines for Developing Quantitative Safety Risk Criteria, American Institute of Chemical Engineers, Center for Chemical Process Safety (2009).

[8] Hart RJ, Morrison DR. The hazard we know: Comparing transportation risk of LPG and LNG. American Institute of Chemical Engineers, 2015 Spring National Meeting and 11th Global Congress on Process Safety, Austin, TX, April 26-30, 2015.

An Inverse-based Methodology for Disturbance Identification of LTI MIMO Systems (# 480120)

Liwen Chen (Joseph)

Advisor: Prof. Qiang Xu

Dan F. Smith Department of Chemical Engineering

Lamar University, Beaumont, TX 77710

2017 AIChE Spring Meeting, Mar 26-30, San Antonio, TX

Overview

- **Background (system inversion, applications)**
- **Previous inverse algorithms**
- **New algorithm (development, convergence criteria)**
- **Optimal design of the feedthrough D matrix (feedback stabilization, Lyapunov equation, least-square solution)**
- **Illustrative examples**
- **Concluding remarks**

2

What is Inversion?

- Given the system G

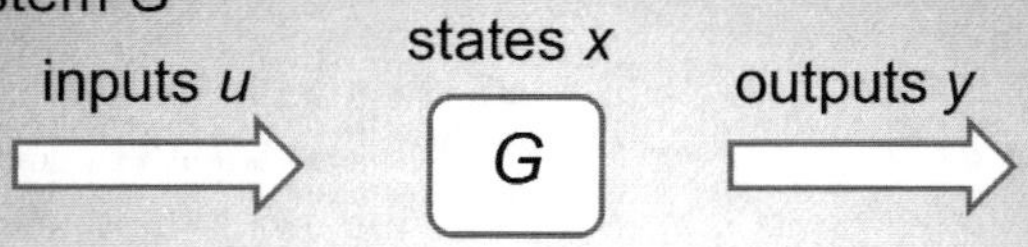

- Left Inverse: Given **measured** y, **recover the estimated** u ($\hat{u} \approx u$)

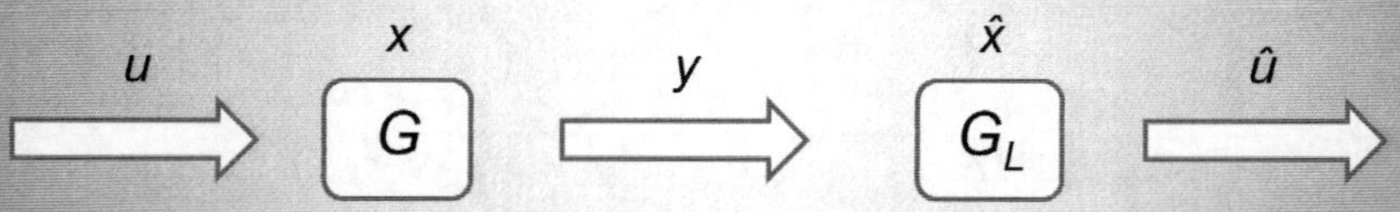

- Right Inverse: Given **reference** y, generate the **required** $\hat{u}$ so that $\hat{y} \approx y$

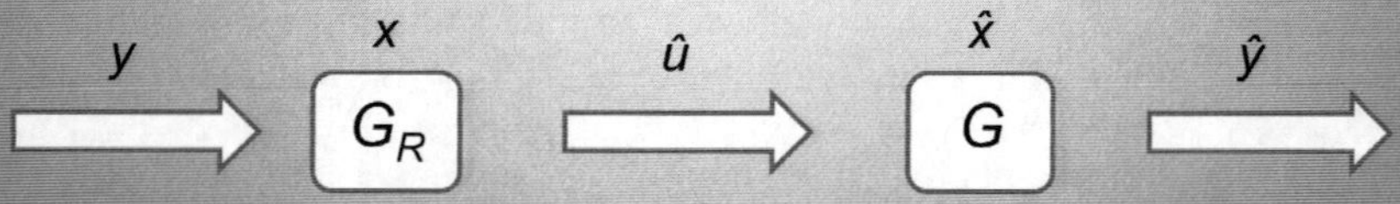

Why Study Inversion?

- Recover critical unknown inputs or states solely from outputs
- Left Inverses
  - Non-interacting control
  - Fault detection and identification (FDI)
- Right Inverses
  - Feed-forward control
  - Fault tolerant control (FTC)

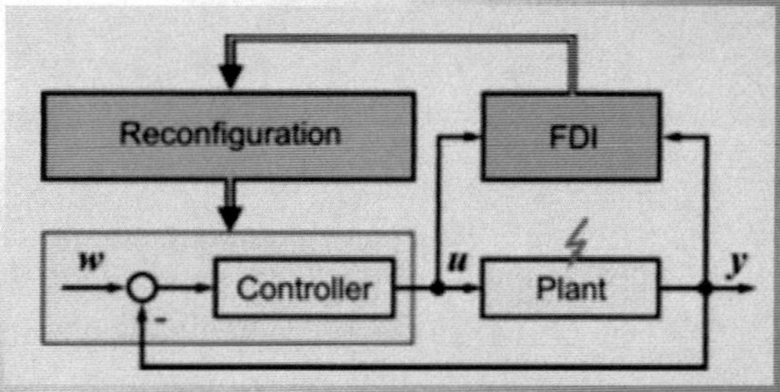

Source: wikipedia "control reconfiguration"

Fault model for residual generation

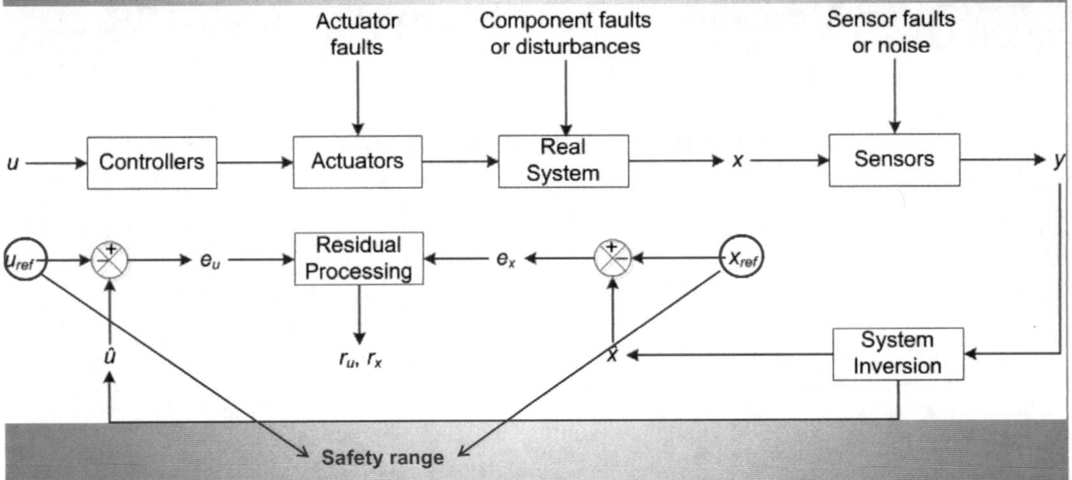

Inverse Problem

- For single-in/single-out (SISO) system

$$y = Gu \quad \Rightarrow \quad u = \frac{1}{G} y$$

- For multi-in/multi-out (MIMO) system

$$\underbrace{\begin{bmatrix} y_1 \\ y_2 \\ \vdots \\ y_p \end{bmatrix}}_{Y} = \underbrace{\begin{bmatrix} G_{11} & G_{12} & \cdots & G_{1m} \\ G_{21} & G_{22} & & G_{2m} \\ \vdots & \vdots & \ddots & \vdots \\ G_{p1} & G_{p2} & \cdots & G_{pm} \end{bmatrix}}_{G} \underbrace{\begin{bmatrix} u_1 \\ u_2 \\ \vdots \\ u_m \end{bmatrix}}_{U}$$

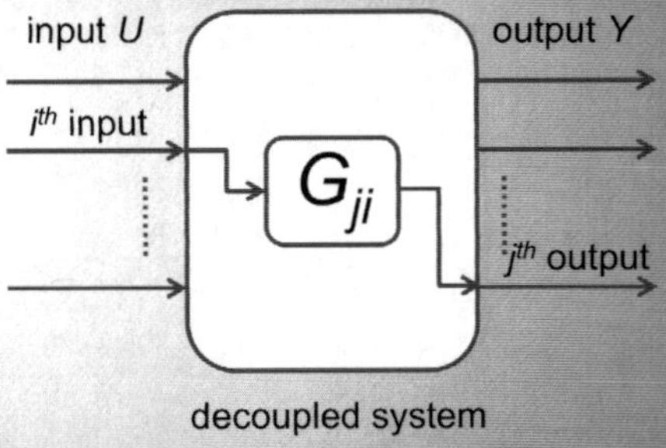

- Left Inverse: find G_L so that $G_L G = I_m$ $(p \geq m)$
- Right Inverse: find G_R so that $G G_R = I_p$ $(m \geq p)$

Inverse Formula

- Consider the continuous-time MIMO system

Feed-through matrix

$$\dot{x} = Ax + Bu, \quad x(t) \in \mathbb{R}^n,\ u(t) \in \mathbb{R}^p \quad (1)$$

$$y = Cx + Du, \quad y(t) \in \mathbb{R}^p \quad (2)$$

- When D is square and nonsingular, D^{-1} can be pre-multiplied on (2) and substitute u into (1) yields the inverse system

$$\dot{x} = \left(A - BD^{-1}C\right)x + BD^{-1}y \quad (3)$$

$$= \hat{A}x + \hat{B}y$$

$$u = -D^{-1}Cx + D^{-1}y \quad (4)$$

$$= \hat{C}x + \hat{D}y$$

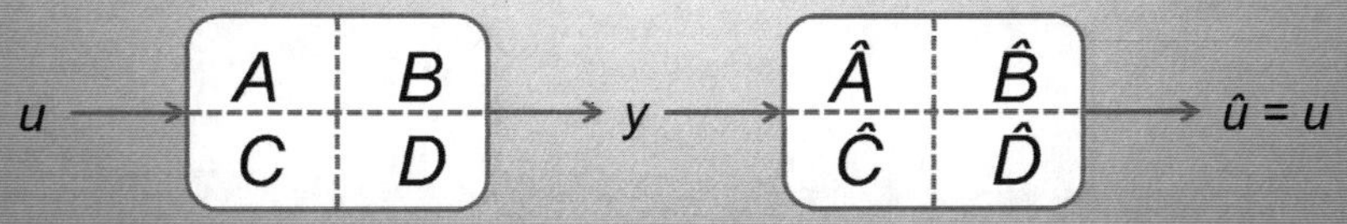

7

Previous Algorithms

- When D is **square** but **singular**
  - Sain & Massey's algorithm[1] (inherent integration L)
  - Silverman's "structure" algorithm[2] (no invertibility check)
- When D is **non-square** but **of full rank**
  - Moylan's stable algorithm[3] (unknown x_0, system order reduction)
- Common features
  - Output y needs to be differentiated over time (dy / dt)
  - In most cases, $D \neq 0$
- **What if $D = 0$ (strictly proper system)?**

Source:
1. Sain, MK., Massey, JL. Invertibility of linear time-invariant dynamical systems, *IEEE Trans. Autom. Control*, 1969; 14:141-149.
2. Silverman, LM. Inversion of multivariable linear systems, *IEEE Trans. Autom. Control*, 1969; 14: 270-276.
3. Moylan, PJ. Stable inversion of linear systems. *IEEE Trans. Autom. Control.* 1977; 22(1):74-78.

8

Incentives

- Strictly proper system ($G(s) = C(sI - A)^{-1}B$)

$$\dot{x} = Ax + Bu \quad (1)$$
$$y = Cx \quad (2)$$

- "Small D" Method (Skogestad & Postlethwaite[1])

$$\dot{\hat{x}} = A\hat{x} + B\hat{u} \quad (3)$$
$$\hat{y} = C\hat{x} + D\hat{u} = y \quad (4)$$

- Problems
  - Existence (if A is stable, D should guarantee $A - BD^{-1}C$ is also stable)
  - How to identify
  - Closeness between u and $\hat{u}$ (reference tracking)

Source: 1. Skogestad, S. & Postlethwaite, I. Multivariable feedback control: analysis and design 2nd ed., Wiley, New York, 2005.

9

Methodology Framework

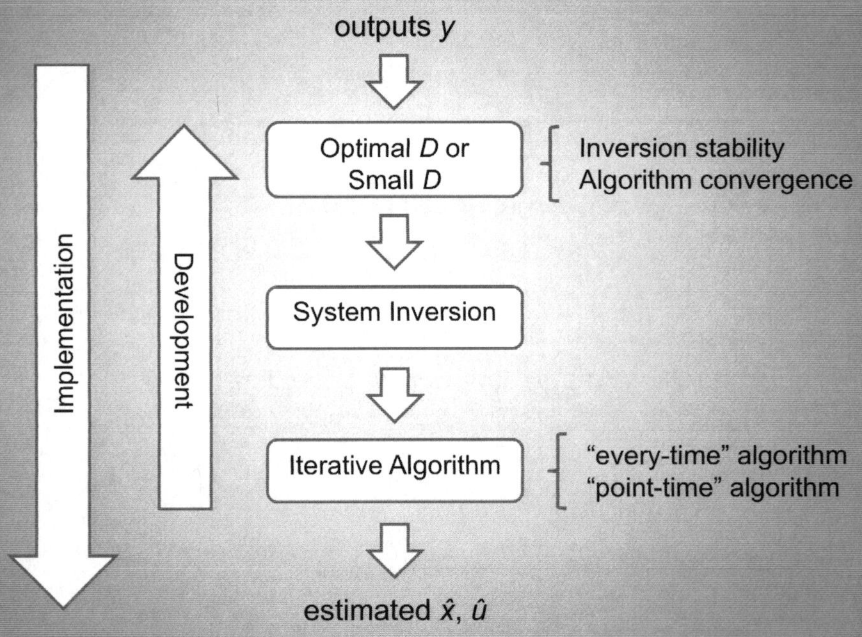

10

Algorithm Development

- Iterative algorithms
  - "every-time" algorithm (ramp, sinusoidal, exponential decay,...)
  - "point-time" algorithm (step)
  - improve on reference tracking

$$\hat{y}_1 = C\hat{x}_1 + D\hat{u}_1 = y$$
$$\hat{y}_2 = C\hat{x}_2 + D\hat{u}_2 = \hat{y}_1 + D\hat{u}_1$$
$$\vdots$$
$$\hat{y}_k = \boxed{C\hat{x}_k + D\hat{u}_k = \hat{y}_1 + D\hat{u}_{k-1}}$$

11

Algorithm Development

- Let $k = k + 1$ recursively,

$$\begin{aligned}\hat{U}_k(s) &= \hat{U}_1(s) + F(s)\left(\hat{U}_1(s) + F(s)\hat{U}_{k-2}(s)\right)\\ &= \left(I + F(s)\right)\hat{U}_1(s) + F^2(s)\hat{U}_{k-2}(s)\\ &= \cdots = \sum_{i=0}^{k-1} F^i(s)\hat{U}_1(s)\end{aligned}$$

- Since $\hat{U}_1(s) = (I - F(s))U(s)$

$$\begin{aligned}\hat{U}_k(s) &= \sum_{i=0}^{k-1} F^i(s)\left(I - F(s)\right)U(s)\\ &= \left(I - F^k(s)\right)U(s) \text{ where } F(s) = G(s)^{-1}D\end{aligned}$$

12

Uniform Convergence

- Define the error vector $E_k(s)$ and iterative step vector $\Delta E_k(s)$

$$E_k(s) = U(s) - \hat{U}_k(s) = F^k(s)U(s)$$
$$\Delta E_k(s) = E_{k-1}(s) - E_k(s) = F^{k-1}(s)G^{-1}(s)Y(s)$$

- The error between real and estimated states is

$$X(s) - \hat{X}_k(s) = (sI - A)^{-1}B(U(s) - \hat{U}_k(s)) = (sI - A)^{-1}BE_k(s)$$

- Obviously, if $||F(s)|| < 1$, $F^k(s) \to 0$, implying* $\hat{U}_k(s) \approx U(s)$, $\hat{X}_k(s) \approx X(s)$.

- Remarks:
  - 1. $||F(s)|| < 1$ is quite restrictive
  - 2. Hard to implement in the time domain

*$U(s)$ is assumed to be bounded in the frequency range.

13

Asymptotic Convergence

- If all poles of $F(s)$ are in the left half-plane (LHP)

$$e_{ss} := \lim_{t\to\infty} e(t) = \lim_{s\to 0} sE(s) = \lim_{s\to 0} sF(s)U(s) \qquad \text{“Final Value Theorem”}$$
$$= F(0)\lim_{s\to 0} sU(s) = \left(I - \hat{C}\hat{A}^{-1}B\right)\lim_{s\to 0} sU(s)$$

- Then, the algorithm converges asymptotically after k^{th} iteration if $||F(0)|| < 1$

- For unknown step inputs, the asymptotic convergence implies the uniform convergence.

14

Recap

- "Every-time" algorithm

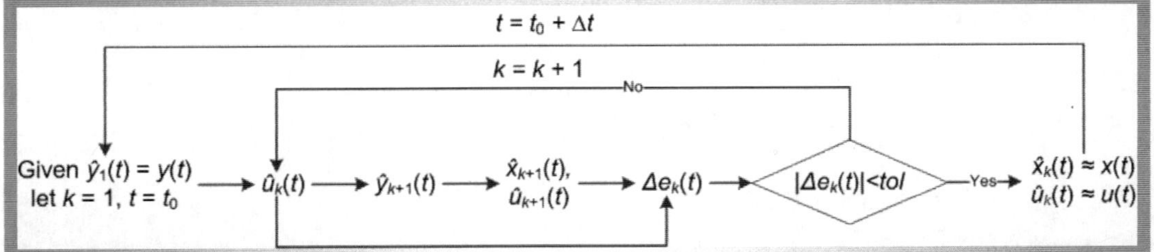

- "Point-time" algorithm

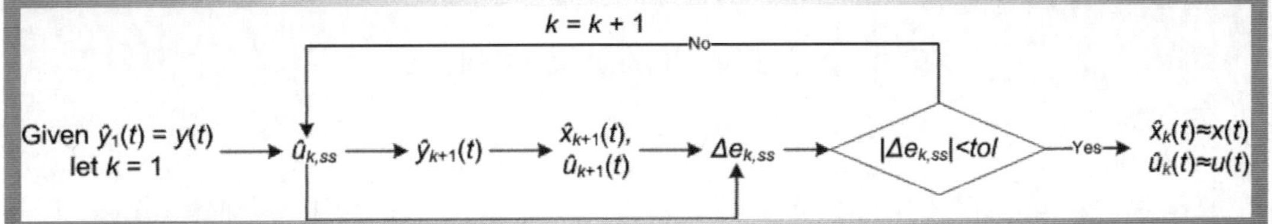

15

Feedback Stabilization

- Lyapunov equation

stability margin

$$(-\mu I - A)W + W(-\mu I - A)^T = -BB^T$$

Let $K = 0.5B^TW^{-1}$, after arrangement,

$$W^{-1}(A - BK) + (A - BK)^T W^{-1} = -2\mu W^{-1}$$

Find $\hat{A}$ so that $\hat{A} = A - BK$

- Therefore, the dummy D can be obtained by

$$D = \tfrac{1}{2}\left(B^T W^{-1} C^+\right)^+ \quad \text{so that } \left\|\hat{A} - (A - BK)\right\|_F \text{ is minimized}$$

where $()^+$ denotes Moore-Penrose pseudoinverse of any matrices and $||\cdot||$ denotes the Frobenious norm.

16

Optimal Design of *D* matrix

- Minimize the spectral abscissa of $\hat{A}$

$$\mu_o = \arg\min_{\mu} \eta_{\hat{A}} \quad \text{s.t.} \quad \begin{cases} \eta_{\hat{A}} < 0 \\ \|F(0)\| = \|I - \hat{C}\hat{A}^{-1}B\| < 1 \end{cases}$$

(spectral abscissa: $\eta_{\hat{A}}$)

- Least-square error (LSE)

$$\text{LSE} = \|D^{-1}C - K\|_F$$

- Plot $\eta_{\hat{A}}$, η_{A-BK}, $\|F(0)\|$, LSE against μ in a suitable range

- One dimensional search for μ_o by Matlab function *fmincon*()

17

Implementations

- Square Systems
  - Controllable
  - Stabilizable (uncontrollable but stable)
- Non-square Systems
  - Tall ($m < p$)
  - Fat ($m > p$)
- Industrial Cases
  - Distillation Tower
  - CSTR

18

Controllable system

- 4th order, 2 inputs, 2 outputs, strictly proper stable system (λ_A = -5.96, -2.72, -2.25, -0.58)

$$\begin{bmatrix}\dot{x}_1\\ \dot{x}_2\\ \dot{x}_3\\ \dot{x}_4\end{bmatrix} = \underbrace{\begin{bmatrix}-1.5 & 1.0 & 0 & 1.0\\ 2.0 & -3.0 & 2.0 & 0\\ 0 & 0.5 & -2.0 & 1.0\\ 1.0 & -1.5 & 0 & -5.0\end{bmatrix}}_{A}\begin{bmatrix}x_1\\ x_2\\ x_3\\ x_4\end{bmatrix} + \underbrace{\begin{bmatrix}1.0 & 0\\ 0 & 0\\ 1.0 & 1.0\\ 0 & 1.8\end{bmatrix}}_{B}\begin{bmatrix}u_1\\ u_2\end{bmatrix} \quad \left(x_0 = \begin{bmatrix}0.2\\ 0.2\\ 0.2\\ 0.2\end{bmatrix}\right)$$

$$\begin{bmatrix}y_1\\ y_2\end{bmatrix} = \underbrace{\begin{bmatrix}0 & 2.4 & -3.1 & 1.0\\ 1.0 & 6.0 & -0.5 & -2.8\end{bmatrix}}_{C}\begin{bmatrix}x_1\\ x_2\\ x_3\\ x_4\end{bmatrix}$$

19

Optimal *D* matrix

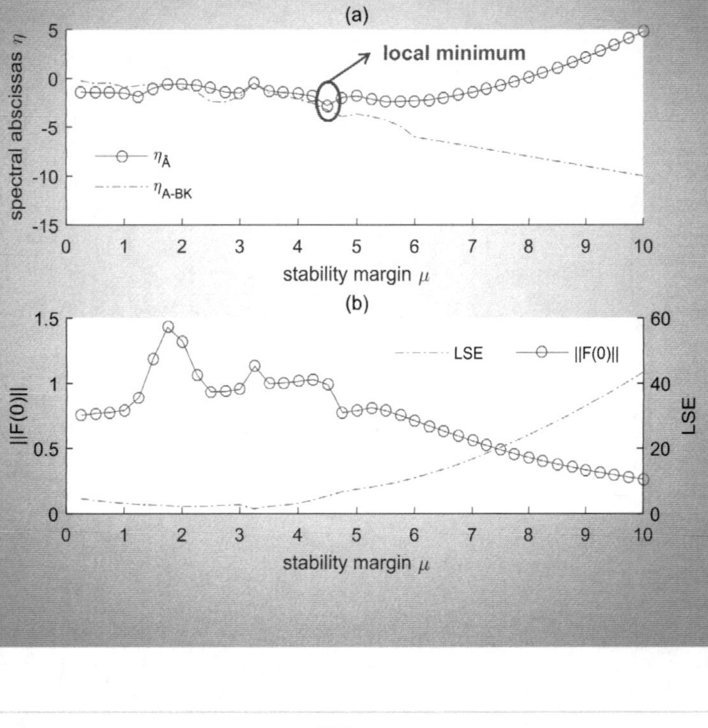

20

Optimal *D* vs. Small *D*

- Integral of Absolute Error (IAE) is used to evaluate reference tracking

$$\text{IAE} = \int_0^{\infty} \left| e(t) \right|_1 dt = \int_0^{\infty} \left| u(t) - \hat{u}(t) \right|_1 dt$$

				IAE	
	small D	optimal D		small D	optimal D
μ	NA	4.50	w/o iteration	13.16	60.91
λ_{A^-}	-0.24±2.03i; -35.90;-61.53	-2.79±0.32i; -2.79±1.60i	w/ iteration	1.06	0.17
η_{A^-}	-0.24	-2.7847			
small *D*	$\begin{bmatrix} -0.1 & 0 \\ 0 & -0.1 \end{bmatrix}$		**optimal *D***	$\begin{bmatrix} 0.18 & -2.75 \\ 5.09 & 3.97 \end{bmatrix}$	

21

Step Disturbance Identification

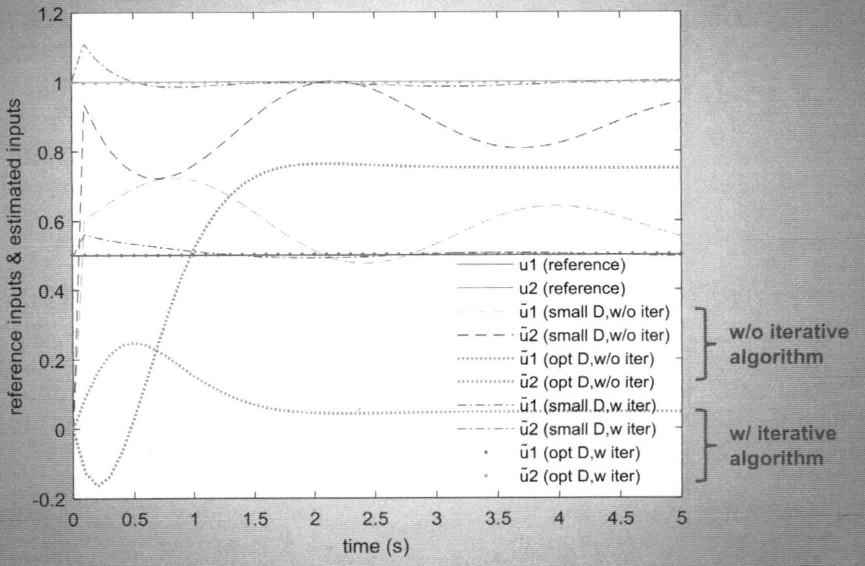

22

Ramp Disturbance Identification

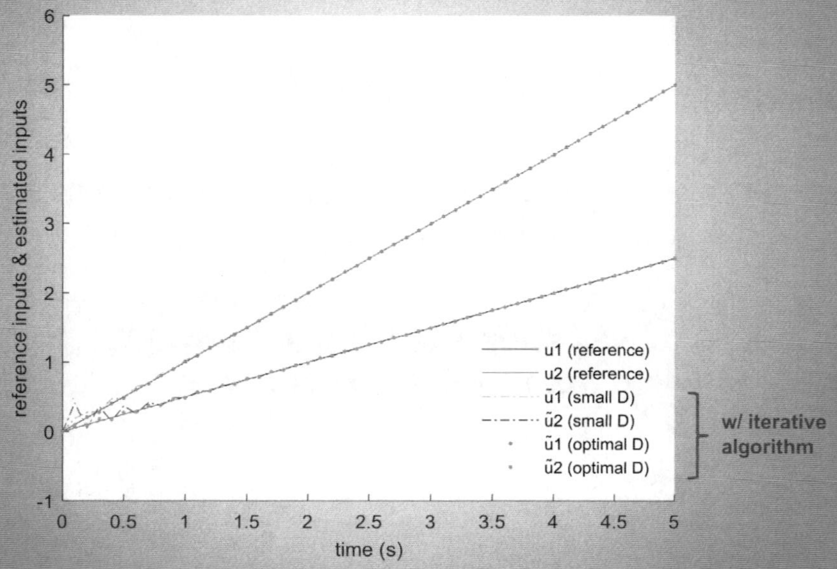

Concluding remarks

- The optimal dummy *D* design
  - provides a systematic way to obtain stable approximate inverse for the strictly proper MIMO system
  - guarantees the asymptotic convergence of the iterative algorithms of most input signals (uniform convergence for step signals)
- Both "every-time" and "point-time" algorithms show significant improvements on reference tracking reflected by smaller IAE
- Current work

  A methodology incorporating the above two methods to be applied for input and state recovery for strictly proper nonlinear MIMO systems

Acknowledgements

- Prof. Mohsen Maesumi from Department of Mathematics at Lamar University;
- Prof. Anthony Michel from Department of Electrical Engineering at University of Notre Dame;
- This work was supported in part by Graduate Scholarship from Lamar University.

25

Questions?

26

Multiscale Based Nonlinear Fault Detection of Industrial Processes

Chiranjivi Botre[a], Jonathan Raftery[a], Majdi Mansouri[b], Mohamed N. Nounou[c], Hazem N. Nounou[b] and M. Nazmul Karim[a]

[a] Artie McFerrin Dept. of Chemical Engineering, Texas A&M University, College Station, Texas 77843, USA
[b] Electrical and Computer Engineering Program, Texas A&M University at Qatar, Doha, QATAR,
[c] Chemical Engineering Department, Texas A&M University at Qatar, Doha, QATAR

1

Process Monitoring

- What is process monitoring?

  Observation of a particular process variables to ensure that it stays within set control limits

- Why carry out process monitoring?

  To ensure safe production

  To maintain product quality

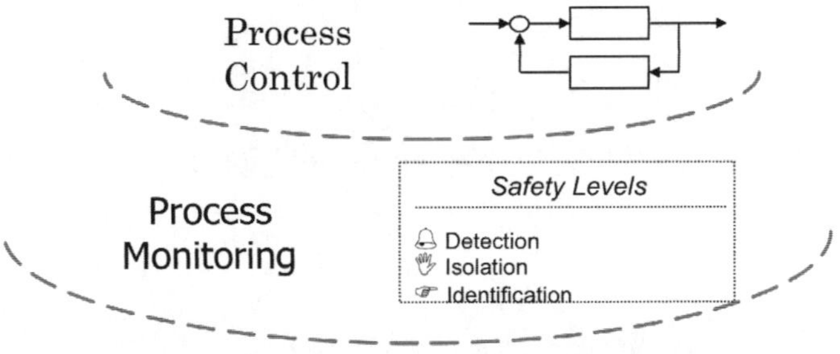

2

Introduction - Faults in Processes

17AIChE Spring Meeting, San Antonio, TX

Artie McFerrin Department of CHEMICAL ENGINEERING TEXAS A&M UNIVERSITY

Abnormal changes in several variables caused by deviation in process

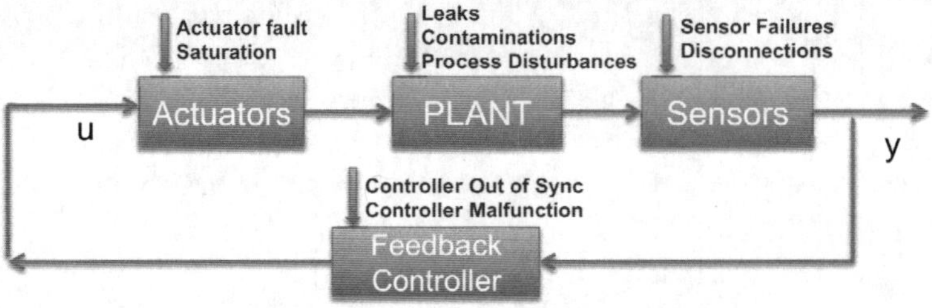

Types of Fault

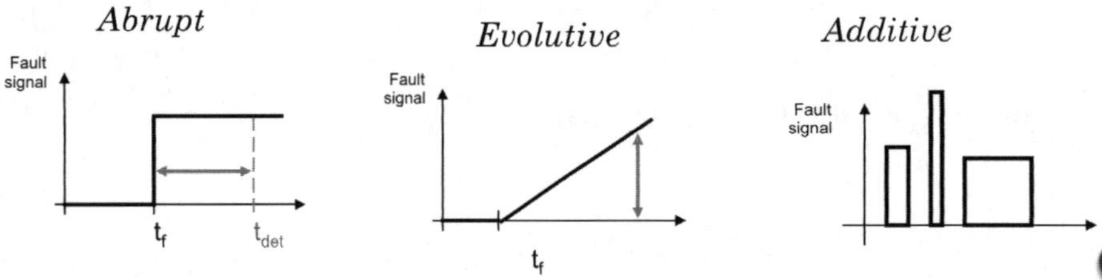

3

Introduction - Fault Detection

17AIChE Spring Meeting, San Antonio, TX

Artie McFerrin Department of CHEMICAL ENGINEERING TEXAS A&M UNIVERSITY

Fault detection techniques can be categorized in three parts

Model based | Knowledge (qualitative) based | Data based

- Obstacles with industrial process data
  - ✓ Complex nonlinearity in data set
  - ✓ Noise in data set
  - ✓ Correlated variables in data set

} Negatively affects fault detection performance

Objective : Develop a method for improving fault detection performance of complex nonlinear industrial processes

Venkatasubramanium et.al. Computers and Chemical Engineering (2003)

4

Introduction - Multivariate Statistical Methods

17 AIChE Spring Meeting, San Antonio, TX

Artie McFerrin Department of CHEMICAL ENGINEERING TEXAS A&M UNIVERSITY

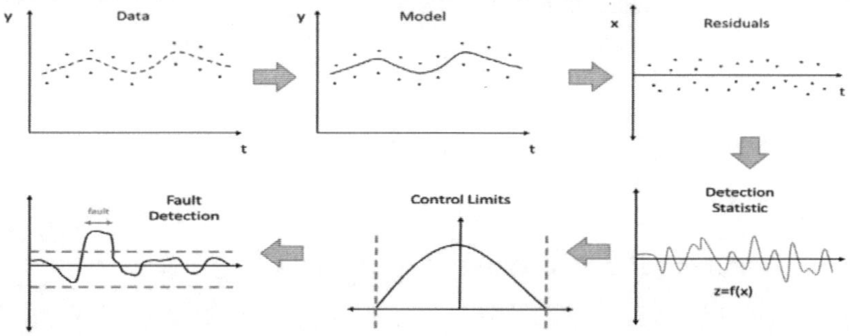

Partial Least Square (PLS):

Process online variables → X

Process quality variables → Y

X ⟷ Y

$$X = t_1 p_1^t + E_1 \qquad Y = u_1 q_1^t + F_1$$

$$Y = BX + \varepsilon$$

Where,

- t_1 and u_1 are the score matrix
- p_1^t and q_1^t are the loading vectors
- B is the regression coefficient

5

De-noising, de-correlating and normalizing data

Wavelet transformation of data

17 AIChE Spring Meeting, San Antonio, TX

Artie McFerrin Department of CHEMICAL ENGINEERING TEXAS A&M UNIVERSITY

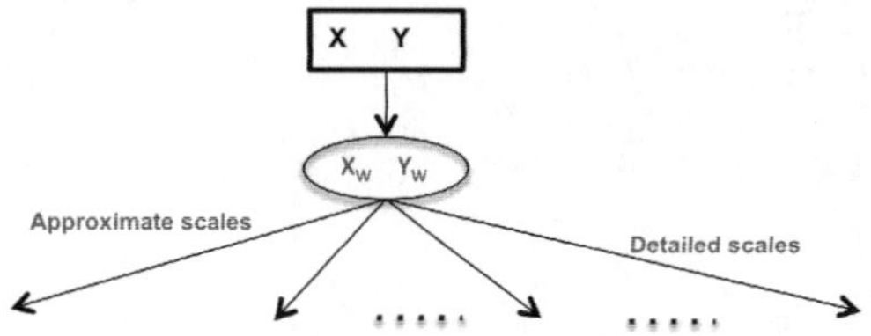

Ability to naturally de-noise signals

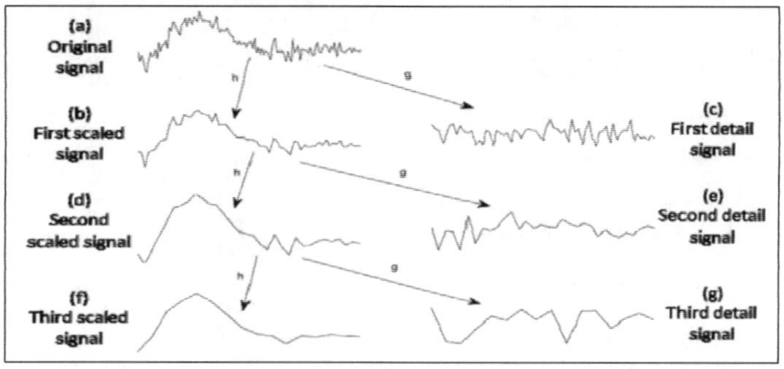

Z. Sheriff et al. 2014 International Conference on Control,Decision and Information Technologies (CoDIT), 2014

6

De-noising, de-correlating and normalizing data

Wavelet transformation of data

17 AIChE Spring Meeting, San Antonio, TX

Artie McFerrin Department of CHEMICAL ENGINEERING TEXAS A&M UNIVERSITY

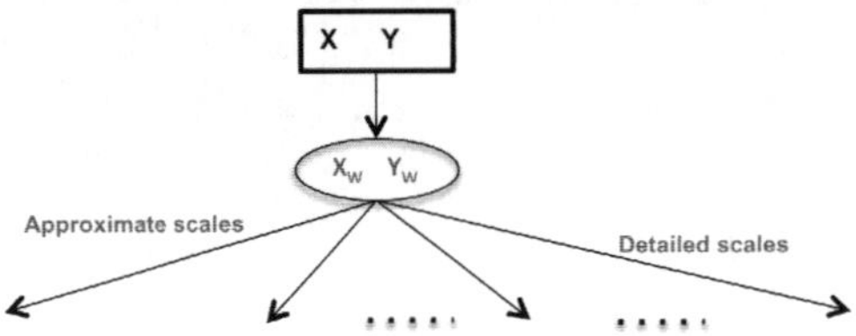

De-correlation of correlated data at multiple scales

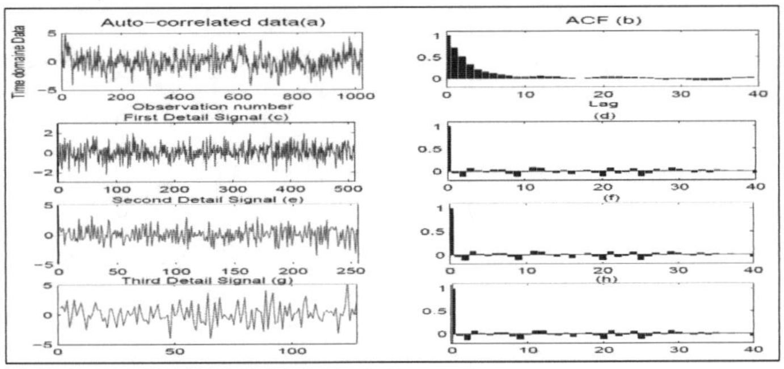

Z. Sheriff et al. 2014 International Conference on Control,Decision and Information Technologies (CoDIT), 2014

7

De-noising, de-correlating and normalizing data

Wavelet transformation of data

17 AIChE Spring Meeting, San Antonio, TX

Artie McFerrin Department of CHEMICAL ENGINEERING TEXAS A&M UNIVERSITY

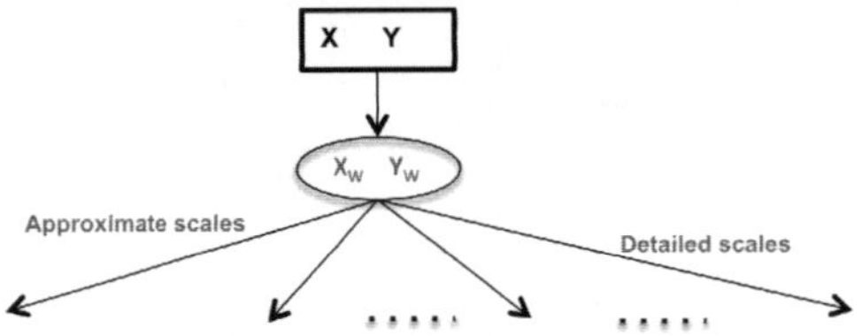

Wavelet coefficients (detail) are Gaussian regardless of distribution of time-domain data

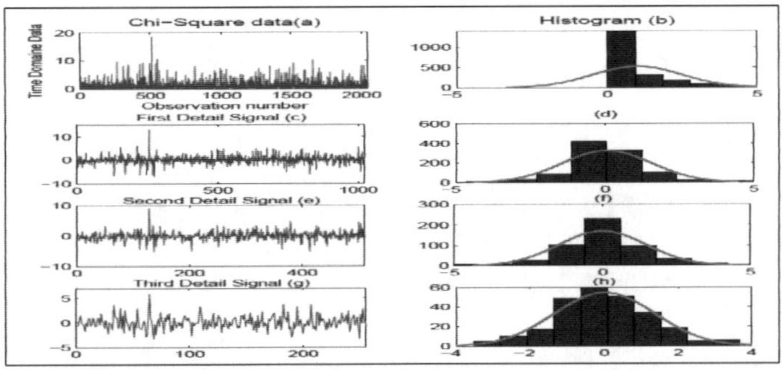

Z. Sheriff et al. 2014 International Conference on Control,Decision and Information Technologies (CoDIT), 2014

8

Nonlinear Multivariate Model

Kernel Partial Least Square

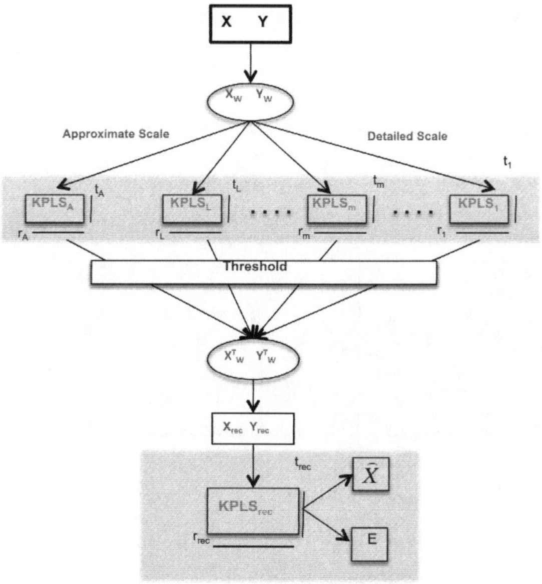

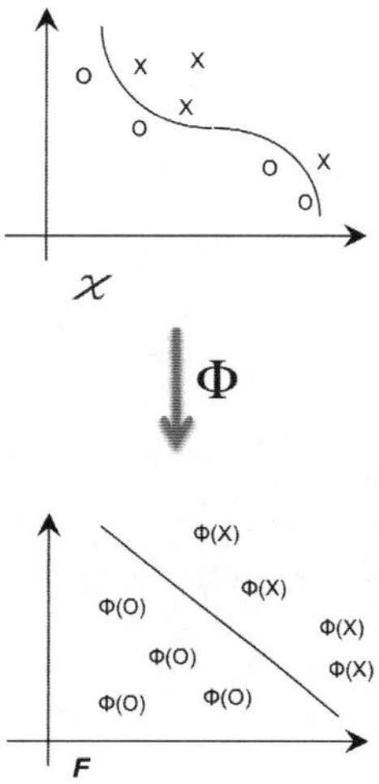

9

Kernel Partial Least Square (KPLS)

- Mercer's theorem states that there exist a mapping into a feature space where a kernel function acts as a dot product.

$$k\left(x_i, x_j\right) = \Phi\left(x_i\right)\Phi\left(x_j\right)$$

- The Gaussian radial basis function (RBF): $K(x,y) = \exp\left(-\frac{\|x-y\|^2}{2h^2}\right)$
- Polynomial Kernel: $K(x,y) = \langle x,y \rangle^d$
- Sigmoid Kernel: $K(x,y) = \tanh\left(\beta_0 \langle x,y \rangle + \beta_1\right)$

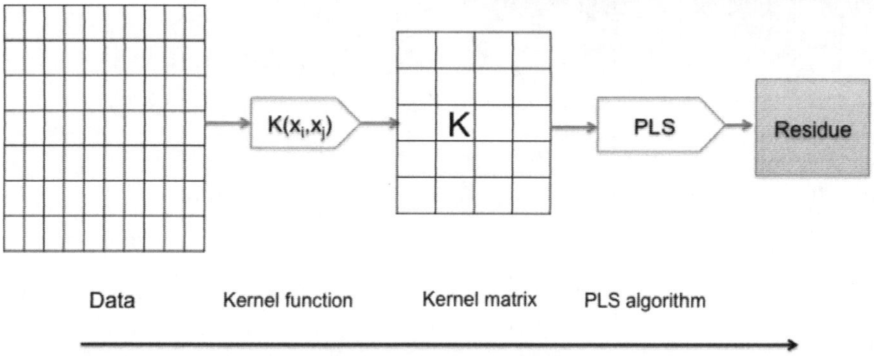

10

Rasipal et. al. The Journal of Machine Learning Research(2001)

Composite FD Statistical Test
Generalized Likelihood Ratio Test (GLRT)

17 AIChE Spring Meeting, San Antonio, TX

Artie McFerrin Department of CHEMICAL ENGINEERING, TEXAS A&M UNIVERSITY

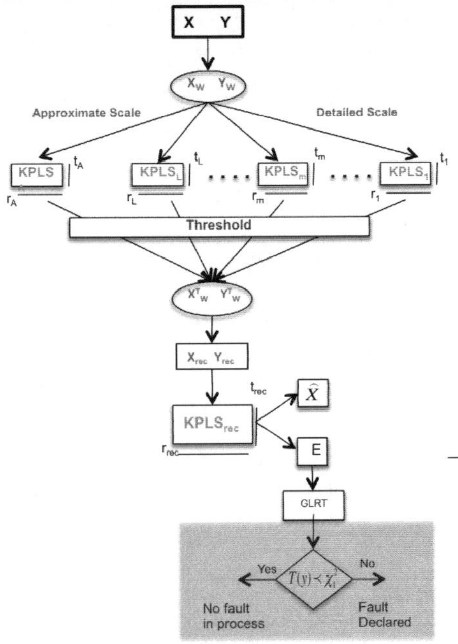

- GLR test - solves composite hypothesis testing problem

$$\begin{cases} H_0 = \{E \sim N(0, \sigma^2 I_N)\} & \text{Null Hypothesis} \\ H_1 = \{E \sim N(\theta, \sigma^2 I_N)\} & \text{Alternative Hypothesis} \end{cases}$$

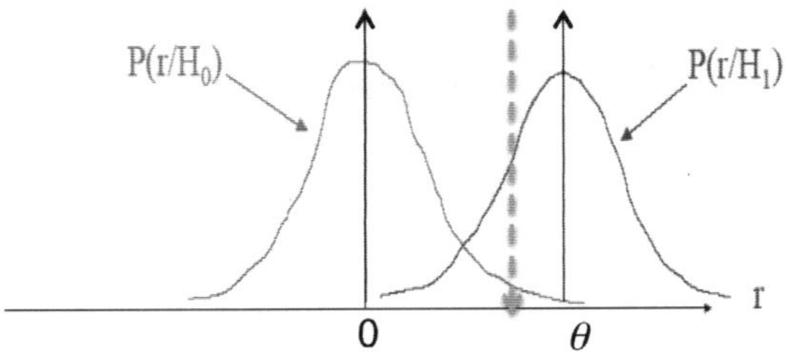

Fault declared:- Statistical mean value (θ) outside threshold limit

11

Botre et al., Journal of Loss Prevention in the Process Industries (2016)

Composite FD Statistical Test
Generalized Likelihood Ratio Test (GLRT)

17 AIChE Spring Meeting, San Antonio, TX

Artie McFerrin Department of CHEMICAL ENGINEERING, TEXAS A&M UNIVERSITY

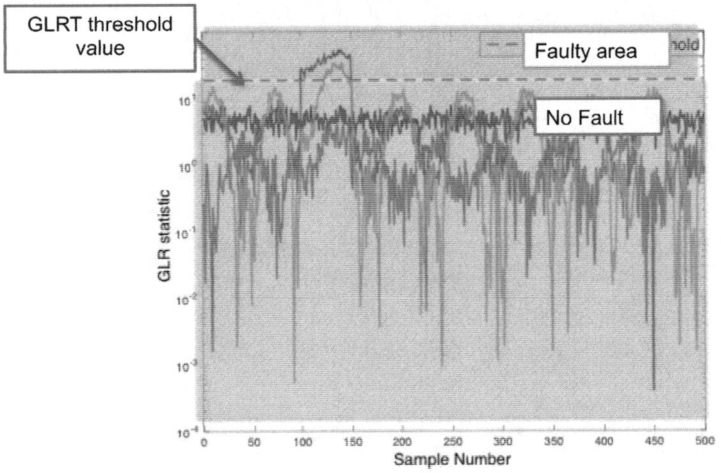

Fault detection performance can be evaluated based on two parameters

1. **Missed detection rate (MDR) (%)** : percentage of faulty observations undetected
2. **False alarm rate (FAR) (%)** : percentage of wrong fault declared in fault free region

12

Kernel Partial Least Square (KPLS)

17 AIChE Spring Meeting, San Antonio, TX

Artie McFerrin Department of CHEMICAL ENGINEERING TEXAS A&M UNIVERSITY

$$K(x,y)=\exp\left(-\frac{\|x-y\|^2}{2\times h^2}\right)$$

Parameter value has an impact on fault detection efficiency.
Value of h will determine the spread of the transformed data

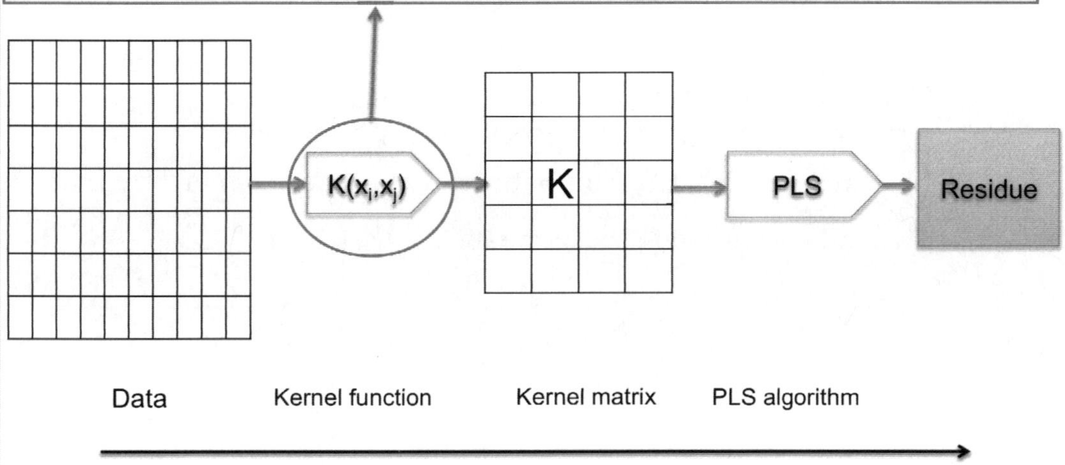

13

Selection of Kernel Parameter

17 AIChE Spring Meeting, San Antonio, TX

Artie McFerrin Department of CHEMICAL ENGINEERING TEXAS A&M UNIVERSITY

$$K(x,y)=\exp\left(-\frac{\|x-y\|^2}{2\times h^2}\right)$$

Parameter value has an impact on fault detection efficiency.
Value of h will determine the spread of the transformed data

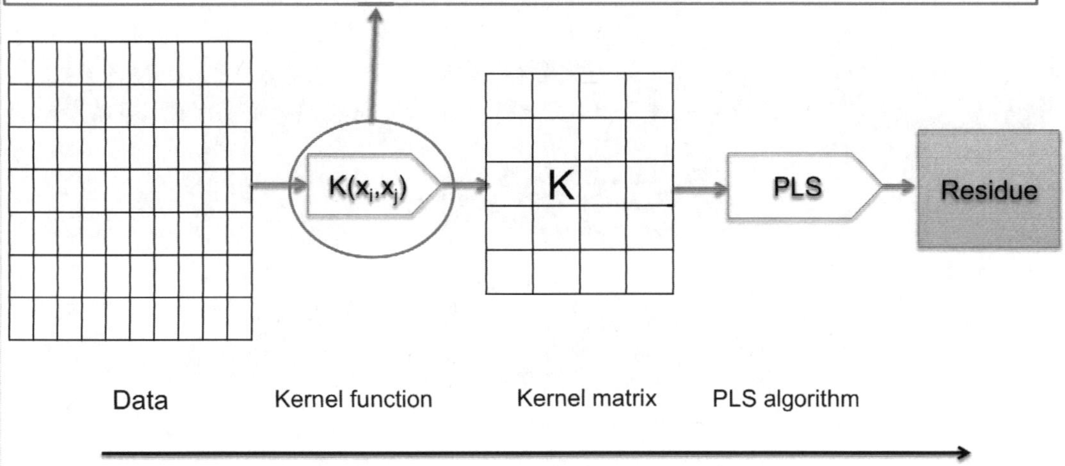

14

Multiobjective Optimization

Pareto Optimality

17AIChE Spring Meeting, San Antonio, TX — Artie McFerrin Department of Chemical Engineering, Texas A&M University

$$\underset{h}{\text{Min}} \quad (\alpha_1 \times \text{Missed detection rate} + \alpha_2 \times \text{False alarm rate})$$

s.t.

Kernel PLS algorithm

GLRT statistic algorithm

$$K(x,y) = \exp\left(-\frac{\|x-y\|^2}{2 \times h^2}\right)$$

- Multi objective genetic algorithm optimization is used to minimize both missed detection rate and false alarm rate.

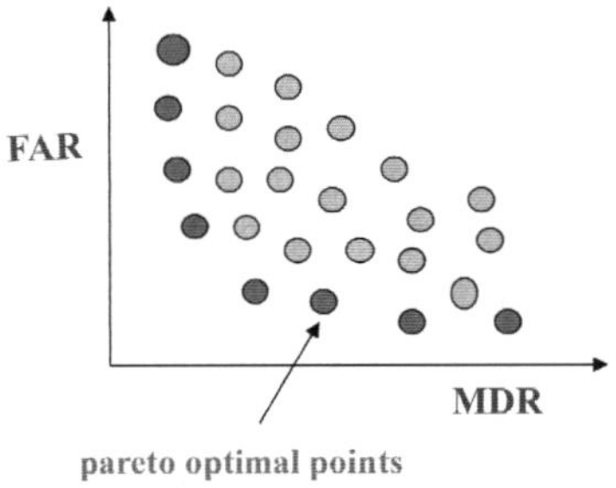

15

Overview of MS-KPLS Algorithm

17AIChE Spring Meeting, San Antonio, TX — Artie McFerrin Department of Chemical Engineering, Texas A&M University

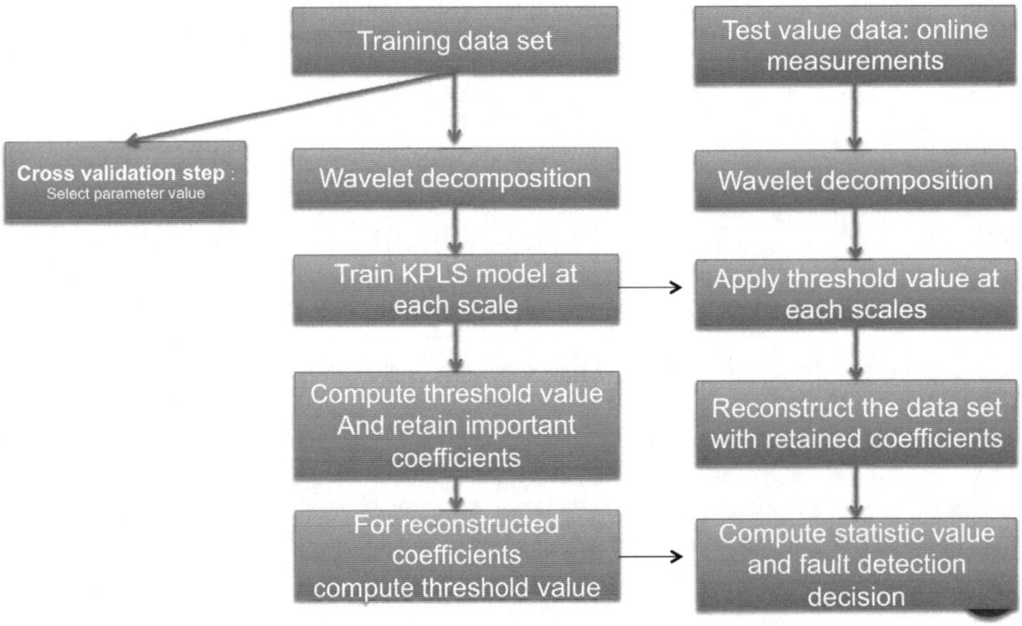

Case Study

A&M ENGINEERING
TEXAS A&M UNIVERSITY

Case Study I: CSTR Simulation

17 AIChE Spring Meeting, San Antonio, TX

Artie McFerrin Department of CHEMICAL ENGINEERING TEXAS A&M UNIVERSITY

The dynamic model for the non-isothermal CSTR

$$\frac{dC_A}{dt} = \frac{F}{V}(C_{A0} - C_A) - K_0 e^{\frac{-E}{RT}} C_A$$

$$\frac{dT}{dt} = \frac{F}{V}(T_0 - T) + \frac{(-\Delta H)k_0}{\rho C_p} e^{\frac{-E}{RT}} C_A - \frac{Q}{V\rho C_p}$$

$$Q = \frac{aF_c^{b+1}}{F_c + \left(\frac{aF_c^b}{2\rho_c C_{pc}}\right)}(T - T_{cin})$$

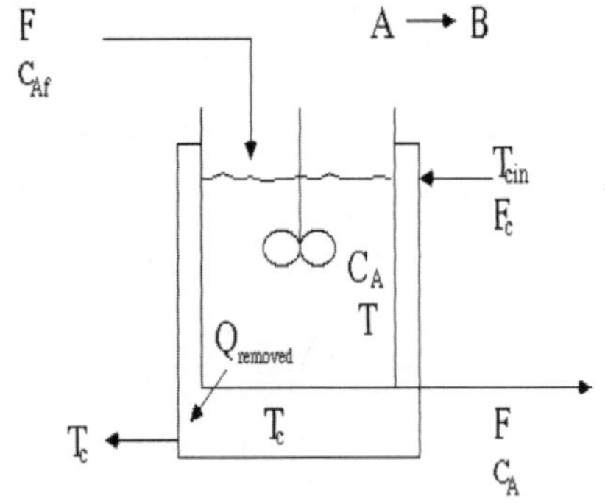

- The outlet temperature (T) and the concentration (CA) are controlled using proportional integral (PI).
- Gaussian noise is introduced in the process variables at random sample time

Mansouri et al., Journal of Loss Prevention in the Process Industries (2015)

18

CSTR :- Step I

Pareto Optimal Front

- Selection of the kernel function parameter

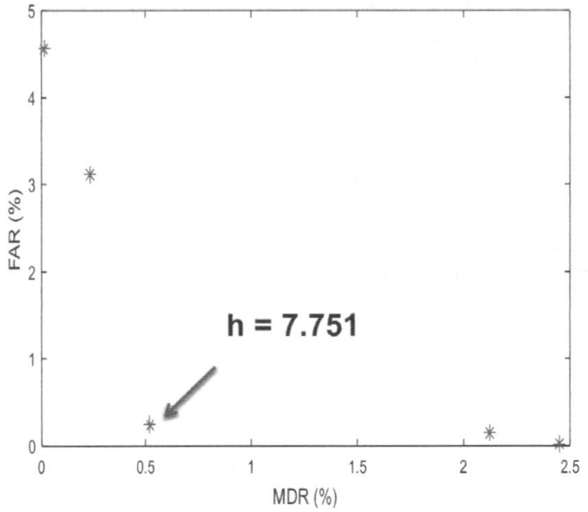

Thus, kernel parameter value (h) = 7.751

19

IDV 14 Fault Results

KPLS based Q^2 statistic

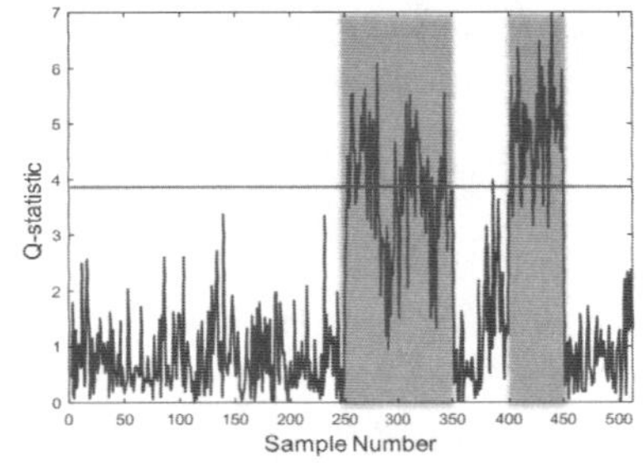

KPLS based GLRT

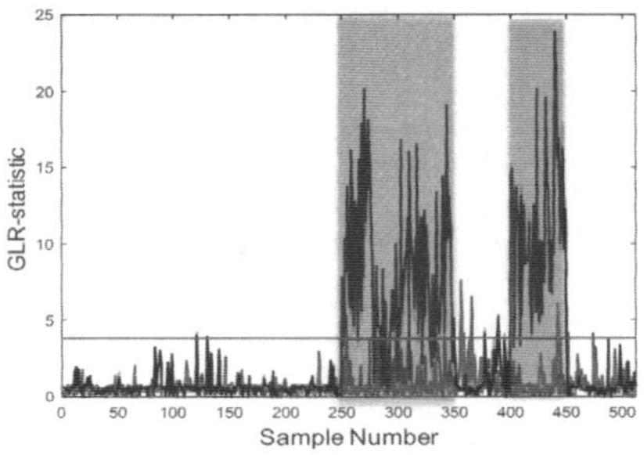

MS-KPLS based GLRT

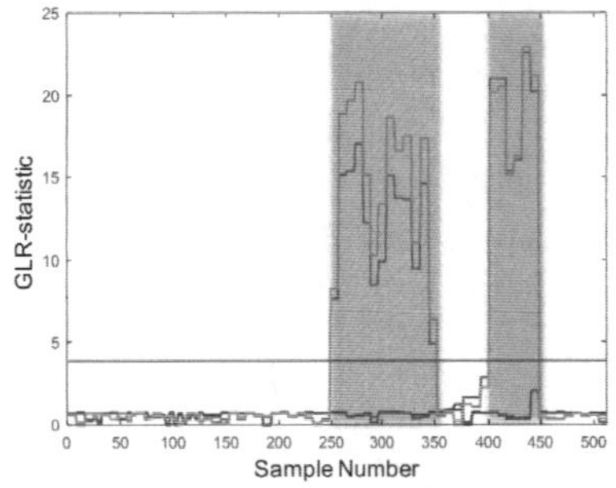

	False alarm (%)	Missed detection (%)
KPLS Q-square	0.28	41.33
KPLS GLRT	3.71	12.23
MS-KPLS GLRT	**0**	**0**

20

Case Study II: Tennessee Eastman Process (TEP) Problem

17AIChE Spring Meeting, San Antonio, TX

Artie McFerrin Department of CHEMICAL ENGINEERING TEXAS A&M UNIVERSITY

- Realistic simulation program of a chemical plant and is widely accepted as a benchmark for control and monitoring studies. It is a continuous process.
- The reaction scheme is given as

$$A(g) + C(g) + D(g) \longrightarrow G(liq)$$

$$A(g) + C(g) + E(g) \longrightarrow H(liq)$$

$$A(g) + E(g) \longrightarrow F(liq)$$

$$3D(g) \longrightarrow 2F(liq)$$

Downs, J. J. and Vogel, Computers and Chemical Engineering (1993)

21

Case Study: Tennessee Eastman Process (TEP) Problem

17AIChE Spring Meeting, San Antonio, TX

Artie McFerrin Department of CHEMICAL ENGINEERING TEXAS A&M UNIVERSITY

TEP Process Flow Diagram

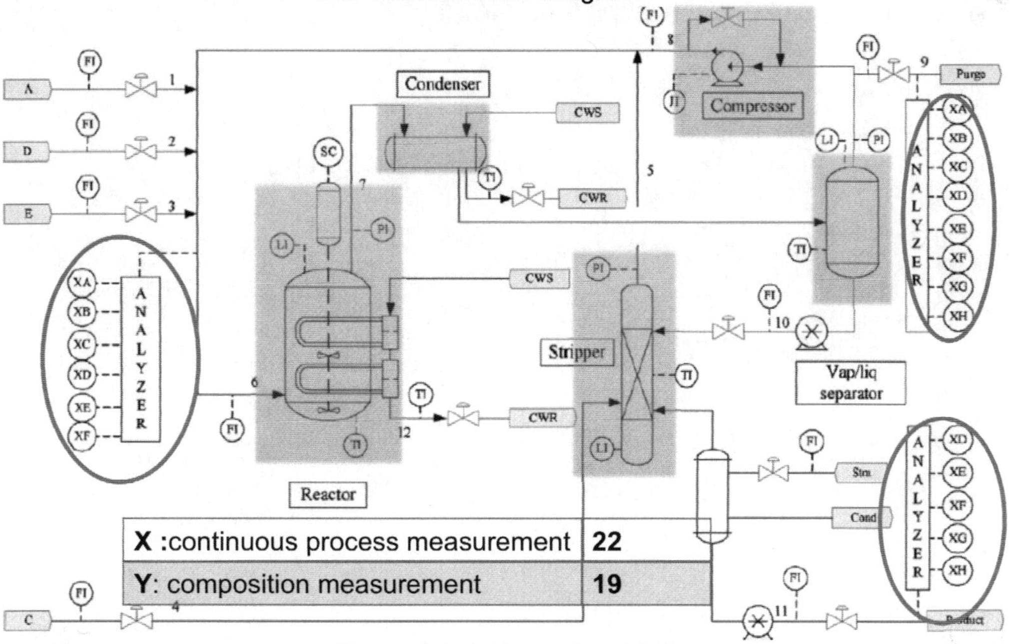

Yin et al., Journal of Process Control (2012)

Faults in Process Variables

17 AIChE Spring Meeting, San Antonio, TX

Artie McFerrin Department of CHEMICAL ENGINEERING TEXAS A&M UNIVERSITY

Fault number	Process variable	Type
IDV(1)	A/C feed ratio, B composition constant	Step
IDV(2)	B composition, A/C ration constant	Step
IDV(3)	D feed temperature	Step
IDV(4)	Reactor cooling water inlet temperature	Step
IDV(5)	Condenser cooling water inlet temperature	Step
IDV(6)	A feed loss	Step
IDV(7)	C header pressure loss-reduced availability	Step
IDV(8)	A, B, and C feed composition	Random variation
IDV(9)	D feed temperature	Random variation
IDV(10)	C feed temperature	Random variation
IDV(11)	Reactor cooling water inlet temperature	Random variation
IDV(12)	Condenser cooling water inlet temperature	Random variation
IDV(13)	Reaction kinetics	Slow drift
IDV(14)	Reactor cooling water valve	Sticking
IDV(15)	Condenser cooling water valve	Sticking
IDV(16)	Unknown	Unknown
IDV(17)	Unknown	Unknown
IDV(18)	Unknown	Unknown
IDV(19)	Unknown	Unknown
IDV(20)	Unknown	Unknown
IDV(21)	The valve fixed at steady state position	Constant position

- IDV 2 and IDV 14 fault case studied
- Total process time T=48 hrs.
- Fault was introduced at t = 225 sample time.

Yin et al., Journal of Process Control (2012)

23

TEP:- Step I

Pareto Optimal Front

17 AIChE Spring Meeting, San Antonio, TX

Artie McFerrin Department of CHEMICAL ENGINEERING TEXAS A&M UNIVERSITY

Selection of the kernel function parameter

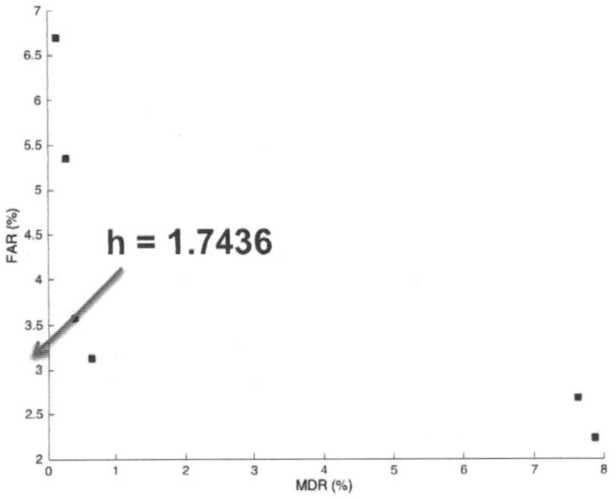

Thus, kernel parameter value (h) = 1.7436

24

IDV 2 Fault Results

17 AIChE Spring Meeting, San Antonio, TX

Artie McFerrin Department of CHEMICAL ENGINEERING TEXAS A&M UNIVERSITY

KPLS based Q^2 statistic

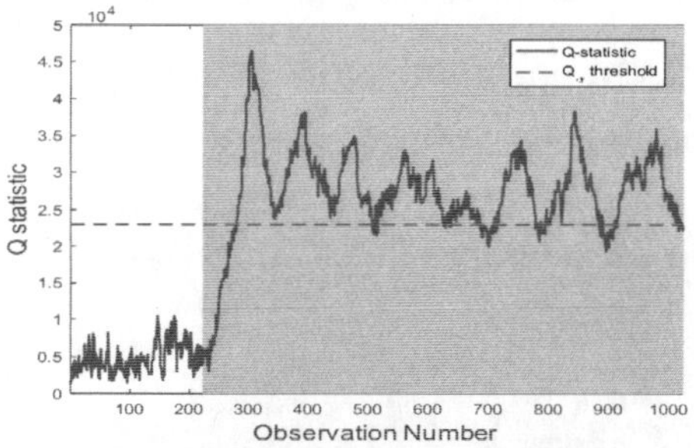

KPLS based GLRT

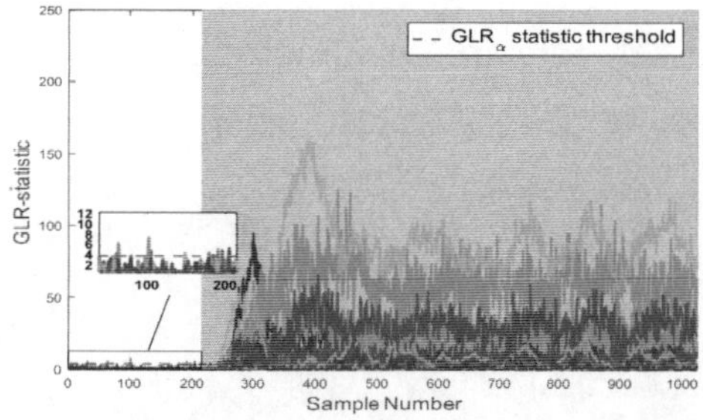

MS-KPLS based GLRT

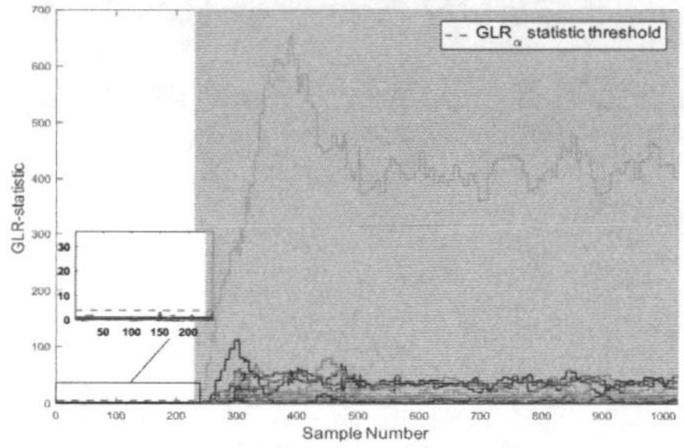

	False alarm (%)	Missed detection (%)
KPLS Q-square	0	17.2150
KPLS GLRT	6.23	2.122
MS-KPLS GLRT	**0**	**0.93**

25

IDV 14 Fault Results

17 AIChE Spring Meeting, San Antonio, TX

Artie McFerrin Department of CHEMICAL ENGINEERING TEXAS A&M UNIVERSITY

KPLS based Q^2 statistic

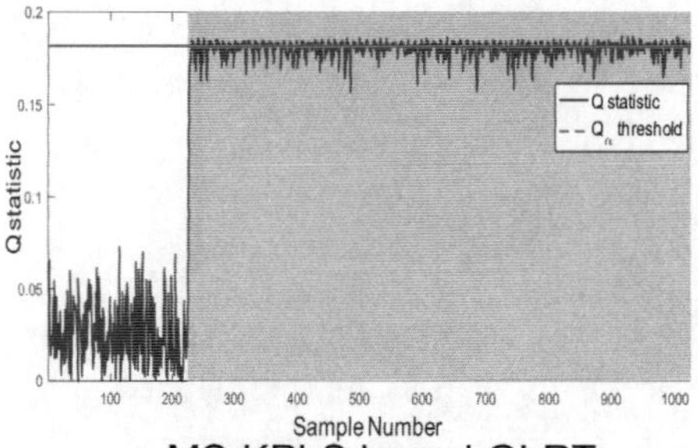

KPLS based GLRT

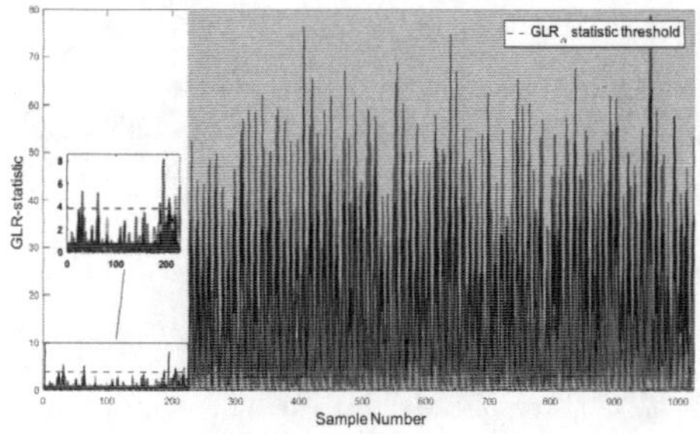

MS-KPLS based GLRT

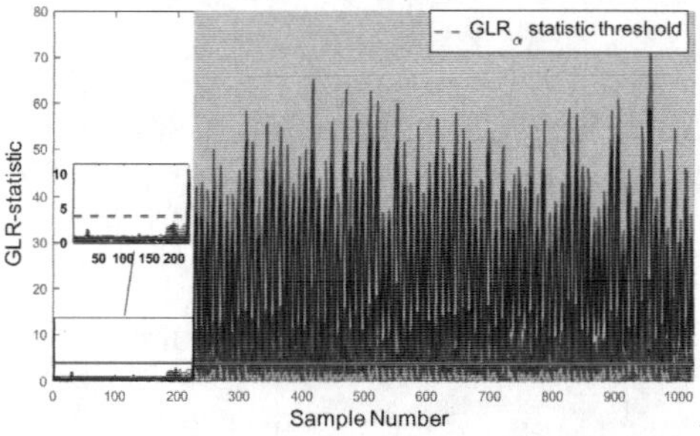

	False alarm (%)	Missed detection (%)
KPLS Q-square	0	42.89
KPLS GLRT	8.23	1.03
MS-KPLS GLRT	**0**	**0**

26

Conclusion

17AIChE Spring Meeting, San Antonio, TX

Artie McFerrin Department of CHEMICAL ENGINEERING TEXAS A&M UNIVERSITY

- Superior fault detection performance of developed MS- KPLS based GLRT with optimized parameter compared to conventional multivariate methods

- Pareto optimality improves efficiency of the kernel partial least square (KPLS) model

- Fault detection results of CSTR and TEP case study demonstrates that developed model can be successfully applied to the processes which have,
  - ✓ Complex nonlinearity in data set
  - ✓ Noise in data set
  - ✓ Auto-correlated variables in data set

27

Acknowledgements

17AIChE Spring Meeting, San Antonio, TX

Artie McFerrin Department of CHEMICAL ENGINEERING TEXAS A&M UNIVERSITY

Artie McFerrin Department of CHEMICAL ENGINEERING TEXAS A&M UNIVERSITY

TEXAS A&M ENERGY INSTITUTE TEXAS A&M UNIVERSITY

الصندوق القطري لرعاية البحث العلمي
Qatar National Research Fund
Member of Qatar Foundation عضو في مؤسسة قطر

- We would like to thank the QNRF (Qatar national research foundation) for their funding of this work.

I would also like to thank Dr. Karim research group for their help in this work.

28

17 AIChE
Spring Meeting, San Antonio, TX

Artie McFerrin Department of
CHEMICAL ENGINEERING

Thank You

Questions?

Recent Advances in MVC Technology

Darwin Logerot
ProSys
2822 O'Neal Lane Bldg A Baton Rouge, LA 70816
Darwin.Logerot@prosys.com

Vidya Kuppuraj
ProSys
2822 O'Neal Lane Bldg A Baton Rouge, LA 70816
Vidya.Kuppuraj@prosys.com

Prepared for Presentation at
American Institute of Chemical Engineers
2017 Spring Meeting and 13th Global Congress on Process Safety
San Antonio, TX
March 26 – 29, 2017

Recent Advances in MVC Technology

Darwin Logerot
ProSys
2822 O'Neal Lane Bldg A Baton Rouge, LA 70816
Darwin.Logerot@prosys.com

Vidya Kuppuraj
ProSys
2822 O'Neal Lane Bldg A Baton Rouge, LA 70816
Vidya.Kuppuraj@prosys.com

Keywords: Multi-Variable Control, MVC

Abstract

The Multi-Variable Controller (MVC) is ubiquitous in refining and petrochemicals industries. The MVC technology in recent years has changed a lot from its start three decades ago. The underlying architecture of MVC is still the same but there have been changes in the modeling methodology. Typical MVC in each cycle execution performs three steps: i) Open loop prediction ii) Steady state optimization and iii) Generates dynamic move plan. A linear programming problem is formulated in steady state optimizer routine to compute optimized targets. A dynamic move plan is generated from the optimized targets and open loop prediction. The move plan from the dynamic optimization is downloaded to the lower level regulatory control. The set points of regulatory control is used as Manipulated Variables (MV) in MVC. The latest in MVC technology is MVC directly manipulating the control valve output (OP) to control the process. The model used for this purpose is called PV (Process Variable) model, where MVs are PVs. The MVC control execution is same as the traditional MVC but the dynamic move plan is computed in terms of PV change and transformed to OP change via OP-PV transformations. This OP change is downloaded to DCS to implement the move. The advantage of this method is that a valve can operate to its saturation limit without losing model fidelity. PV Models also increase a Plant's Capacity by increasing the feasible operating region of the process.

1 Introduction

Regulatory control has been implemented and running successfully in all process industries for decades. Regulatory control is a SISO (Single Input Single Output) model. Modern process plants are thermally integrated for energy optimization purposes and any change in process flow has an effect downstream of the process. This thermal integration makes modern process plants very complex and regulatory control loops are insufficient to achieve the desired performance required by the process engineers. In reality, modern chemical plants are multi-variable in nature and require a MIMO (Multi Input and Multi Output) model to control the process effectively. The Multi-Variable Control (MVC) which exploits the multi-variable nature of the process is best suited for the modern complex chemical plants. MVC was developed three decades ago and has been successfully deployed in numerous process plants. Each MVC installation can have benefits of several millions of dollars.

MVC control uses a multi-variable model to predict the responses of the plant. Based on a performance measure this model computes the optimal targets depending on current state and computes a dynamic move plan to achieve the optimal target value. MVC technology has changed a lot in recent years from its start three decades ago. The underlying architecture of MVC is still the same but there have been changes in the modeling methodology. Typical MVC performs three steps in each cycle of execution: i) Open loop prediction ii) Steady state optimization and iii) Generates dynamic move plan. A linear programming problem is formulated in steady state optimizer routine to compute optimized targets. A dynamic move plan is generated from the optimized targets and open loop prediction. The move plan from the dynamic optimization is downloaded to the lower level regulatory control. This article will discuss the recent advances in the MVC technology.

2 Dynamic Matrix Control (DMC)

The heart of the Multi-Variable Controller is the dynamic model of the process used by MVC for control purpose. The accurate process dynamics is normally represented by step response models. The model matrix which consists of model responses for dependents/controlled variables (CV) to each independent/manipulated variable (MV) is represented by step response model. A CV can be affected by several MVs and these responses are recorded in the model matrix of the MVC controller. The step response model captures the dynamic response and the steady state gain associated for CV response for each MV. The step response coefficients of this MIMO model is represented in a matrix form which is called Dynamic Matrix and the controller which uses this dynamic model for control is called a Dynamic Matrix Controller. In the model matrix, manipulated variables are usually setpoints of the regulatory control. In some cases, the valve output can also be a manipulated variable. The controlled variables are those which are affected by the movement of the manipulated variables. There is a subset of controlled variables which have to be maintained within certain limits for safe and efficient operation.

3 MVC Overview

All commercially available MVC technologies use a dynamic model matrix to represent the cause and effect nature of the process. The dynamic models are normally step response models between manipulated variables and dependent variables. The MVC controller uses the process model to compute optimal target values based on a performance measure. The performance measure is a cost criterion which uses the MV costs to compute an optimum target value. The MVC controller cycle consists of three steps: i) Open loop prediction ii) Steady state optimization and iii) Generate dynamic move plan.

Open loop prediction uses the process model and the previous MV moves to compute an open loop prediction of the process. MVC controllers usually have more control variables than manipulated variables. This non square nature of the MVC controller is squared-up in the steady state optimization. The steady state optimizer uses the current value, MV costs, and the operator limits to compute steady state targets. The dependent variable limits are posed in terms of manipulated variables by using steady state gain of the process model. The decision variables in the steady state optimization are manipulated variables. A linear programming problem is formulated and solved for optimal targets. The MV and CV operator limits form the feasible region of the linear program. The number of active constrains is equal to the number of manipulated variables. In the dynamic move plan step, a move plan is generated for manipulated variables using an error vector which is computed as the difference between steady state target and open loop prediction. The optimal moves computed are downloaded to the DCS for implementation.

4 Recent Advances in MVC

The Dynamic Model Matrix structure in an MVC controller is entirely dependent on the lower level regulatory configuration. The models in the MVC controller change when there are changes in the tuning of a regulatory controller. This is because the dynamics of the regulatory controller gets altered by changing the PID controller tuning settings. In reality, dynamic models in an MVC controller have to be updated for PID tuning changes. Since the model changes can be tolerated using feedback, MVC models are not updated for PID tuning changes. When there is a change in the mode of the regulatory control, it necessitates changes in the model structure. For example, if a PID controller is changed from automatic mode to manual mode, it will trigger changes in the downstream of the PID controller and the MVC models have to be updated to mimic the process accurately. In this case, the model update is not a straightforward procedure. It necessitates retesting of the process for the new PID configuration which involves additional cost and engineering time. Recent advances in the modeling techniques allow computing an updated model matrix when there is a change in the PID mode configuration. This is performed by pivoting the model matrix and by swapping MVs and CVs. There is no need for additional testing to update the model matrix. The accuracy of the updated models depends on the accuracy of the original model matrix.

The performance of a MVC depends on the accuracy of the model matrix used for control purposes. There are some situations where the model matrix no longer accurately depicts the process. For example, the flooding of a distillation column invalidates the model used in the MVC controller. This is because tower flooding is not represented by the process model in that differential pressure of the tower is a steady state variable during the normal operation becomes a ramp variable during tower flooding scenario. This change in the behavior of the tower flooding is not represented by the models and hence models become invalid. The same situation happens when the valve output is saturated. The valve saturation also invalidates the models in the MVC controller. For the same reason as tower flooding, in some cases, the valve saturation can lead to a ramp like behavior for some variables which normally are a steady state variable. This change in the behavior is not adequately captured by the model matrix. Therefore, it is important to keep all the valve outputs in control so that the process model represent the actual process behavior. The control of normally saturating valves results in giving up some capacity of the process and lesser throughput because these valves are to be operated close to their saturation limit. One of the latest advances in MVC technology is direct manipulation of the control valve output (OP) to control the process. The model used for this purpose is called PV (Process Variable) model, where MVs are PVs.

In the PV model, PVs (process variables) are the manipulated variables in the MVC controller. The dynamic move calculation computes the PV change necessary to achieve the SS targets. The PV change is transformed to OP (valve output) change using valve transformations. The valve transformations used are updated regularly and OP change computed is downloaded to the DCS to implement the move. The advantage of using PV models is that the valve saturation no longer invalidates the models and there is no need to sacrifice the process throughput to keep the control valves in control. PV Models also increase a Plant's Capacity by increasing the feasible operating region of the process. The advantage of this method is that valves can operate to their saturation limit without losing model fidelity.

480375 Model Predictive Control for Ethylene Oxide Explosion Limits Under Plant Upset Conditions

Abstract:

In this paper, a model predictive controller is designed to for an ethylene oxide plant using industry-standard distributed control system and dynamic process simulation in tandem. The controller is tested for ability to remain within explosion limits and ability to maintain optimal operation parameters under normal operation and plant upset conditions, including sudden failure in the lean cycle water stripping column and cooling duty availability in the reactor cooling systems.

PSE @ LSU

PROCESS SYSTEMS ENGINEERING AT LOUISIANA STATE UNIVERSITY

LSU

SMATDAP NSF

The Smart MATerials Design Analysis and Processing Consortium

REAL TIME OPTIMAL CONTROL AND STATE ESTIMATION IN SEMI-BATCH FREE-RADICAL POLYMERIZATION REACTORS

Santiago D. Salas, Navid Ghadipasha, Wenbo Zhu, Jose A. Romagnoli

Louisiana State University, Cain Department of Chemical Engineering

Terry Mcafee, Wayne F. Reed

Tulane University, Department of Physics and Engineering Physics

Presentation Outline

- Introduction
- Framework
    - *Modelling*
    - *Dynamic optimization*
    - *Dynamic state estimation*
    - *Dynamic state estimation & feedback control*
- Results
- Conclusions and Future Work

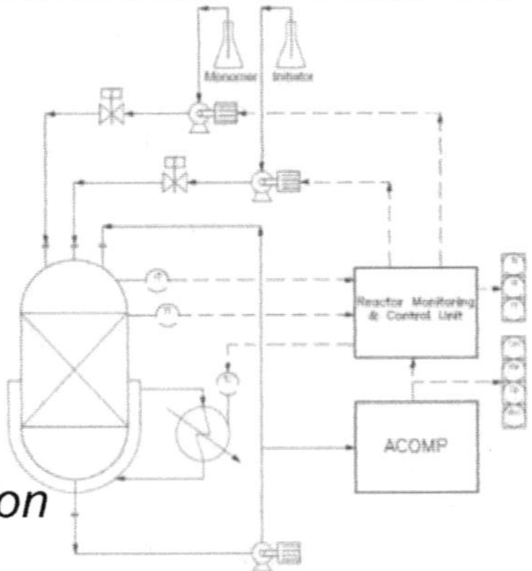

2

Introduction

- **Model-based control** is potentially important in **polymer processing** to achieve desired end-product characteristics.
  - Typically this is done offline to achieve optimal operating conditions/trajectories that define end-product quality.
  - A shortcoming of using advanced operating and control strategies in polymerization processes is the **unavailability of sensors** to determine online process status.

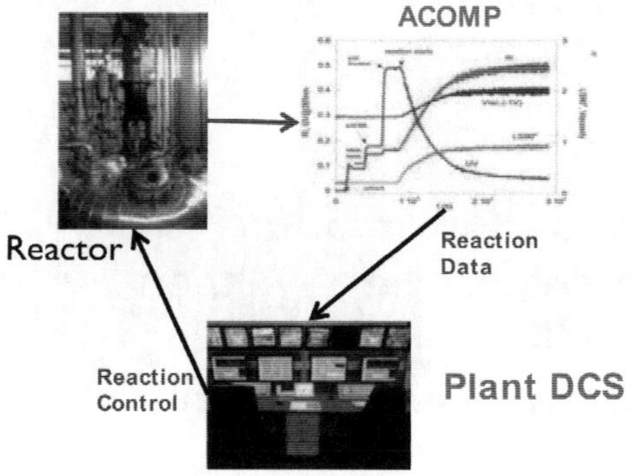

Automatic Continuous Online Monitoring of Polymerization Reactions (ACOMP)

- Florenzano, F. H.; Strelitzki, R.; Reed, W. F. Absolute, on-line monitoring of molar mass during polymerization reactions. *Macromolecules* **1998**, *31*, 7226-7238.
- Wayne F. Reed, A. M. A. Monitoring polymerization reactions: From fundamentals to applications. **2013**.
- McAfee, T.; Leonardi, N.; Montgomery, R.; Siqueira, J.; Zekoski, T.; Drenski, M. F.; Reed, W. F. Automatic Control of Polymer Molecular Weight during Synthesis. *Macromolecules* **2016**, *49* (19), 7170-7183.

Introduction

- In this work, a **dynamic filter for nonlinear systems** and a feedback controller are implemented for:
  - Online state estimation,
  - Data reconciliation,
  - Molar mass distribution monitoring, and
  - Control of polymerization processes.

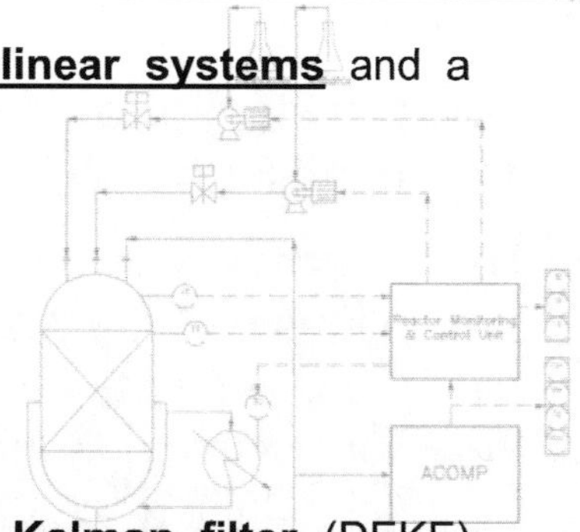

- Specifically, a **discrete-time extended Kalman filter** (DEKF) and **feedback controller** are designed and blended for advance control and monitoring of a free-radical polymerization reaction that synthetizes polyacrylamide using potassium persulfate as initiator in a semi-batch reactor.

- For improving the **robustness** of the filter, its free parameters are tuned using a metaheuristic technique.

Introduction

- **State estimation permits**:
  - **Estimate polymer properties** that cannot be measured directly.
  - **Reconciling measured values** and reducing their noise for real-time control application.

- **A tailor made filter/controller module** is formulated and implemented in Python environment supporting multiple programming paradigms and evaluated in real-time experiments.

Modelling

- As a first step towards, a flexible **mathematical model** was developed for **batch** and **semi-batch operations**.
- The model was then **validated** by comparing the simulations with literature data, and data of batch and semi-batch experiments performed at the Tulane Center for Polymer Reaction Monitoring and Characterization.
- The model was then **used** to:
  - Perform **parameter estimation** and obtain updated kinetic parameters for the system.
  - Dynamic process **optimization** to find optimal recipes for given targets
  - Perform filtering and control studies

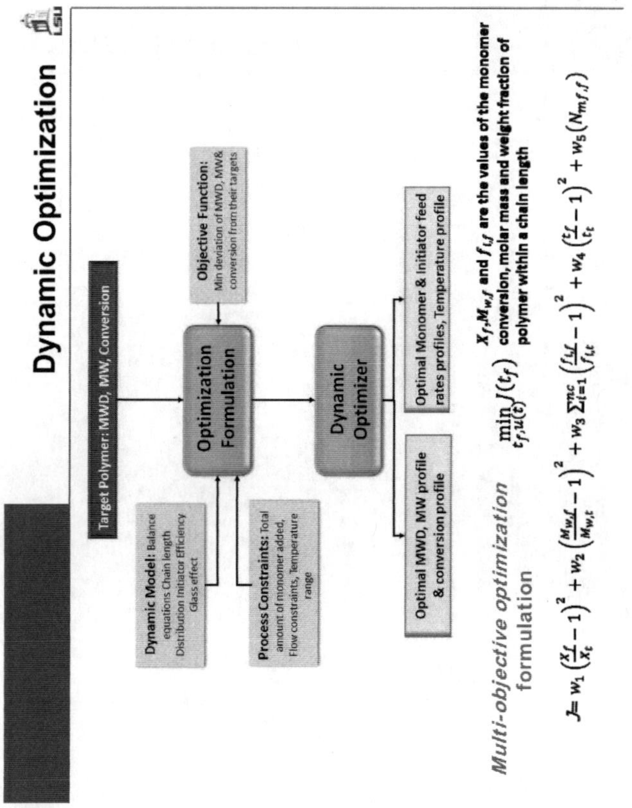

[redacted]

Dynamic Optimization

Increasing Mw

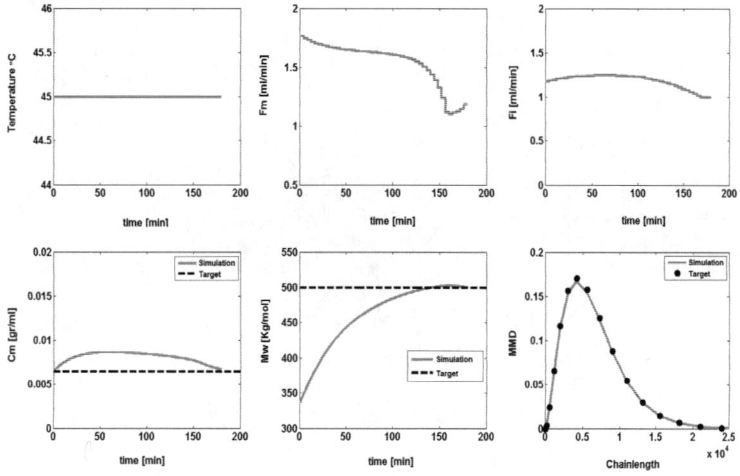

Manipulated variables

Controlled variables (targets)

$$J = w_1\left(\frac{X_f}{X_t}-1\right)^2 + w_2\left(\frac{M_{w,f}}{M_{w,t}}-1\right)^2 + w_3\sum_{i=1}^{nc}\left(\frac{f_{i,f}}{f_{i,t}}-1\right)^2$$

8

[redacted]

Dynamic Optimization

Constant Mw

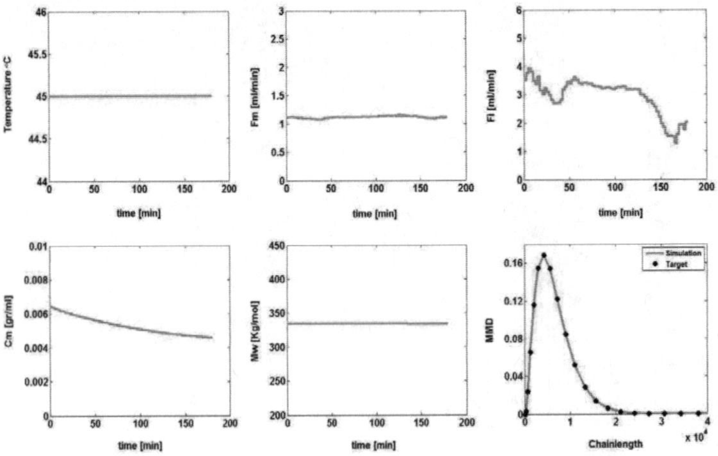

Manipulated variables

Controlled variables (targets)

$$J = w_1 \int_0^{t_f} \left| \frac{\partial M_w}{\partial t} \right| dt + w_2 \sum_{i=1}^{nc} \left(\frac{f_{i,f}}{f_{i,t}} - 1 \right)^2 + w_3 (N_{mf,f})$$

9

[redacted]

Dynamic Optimization

Decreasing Mw

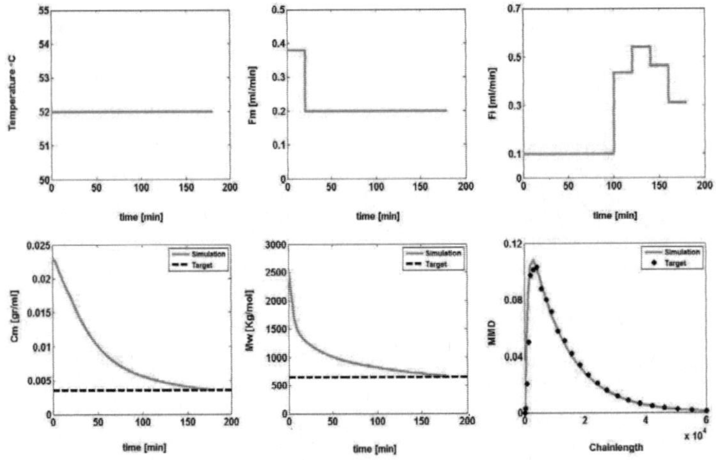

Manipulated variables

Controlled variables (targets)

$$J = w_1 \left(\frac{X_f}{X_t} - 1 \right)^2 + w_2 \left(\frac{M_{w,f}}{M_{w,t}} - 1 \right)^2 + w_3 \sum_{i=1}^{nc} \left(\frac{f_{i,f}}{f_{i,t}} - 1 \right)^2 + w_5 (N_{mf,f})$$

10

Dynamic Optimization

Experimental conditions in all trajectories

Trajectory Description	Initial N_m [mol]	Initial N_s [mol]	Initial N_i [mol]	C_{if} [mol/m3]	C_{mf} [mol/m3]
Increasing M_w optimal trajectory	0.05	30.0	0.008	3.69932	1406.866
Constant M_w optimal trajectory	0.05	30.0	0.008	3.69932	1406.866
Decreasing M_w optimal trajectory	0.10	35.0	0.008	3.69932	1406.866

Ghadipasha, N.; Geraili, A.; Romagnoli, J. A.; Castor, C. A.; Drenski, M. F.; Reed, W. F. Combining On-Line Characterization Tools with Modern Software Environments for Optimal Operation of Polymerization Processes. *Processes* **2016**, *4* (1), 5.

11

Dynamic State Estimation

Model: $\boldsymbol{x_k} = f_{k-1}(\boldsymbol{x_{k-1}}, \boldsymbol{u_{k-1}}) + \boldsymbol{w_k}, \quad \boldsymbol{w_k} \sim N[0, \boldsymbol{Q_k}]$

$\boldsymbol{y}_k = h_k(\boldsymbol{x}_k) + \boldsymbol{v}_k, \quad \boldsymbol{v}_k \sim \mathrm{N}[0, \boldsymbol{R_k}]$

DEKF: $minE[\tilde{\boldsymbol{x}}_k^T \tilde{\boldsymbol{x}}_k], \; \tilde{\boldsymbol{x}}_k = \boldsymbol{x}_k - \hat{\boldsymbol{x}}_k$

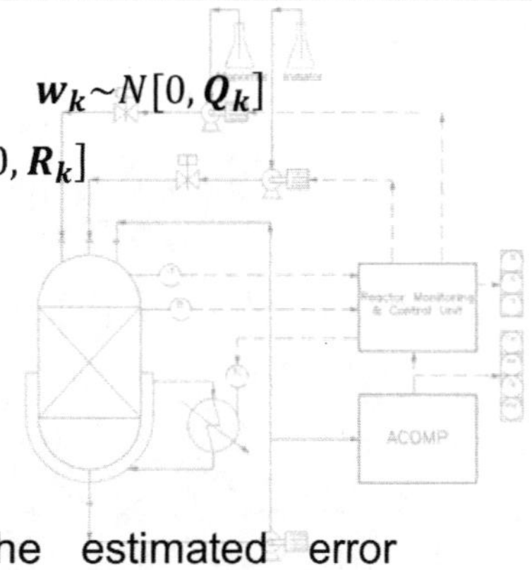

$$\boldsymbol{P}_k^- = \mathbf{F}_{k-1}\boldsymbol{P}_{k-1}^+\mathbf{F}_{k-1}^T + \mathbf{Q}$$

$$\hat{\mathbf{x}}_k^- = f_{k-1}(\hat{\mathbf{x}}_{k-1}^+, \boldsymbol{u}_{k-1}, 0)$$

$$\boldsymbol{K}_k = \boldsymbol{P}_k^- \boldsymbol{H}_k^T\left(\boldsymbol{H}_k \boldsymbol{P}_k^- \boldsymbol{H}_k^T + \boldsymbol{R}\right)^{-1}$$

$$\hat{\mathbf{x}}_k^+ = \hat{\mathbf{x}}_k^- + \mathbf{K}_k[\boldsymbol{y}_k - h_k(\hat{\mathbf{x}}_k^-, 0)]$$

$$\boldsymbol{P}_k^+ = (\mathbf{I} - \mathbf{K}_k\mathbf{H}_k)\boldsymbol{P}_k^-$$

P is the estimated error covariance matrix and it is an indicator of the estimation accuracy and **K** is the Kalman gain matrix.

Simon, D. Optimal state estimation: Kalman, H infinity, and nonlinear approaches. *John Wiley & Sons* **2006**.

12

Dynamic State Estimation

- Stability properties of the estimation error dynamics can be analyzed through the eigenvalues of the matrix $\mathbf{F} - \boldsymbol{K}\mathbf{H}$.
- Covariance matrices ***P***(0), ***Q*** and ***R*** contain the DEKF's **free parameters**.
- State estimation accuracy and data reconciliation ability of the DEKF depend largely on the quality of these parameters and their tuning can be a tedious task for large systems.
- A metaheuristic methodology for tuning them is proposed for an intelligent search of optimal parameters.

13

Dynamic state estimation & feedback control

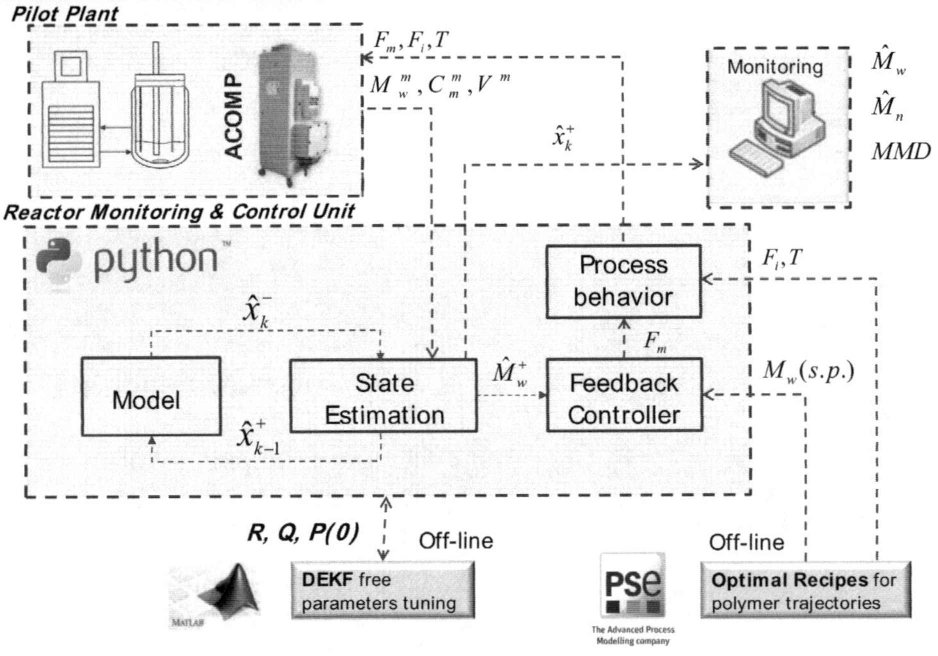

14

Dynamic state estimation & feedback control

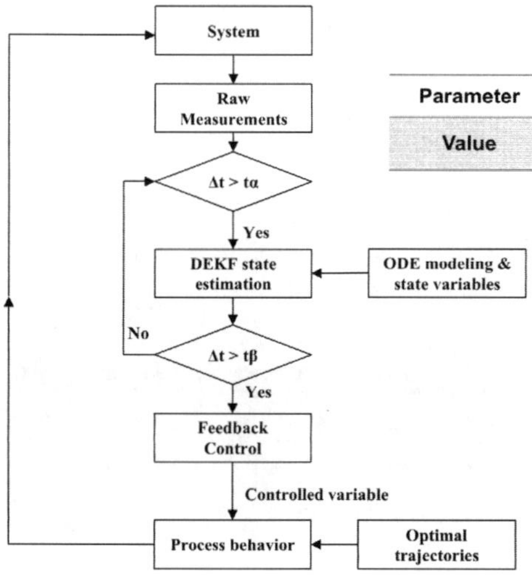

Parameter	k_p	k_i	k_d	t_α	t_β
Value	2E-08	6.667E-10	2E-08	15 sec	2.5 min

The **DEKF** is set in the inner loop at a **higher frequency** than the outer loop where the control action is allocated. Although the DEKF is set in the high-frequency loop, it is still considered as a discrete system.

Time interval set for state estimation is denoted as t_α and for control action t_β.

For operating the module:
$1\,sec < t_\alpha < t_\beta$.

15

Results

DEKF Free parameters

Parameter	Value
P(0)	diag([6.2868E-03, 1.0E-06, 1.0E-04, 71.7574, 1.7437E-02, 9.8167])
Q	diag([6.1502E-03, 3.56E-07, 8.21E-05, 8.697, 3.9721E-03, 4.1485E-03])
R	diag([1.0E-02, 3.3417, 2.9621E-11])

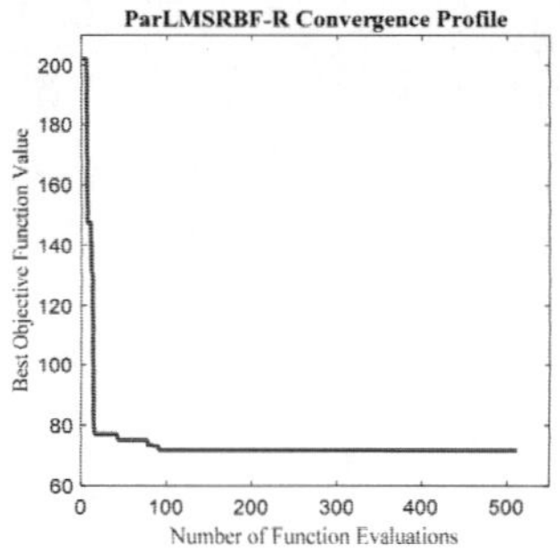

Convergence profile when tuning DEKF free parameters using *ParLMSRBF*-R

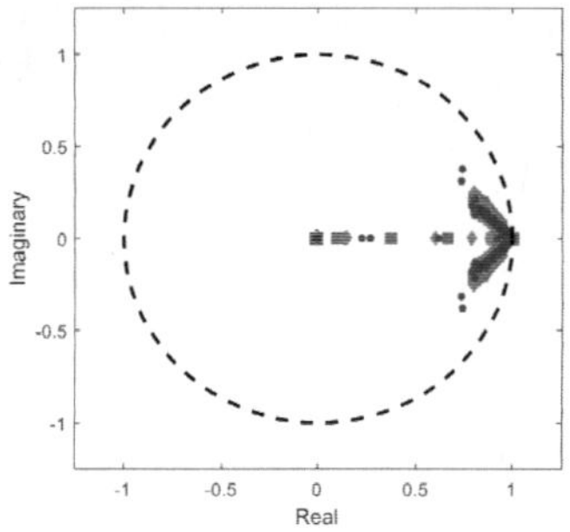

Eigenvalues of the estimation error for the three OL experiments

16

Open Loop - Results

Open Loop <u>increasing</u> Mw trajectory.

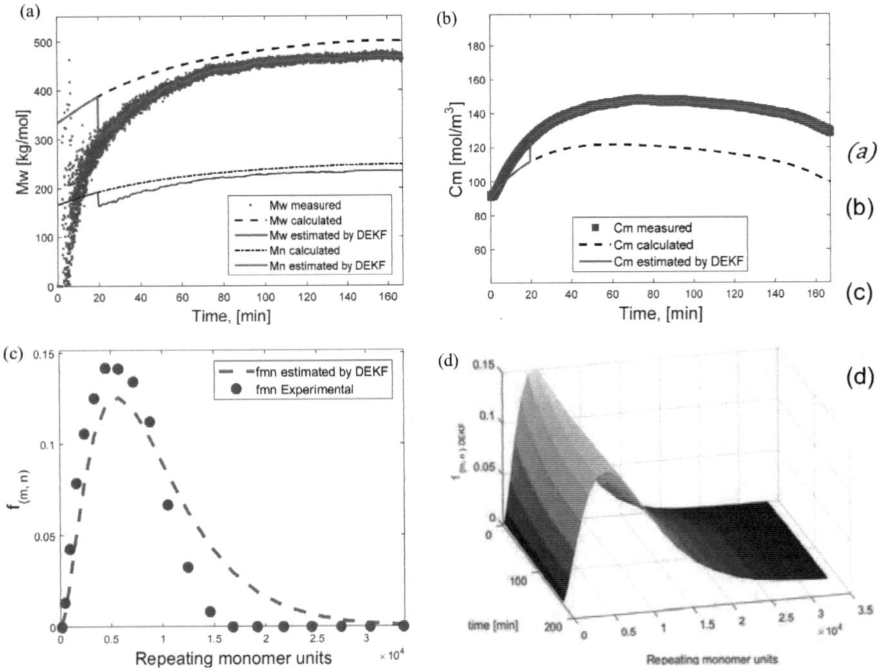

(a) M_w and M_n **trajectories**,
(b) Monomer concentration **trajectories**,
(c) Final chain length distribution by **DEKF** and **experimental**,
(d) Chain length distribution estimated by DEKF **along reaction**.

17

480015 Advanced Control Implementation By Dynamic Matrix Control Pyrolysis Gasoline Hydrogenation Units in Braskem Plant

Abstract:

Advanced Control Implementation by Dynamic Matrix Control Pyrolysis Gasoline Hydrogenation units in Braskem Plant

Authors: Ribeiro, Pleycienne Trajano; Reis, Ivã das Neves; Paim, Anderson; Ribeiro, Joana Arnaldo; Amaral, Francisco Franco.

E-mail:pleycienne.ribeiro@braskem.com;
iva.reis@braskem.com; anderson.paim@braskem.com; joana.ribeiro@braskem.com; francisco.amaral@braskem.com

Keywords: Dynamic Matrix Control; Pyrolysis Gasoline Hydrogenation; Optimization

Advanced controllers based models are being increasingly employed in industrial units, especially in the petrochemical sector due to the advancement of technologies that facilitate the development and the application of these controllers. This type of application provides adaptability and robustness in the operability of the unit, and provide models capable of driving the process for optimum operating conditions.

All phases leading to the development of an advanced control necessarily undergo a comprehensive review of the basic regulatory control, instrumentation, process limits and operational restrictions

Thus, this study aims to describe the process of deploying an advanced process control based on dynamic array in a Pyrolysis Gasoline Hydrogenation (PGH) unit located in basic petrochemicals complex at Braskem, in Camaçari, Bahia. Control strategies have been developed with the objectives to stabilize the variations of the unit feed flow, to avoid instabilities, minimize the consumption of energy and maximize the production of streams with higher added value.

The PGH units has high flow rates, high energy consumption and has the function of di-hydrogenated gasoline pyrolysis produced in naphtha crackers, for the production of BTX (benzene, toluene and xylenes) and other byproducts. In this unit configuration, the first equipment is a depentanizer column that has the function of recovering raw C5 of the heavy pygas. This stream is sent to the isoprene production unit, which also produces piperylenes, DCPD and C5 raffinate. In sequence, there is a column that removes the C10+ compounds from the Pygas . This stream is sent to the unit fuel oil pool, and is burned in others boilers and furnaces of the complex. Subsequently occurs and the selective catalytic hydrogenation of diolefins, olefins and sulfur compounds in two reactors. After the reactors, the lights compounds and the hydrogen sulfide are removed in a striper colmun and the last main equipment is a deoctanizer column, where occurs the separation between dihydrogenated C9 stream and C6-C8 stream. The first stream is directed to automotive gasoline production and the second one is sent to the aromatics extraction units for benzene, toluene and xylenes production.

The developed control strategy provided a reduction in the unit's energy consumption by 10.0%, and also improved the recovery from the main products streams, without compromising its specifications as follows: increase of raw C5 stream by decreasing the C5 content in the depentanizer bottom stream and also decreasing the C6 content in the raw C5 stream; lower of production of fuel oil by increasing the production of automotive gasoline and a higher C6-C8 production by lowering the Aromatics C8 content in the automotive gasoline stream and promoted a gain of US$ 2,109,000.00 per year. The results of the new advanced process control proved the effectiveness of this type of implementation in industrial units, resulting in measurable financial gains and providing qualitative gains in the form of unit operation.

480968 From Field to Fuel: The Effect of Biomass Feedstock Composition on Fast Pyrolysis, Vapor Phase Upgrading, and Hydrotreatment

Abstract:

Fast pyrolysis of biomass followed by hydrotreating of the resulting pyrolysis liquid can produce a low-oxygen hydrocarbon liquid which holds promise as a refinery blendstock that could be used to produce transportation fuels. However many questions remain. The biomass feedstock affects the yield of pyrolysis oil and its physical and chemical properties. Catalytic upgrading of the pyrolysis vapors produces a higher-quality pyrolysis oil but in lower yield than standard uncatalyzed fast pyrolysis. The quality of the oil affects the yield and operability of the hydrotreating, including the number of stages required to produce a satisfactory intermediate or finished blendstock. Thus the best and most cost-effective path is still in question.

Therefore the Pacific Northwest National Laboratory (PNNL), Idaho National Laboratory (INL), and National Renewable Energy Laboratory (NREL), have collaborated to study the effect of feedstock properties and vapor-phase upgrading (VPU) on the yield and quality of the pyrolysis and hydrotreated products. Feedstocks were prepared and characterized at INL, fast pyrolysis and vapor phase upgrading on select samples was performed at NREL, and upgrading of the bio-oils to a finished fuel blendstock was performed at PNNL. Multiple linear regression models were used to correlate feedstock composition to bio-oil yields and qualities. Within experimental error, oil yields from fast pyrolysis and upgrading are shown to be linear functions of the blend components. The water content of fast pyrolysis oils was also observed to increase linearly with the summed content of potassium and sodium, ranging from 21% for clean pine to 37% for switchgrass. Pyrolysis oil yield on a dry basis (unit mass of dry oil per unit mass of dry feedstock) was observed to be affected primarily by three components (predictor variables) in the feedstock: volatile matter, total lignin, and the summed content of potassium and sodium. Importantly, the predictor variables are correlated, primarily because the herbaceous materials have high lignin and alkali metal contents and low volatile matter contents and also exhibit lower fast pyrolysis oil yields than the woody materials. Ongoing work is extending the model to predict the yield through the complete process to hydrocarbons.

BURNS & McDONNELL

Understanding California's Low Carbon Fuel Standard (LCFS)

March 2017

Agenda

- Overview
- Renewable Fuel Standard
- California Carbon Cap and Trade
- California Low Carbon Fuel Standard
- LCFS vs RFS

Overview

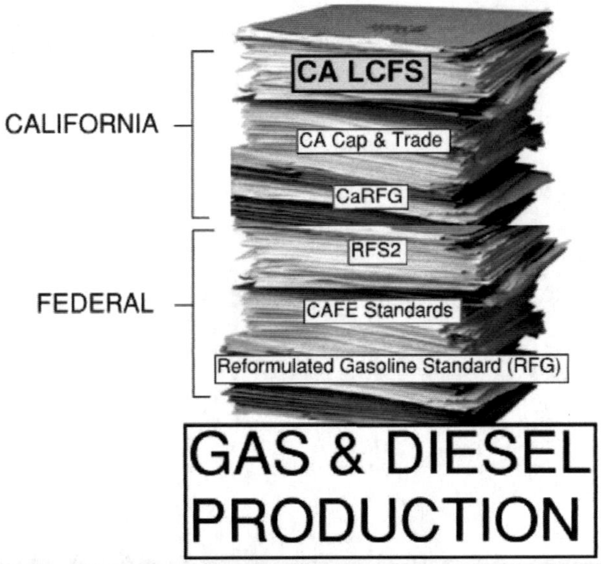

Renewable Fuel Standard (RFS)

- Federal program administered by the EPA mandating a minimum volume of renewable fuel in transportation fuel
  - Energy Policy Act of 2005 (RFS1)
  - Energy Independence and Security Act of 2007 (RFS2)
- Increasing mandated volumes up to 36 billion gallons in 2022
- Mandated volumes are updated regularly to attempt correction for demand and availability

Renewable Fuel Standard- Original

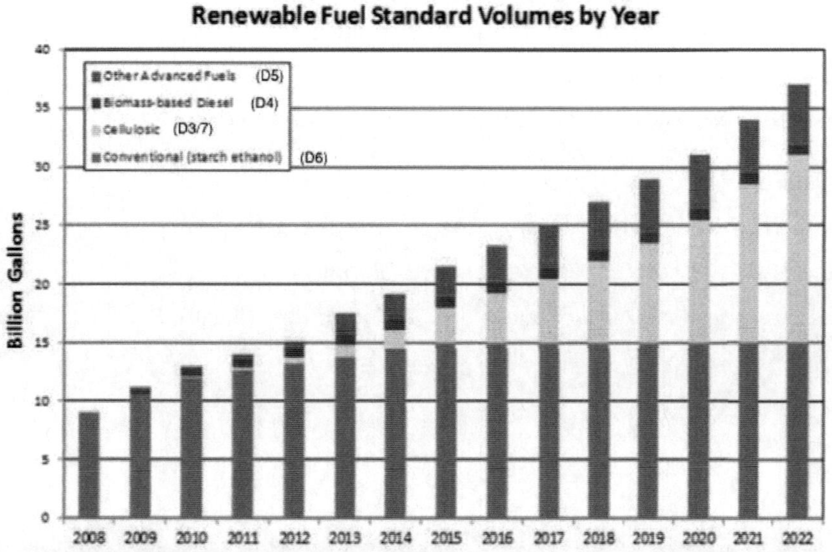

BURNS & McDONNELL

Renewable Fuel Standard- With Actuals

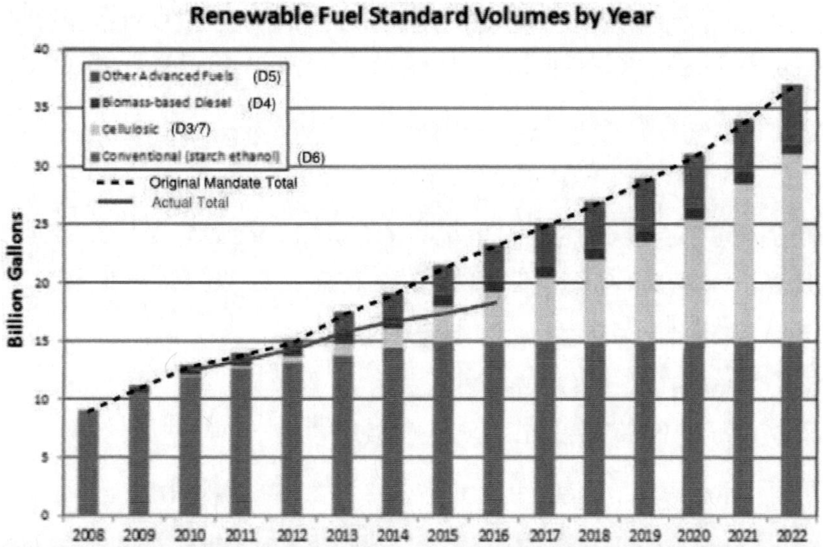

BURNS & McDONNELL

Renewable Fuel Standard Applied

- Obligations
  - Obligated Parties: Transportation fuel refiners or importers
  - Based on fuel sales - Renewable Volume Obligation (RVO)
- Compliance
  - Tracked by Renewable Identification Numbers (RIN)
  - Generated with biofuel and released at blending
  - Produce/purchase biofuel and blend OR purchase RINs in market

BURNS & McDONNELL 7

California Green House Gas Regulations

- AB 32 - 2006 Global Warming Solutions Act
- Cap & Trade is one of several elements resulting from this broad climate change initiative. Others
  - Renewable Portfolio Standard, RPS
  - Low Carbon Fuel Standard, LCFS
- Market based regulation designed to reduce greenhouse gases (GHGs) from multiple sources representing 85% of California GHGs to:
  - 1990 levels by 2020
  - 80% of 1990 levels by 2050
- The cap declines about 3%/yr beginning in 2013.

BURNS & McDONNELL

Cap & Trade GHG Inventory and Projections

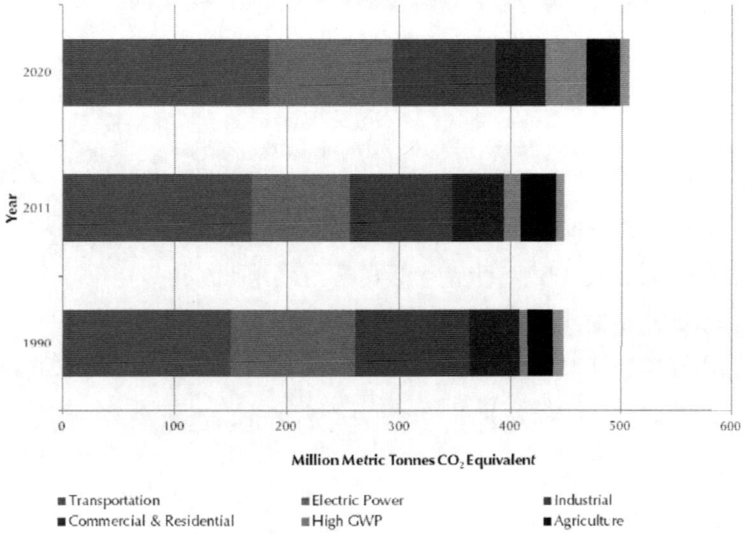

BURNS & McDONNELL 9

California Green House Gas Regulations

- AB 32 - 2006 Global Warming Solutions Act
- Cap & Trade is one of several elements resulting from this broad climate change initiative. Others
  - Renewable Portfolio Standard, RPS
  - Low Carbon Fuel Standard, LCFS
- Market based regulation designed to reduce greenhouse gases (GHGs) from multiple sources representing 85% of California GHGs to:
  - 1990 levels by 2020
  - 80% of 1990 levels by 2050
- The cap declines about 3%/yr beginning in 2013.

BURNS & McDONNELL

California Low Carbon Fuel Standard

- State Program Administered by the CA EPA Air Resources Board (ARB)
  - Executive Order S-1-07 (2007)→ ARB Initial Imp. (2011) →Re-adoption (2015)
- Encourage low carbon transportation fuels (gasoline/diesel) to reduce GHG emissions, diversify fuel pool, reduce petroleum use, reduce air pollutants
- Target is 10% reduction in "Carbon Intensity" by 2020 (2010 basis)

California Environmental Protection Agency
AIR RESOURCES BOARD

BURNS & McDONNELL

California Low Carbon Fuel Standard

- Carbon Intensity
  - Amount of GHG emissions per unit of fuel energy measured in grams of CO2 equivalence per MJ
  - Models are used to calculate the carbon intensity based on CA life cycle analysis rules (CA GREET 2.0)
- Deficits: Gasoline and Diesel producers/importers
- Credits: Biofuels, Alt Fuels (Nat Gas)
- Exempt: small volume fuels, propane, military, locomotive, oceanic, air craft

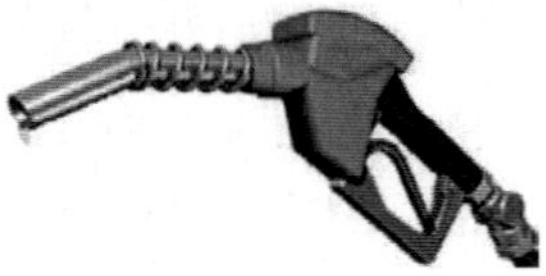

BURNS & McDONNELL

California Low Carbon Fuel Standard

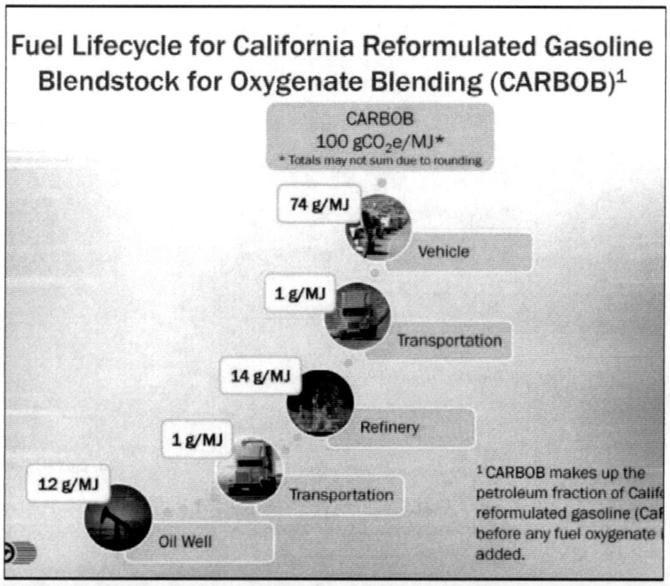

California Low Carbon Fuel Standard

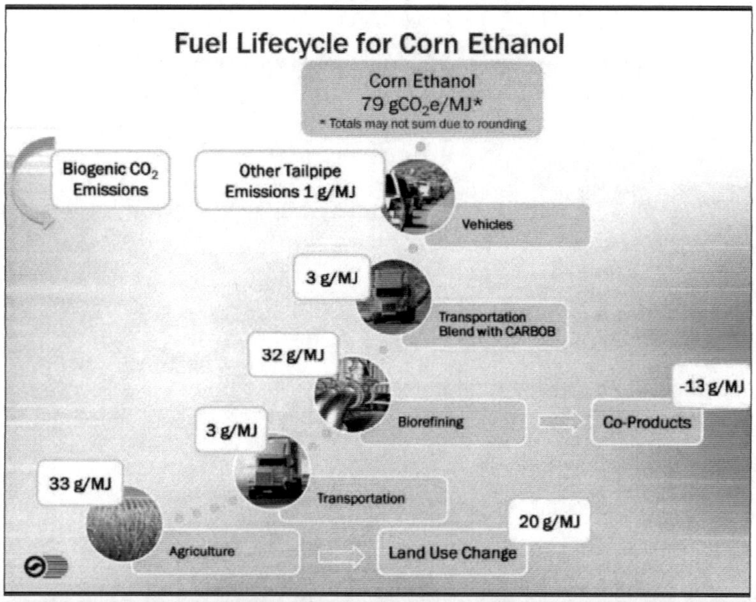

California Low Carbon Fuel Standard

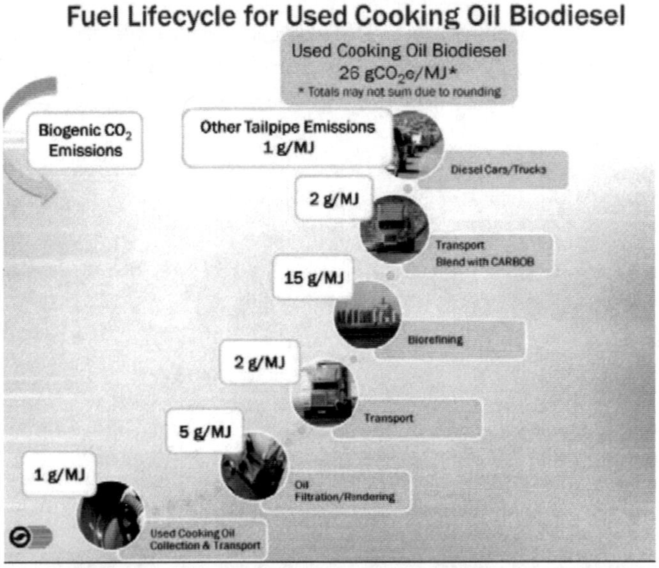

BURNS & McDONNELL 15

California Low Carbon Fuel Standard

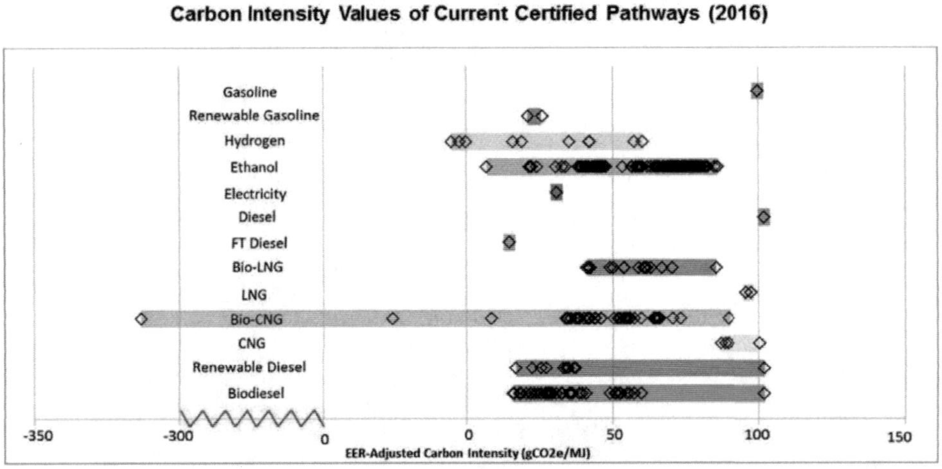

Last Updated 12/22/2016

BURNS & McDONNELL 16

California Low Carbon Fuel Standard

- Credit is metric ton CO2 equivalent (below the line)

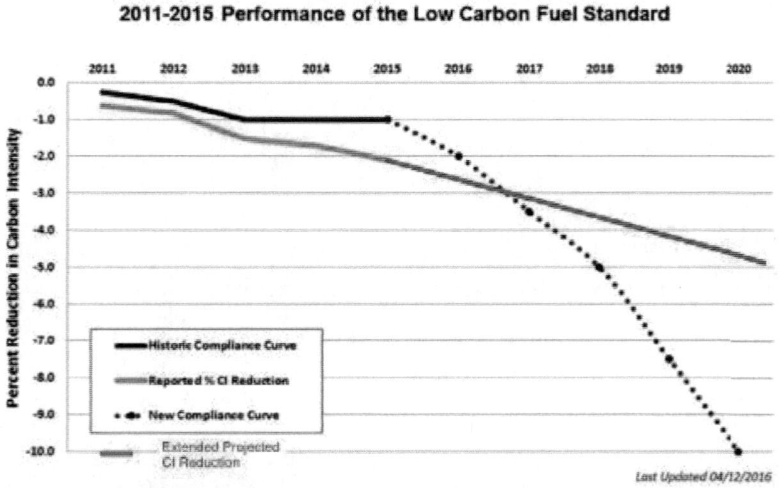

BURNS & McDONNELL 17

California Low Carbon Fuel Standard

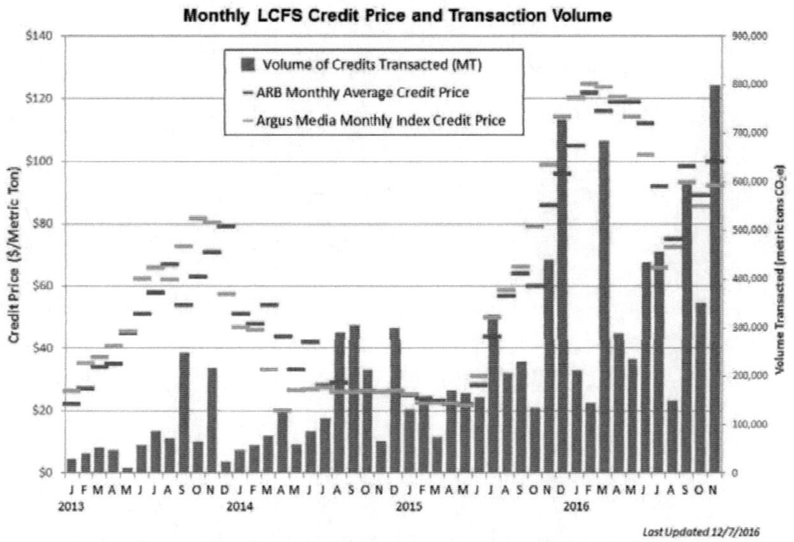

BURNS & McDONNELL 18

California Low Carbon Fuel Standard

- Obligated parties track compliance and report to ARB
  - Responsible for Compliance and Credits Authenticity
- Credit Clearance Market/Cap
  - $200/credit (MTCO2e) in 2016
  - 5% interest rate
  - Can't choose purchased credits
- Pacific Coast Collaborative- CA, OR, BC (WA)

BURNS & McDONNELL

19

California Low Carbon Fuel Standard

- Future
  - SB 32/ AB 197→ 40% GHG reduction by 2030
    - LCFS Part of Plan→ 18-25% CI Reduction
  - Co-processing, Renewable Jet Fuel, 3rd Party Verification, Unique ID, Modifications (all)
- Projected Credit Deficit:
  - First annual deficit (2017)
  - Overall deficit (2019)
  - 9.5-11 million credit deficit (2020)

BURNS & McDONNELL

20

RFS vs LCFS

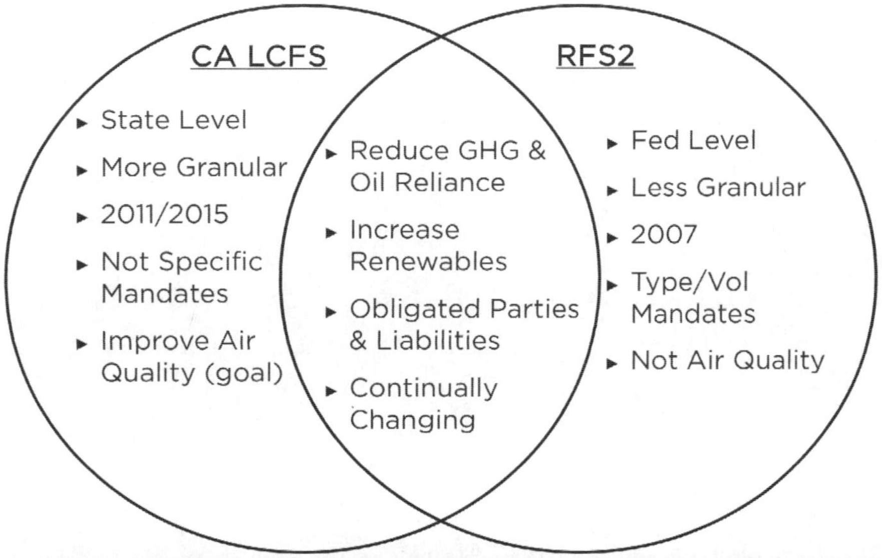

Summary

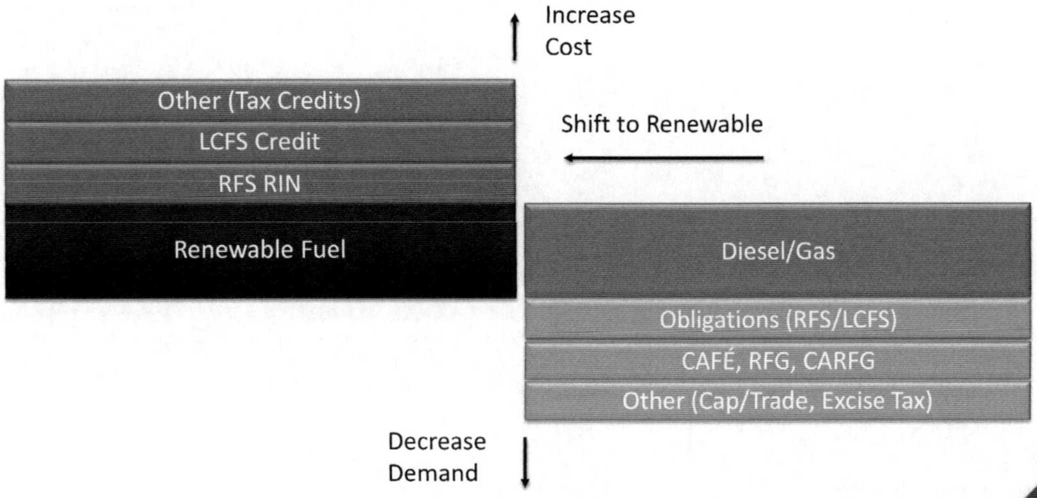

BURNS & McDONNELL℠

480316 Biojet Fuel Production from Ethanol and Isobutanol

Abstract:

Biojet fuels production from renewable ethanol and isobutanol will significantly reduce the greenhouse gas emissions from the aviation industry. In this study, ethanol and isobutanol were converted to jet fuel boiling range hydrocarbons in a fixed-bed catalytic reactor. Figure 1 shows the block flow diagram. The catalysts evaluated in the tests include: ZSM-5, MgO/alumina, and a large pore size zeolite. Different reaction temperatures, 300, 350, and 400°C, were compared. The conversion yields were between 40-80%. High selectivity for aromatics such as mesitylenes, xylenes, and toluene was obtained for ZSM-5. An economic analysis will be presented.

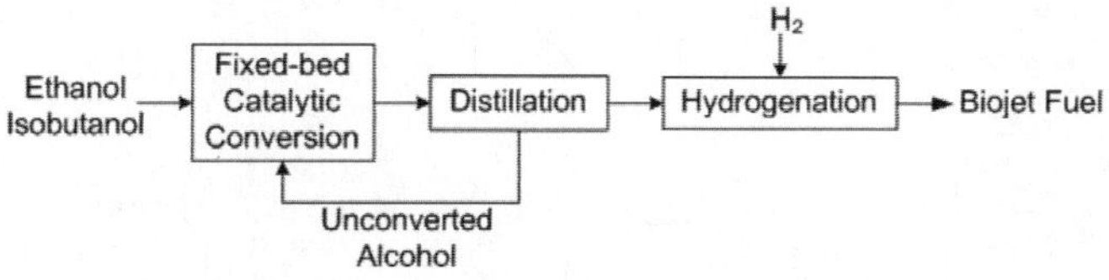

Figure 1. The Block Flow Diagram for the Alcohols to Jet Fuels Process

13TH GLOBAL CONGRESS ON PROCESS SAFETY

Effectively Addressing the Results of Damage Mechanism Reviews in a Process Hazard Analysis: A Checklist Approach

Tazim S. Rehmat, ABSG Consulting Inc., 300 Commerce Drive, Suite 200, Irvine, CA, 92602 (714) 734-2526 TRehmat@absconsulting.com

Gary Carrithers, ABSG Consulting Inc., 10301 Technology Drive, Knoxville, TN, 37932 (502) 592-9386 GCarrithers@absconsulting.com

Randal Montgomery, ABSG Consulting Inc., 16855 Northchase Drive, Houston, TX 77060 (865) 671-5832 RMontgomery@absconsulting.com

David Whittle, ABSG Consulting Inc., 10301 Technology Drive, Knoxville, TN, 37932 (865) 671-5819 DWhittle@absconsulting.com

Sayed Termah, ABSG Consulting Inc., 16855 Northchase Drive, Houston, TX 77060 (281) 994-0859 STermah@abs-group.com

Prepared for Presentation at
American Institute of Chemical Engineers
2017 Spring Meeting and 13th Global Congress on Process Safety
San Antonio, TX
March 26 – 29, 2017

Effectively Addressing the Results of Damage Mechanism Reviews in a Process Hazard Analysis: A Checklist Approach

Tazim S. Rehmat, ABSG Consulting Inc., 300 Commerce Drive, Suite 200, Irvine, CA, 92602 (714) 734-2526 TRehmat@absconsulting.com

Gary Carrithers, ABSG Consulting Inc., 10301 Technology Drive, Knoxville, TN, 37932 (502) 592-9386 GCarrithers@absconsulting.com

Randal Montgomery, ABSG Consulting Inc., 16855 Northchase Drive, Houston, TX 77060 (865) 671-5832 RMontgomery@absconsulting.com

David Whittle, ABSG Consulting Inc., 10301 Technology Drive, Knoxville, TN, 37932 (865) 671-5819 DWhittle@absconsulting.com

Sayed Termah, ABSG Consulting Inc., 16855 Northchase Drive, Houston, TX 77060 (281) 994-0859 STermah@abs-group.com

Keywords: Damage mechanism review, process hazard analysis, mechanical integrity, checklist, Cal/OSHA

Abstract

Performing damage mechanism reviews (DMRs) to determine credible degradation modes and susceptibilities of processing equipment has become an important step in developing effective mechanical integrity plans in the refining and chemical processing industries. DMRs aid in formulating inspection plans to mitigate risk from loss of containment and/or unplanned outages and are the subject of the corresponding API Recommended Practice (RP) 571. In addition, California OSHA's (Cal/OSHA's) recently revised *Process Safety Management for Petroleum Refineries* regulation now requires that DMRs be performed and process hazard analyses (PHAs) address DMR reports that are applicable to the process unit being analyzed.

A refinery PHA team subject to that regulation will now need to consult with individuals with expertise in damage mechanisms when performing the review of applicable DMR reports. To help effectively address this requirement, the primary PHA methodology can be supplemented with a checklist to help the PHA team identify DMR issues that may be applicable to the process unit. This paper describes the application and contents of a checklist the PHA team can use to help identify DMR issues such as (1) identifying and demonstrating that known corrosion and damage mechanisms (high temperature hydrogen attack, chloride stress corrosion cracking, etc.) are understood and controlled, (2) confirming a DMR report appropriately documents applicable damage mechanisms to the process unit, and (3) confirming a DMR is performed for incident investigations where a damage mechanism is identified as a contributing factor.

1. Introduction and Regulatory Background

In 2012, the governor of California formed an Interagency Working Group on refinery safety to examine ways to improve public safety and worker safety at petroleum refineries. The objective of the group was to identify a means to improve refinery and agency performances. The Interagency Working Group published a report in February 2014 that identified (1) regulatory gaps, (2) areas of overlap between regulatory agencies, and (3) areas in need of improved coordination.

Subsequent to the findings of the Interagency Working Group, California announced it would implement changes to the Cal/OSHA process safety management (PSM) regulation and strengthen the California Accidental Release Prevention program (CalARP) regulations as they apply to petroleum refineries.

2. Regulatory Revisions

As a result of the above initiative, a new section, 5189.1 *Process Safety Management for Petroleum Refineries*) [1], was added to the Cal/OSHA regulation that specifically targets petroleum refineries. The latest version of the change to the Cal/OSHA regulation was published in July 2016 and is expected to be finalized in 2017.

The regulation includes revision of existing PSM elements and addition of new PSM elements and topics (see Figure 1). Damage mechanism review (DMR) is a new process safety element that has specific requirements and needs to be considered and incorporated into other PSM elements (see Figure 2). This paper focuses on implementing the DMR requirement into a process hazards analysis (PHA).

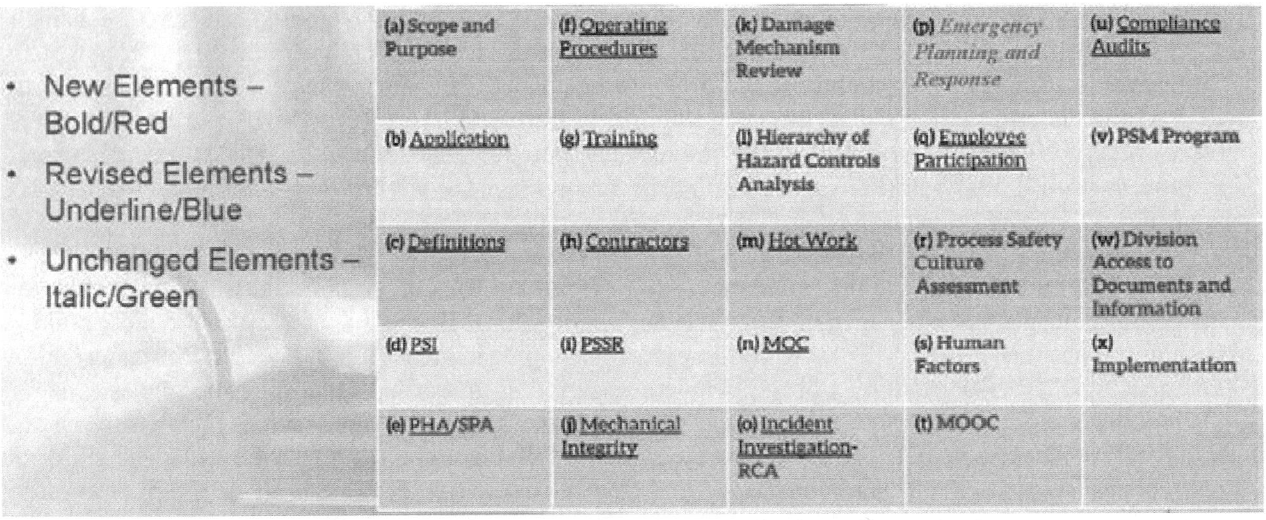

Figure 1. Regulatory Revisions

Figure 2. Incorporating DMR into Other PSM Elements

3. What Is a DMR?

Cal/OSHA's new regulation, *Process Safety Management for Petroleum Refineries,* defines a damage mechanism as *a mechanical, chemical, physical, or other process that results in equipment degradation.*

A DMR is an analysis process designed to determine credible damage mechanism susceptibilities of fixed equipment. Such equipment includes process vessels, heat exchangers, process piping, storage tanks, and process heaters. The analysis process is performed circuit-by-circuit by a multi-disciplinary team that typically includes operations, maintenance, mechanical, materials, and inspection personnel. The results from a DMR can be used to formulate inspection plans to mitigate the risk of a loss of containment and/or unplanned outages.

4. Importance of Performing a DMR

Many refining and chemical processes are not always steady state operations. This may be due to process unit feedstock variations and/or upsets in operation. When processes and operating parameters change, new damage mechanisms may be introduced as well as acceleration of existing damage mechanisms. Thus, an understanding of the process/operating conditions and the potential resulting damage mechanisms is often needed in order to establish and maintain a robust mechanical integrity program (see NACE MR0103, API 510, API 570, API RP 580, and API RP 581) [4, 6, 7, 8, 9].

Understanding damage mechanisms is important for several reasons, including:

- Selecting appropriate equipment inspection intervals/due dates, locations, and techniques,
- Making decisions (e.g., modifications to a process, materials selection, monitoring frequencies) that can help eliminate or reduce the probability of a specific damage mechanism, and
- Determining the type of failures that could occur.

5. How to Perform a DMR

DMRs are typically performed using a systematic approach (see Figure 3). The six steps are (1) obtain data, (2) develop information, (3) conduct DMR, (4) identify DMs, (5) document DMR, and (6) use DMR results.

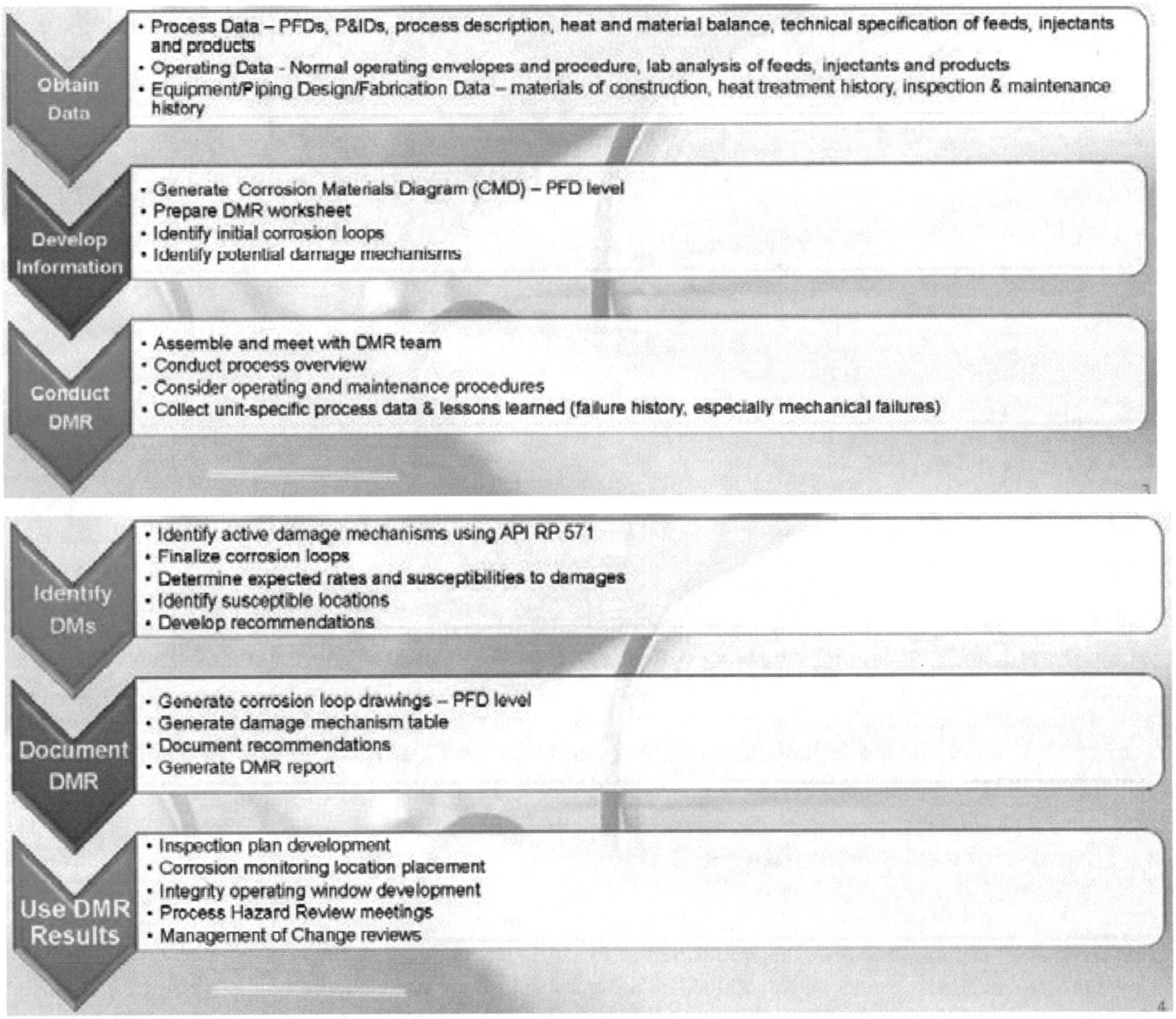

Figure 3. How to Perform a DMR

1. Obtain Data

Data required in performing a DMR include process data, operating data, and equipment/piping design/fabrication data. Site-specific information associated with risk-based inspection studies, preventive maintenance inspections, results from inspections, and failure history should be available electronically for reference during the study.

2. Develop Information

A DMR includes (1) generating a corrosion materials diagram, (2) preparing a DMR worksheet, (3) identifying initial corrosion loops, and (4) identifying potential damage mechanisms.

Developing corrosion loops is an important step in performing a DMR. A corrosion loop is defined as a section of a process that shares common damage mechanism susceptibilities with similar anticipated rates of damage. A corrosion loop is typically comprised of similar materials exposed to similar process/environmental conditions and asset characteristics (materials of construction, insulation type, coatings, heat tracing present, buried, etc.).

Corrosion loops are developed using the following guidelines:

- Maintain a single process chemical (for example, a major indication of a loop boundary is when the service that is flowing through the piping undergoes a change that greatly affects the composition),
- Maintain a single phase, and
- Use the inlet/outlet of a major piece of equipment (pressure vessel, separator, heat exchanger, column, etc.) as a loop boundary.

3. Conduct DMR

A DMR team typically includes a corrosion engineer/specialist, unit inspector, unit process engineer, unit operations representatives, unit mechanical/reliability engineer, unit subject matter experts for licensed technologies (if needed), and a facilitator knowledgeable in the DMR process. The team (1) conducts a process overview, considering operating and maintenance procedures, and (2) collects unit-specific process data and discusses mechanical failure history.

The steps in conducting a DMR include:

1. Developing color-coded process flow diagrams (PFDs) to illustrate various corrosion loops in each process. PFDs, P&IDs, piping isometric drawings, and process stream heat and material balances are normally used to generate color-coded PFDs.
2. Collecting materials of construction and fabrication records to be added to the PFDs so corrosion materials diagrams (CMDs) can be developed. The CMD is a modified process flow diagram or database containing relevant equipment and piping corrosion mechanisms, operating conditions, materials of construction, and corrosion circuits.

4. Identify Damage Mechanisms

Damage mechanisms are identified from a review of established corrosion/materials literature, company-specific studies, and industry guidance documents such as API RP 571 [3].

API RP 571 classifies damage mechanisms into the following major groups:

- Mechanical and Metallurgical Failure Mechanisms,
- Uniform or Localized Loss of Thickness,
- High Temperature Corrosion (>400°F [204°C]), and
- Environment – Assisted Cracking.

Expected damage rates, susceptible locations, and susceptibilities to damages are documented, and recommendations are developed as needed.

5. Document DMR

A DMR report contains the information needed to understand materials damage issues in a specific type of operating process unit at a plant site. The detailed DMR report includes corrosion loop drawings, corrosion damage mechanism tables, and documented findings and recommendations from the DMR. It is recommended that the report include the team member composition, the DMR approach, and data used in performing the review.

6. Use DMR Results

Once completed, the DMR report can be used in a variety of ways, including (1) inspection plan development, (2) corrosion monitoring location (CML) placement, (3) integrity operating window development, (4) PHA meetings, (5) management of change reviews, and (6) incident investigations.

6. How to Address the Results of a DMR in the PHA

Cal/OSHA's new regulation states that ..."*the PHA shall address applicable DMR reports*." Thus, a PHA conducted to meet that regulation now requires consultation with personnel (can be a part-time team member) with expertise in damage mechanisms during PHA meeting sessions. The DMR report serves as additional process safety information for reference during the PHA. To help effectively address this requirement, the primary PHA methodology (e.g., HAZOP and/or what-if analysis) may be supplemented with a DMR-specific checklist to help the PHA team identify hazardous scenarios, while considering the results of any DMR report(s) applicable to the process unit. The checklist discussion during a PHA can serve as an effective transfer of knowledge of unit-specific damage mechanisms to unit personnel.

It should be noted that some damage mechanisms may be identified using the primary hazard evaluation technique (HAZOP and/or what-if analysis) during the PHA. However, supplementing the primary hazard evaluation with a specific checklist allows the PHA team to further address damage mechanisms in a systematic fashion to ensure more thorough coverage than that afforded by the primary analysis technique(s) alone.

The checklist approach also helps the PHA team (1) discuss the contents of the DMR report(s) prepared for the process unit and (2) communicate how the damage mechanisms are being managed by the mechanical integrity group for the process unit. The checklist can also guide the discussion of hazards and potential major incidents associated with any identified damage mechanisms.

Table 1 illustrates a suggested corrosion and materials DMR checklist that can be used by a PHA team that has expertise in damage mechanisms. The application and contents of the checklist can help the PHA team confirm items such as whether (1) known corrosion and damage mechanisms (e.g., high temperature hydrogen attack, chloride stress corrosion cracking) are understood and controlled, (2) the existing DMR report appropriately documents applicable damage mechanisms in the process unit, (3) a DMR is performed for incident investigations where a damage mechanism is identified as a contributing factor, and (4) a process unit change (i.e., management of change) has an effect on damage mechanisms or integrity operating windows.

The contents of the checklist include a set of questions that allows the PHA team to confirm that the DMR requirements of Cal/OSHA's regulation as it applies to the PHA element are appropriately met along with questions that deal with the specific damage mechanisms known to occur in petroleum refineries. The questions are related to specific types of damage mechanisms (e.g., high temperature, low temperature corrosion and stress corrosion cracking, external damage, mechanical damage, and other) and include questions developed from industry standards as well as incidents related to mechanical integrity issues [2,4,5].

The checklist is constructed to include topic-specific items and questions to help a PHA team leader elicit responses from the team. The intent of the questions is to allow a PHA team to brainstorm and discuss the damage mechanism report prepared for the process unit being analyzed, along with ensuring that the report was prepared to include damage mechanisms applicable for all potential operating modes (e.g., normal startup and shutdown). This is similar to a PHA team using a facility siting checklist in a PHA to help identify any facility siting issues while considering the results of any separate facility siting study report conducted for the process unit (per API RP 752 and/or API RP 753) [10, 11].

Potential responses to the questions could include an acknowledgement that the PHA team understands which specific damage mechanisms may be present in the process unit and a discussion of how the damage mechanisms are managed (i.e., inspection programs, defined integrity operating windows, etc.). If a corrosion mechanism listed in the checklist is not present in the process unit being analyzed, the team can respond by indicating it is not applicable.

The checklist discussion may also help to (1) identify other damage mechanisms that may not have been identified during preparation of the DMR report that could occur from a process parameter deviation (flow, temperature, pressure) and (2) ensure mechanical integrity plans are robust and have taken into consideration any process changes identified by the PHA team.

7. PHA Documentation Tips

To confirm that a DMR was adequately addressed and was consistent with Cal/OSHA's expectations, consider using these guidelines when preparing the PHA report:

- Explicitly describe "how" the PHA team addressed/considered damage mechanisms in the process unit being analyzed,
- Reference the damage mechanism report prepared for the process unit for which the PHA is being performed,
- List the team members, including any individuals with expertise in damage mechanism, in the PHA report,
- Include the DMR checklist used for the evaluation,

- Use language in the PHA documentation that shows damage mechanisms were considered during the meeting(s), and
- When applicable, write recommendations to help mitigate any hazards related to damage mechanisms identified by the PHA team.

8. Conclusion

While conducting DMRs and incorporating their results in PHAs and other elements of a process safety management framework are not currently required by most regulatory agencies, state regulators in California have begun requiring them for refineries. As industry continues to recognize the value of conducting and documenting DMRs (per guidance in API RP 571) to enhance mechanical integrity programs and mitigate the risk of loss of containment events, the results of these DMRs can also provide additional relevant information for assisting PHA (and management of change) review teams in identifying potential hazardous scenarios for processes. A checklist, such as the one presented in Table 1, can be an effective method for PHA teams to efficiently identify and discuss DMR issues identified in separate, detailed DMR reports when conducting their hazard evaluations.

9. References

1. Title 8, Division 1, Chapter 4 Subchapter 7. General Industry Safety Orders Group 16. Control of Hazardous Substances Article 109. Hazardous Substances and Processes, §5189.1. *Process Safety Management for Petroleum Refineries* (Draft Regulation), July 2016.
2. *Corrosion Control Documents*, Downstream Segment, API Recommended Practice 970, Proposed First Edition, March 2016.
3. *Damage Mechanisms Affecting Fixed Equipment in the Refining Industry*, Downstream Segment, API Recommended Practice 571, Second Edition, April 2011.
4. NACE Standard MR0103-2003, *Materials Resistant to Sulfide Stress Cracking in Corrosive Petroleum Refining Environments*, Houston, Texas.
5. ABSG Consulting Inc. Course 110 *Process Hazard Analysis Leadership Training*, Resource Material Tab Mechanical Integrity Issues Checklist.
6. *Pressure Vessel Inspection Code: In-service Inspection, Rating, Repair, and Alteration,* API 510, 10th Edition, May 2014.
7. *Piping Inspection Code: In-service Inspection, Repair, and Alteration of Piping Systems,* API 570, 4th Edition, February 2016.
8. *Risk-Based Inspection,* API RP 580, 3rd Edition, February 2016.
9. *Risk-Based Inspection Methodology*, API RP 581, 3rd Edition, April 2016.
10. *Management of Hazards Associated with Location of Process Plant Permanent Buildings*, API RP 752, 3rd Edition, December 2009.
11. *Management of Hazards Associated with Location of Process Plant Portable Buildings*, API RP 753, 1st Edition, June 2007.

Table 1. Process Hazards Analysis Corrosion and Materials Damage Mechanism Review (DMR) Checklist

Item No.	Question	Response	Recommendations
I. General			
1	Does a DMR report exist that addresses the applicable damage mechanisms for the process unit? If not, was there a determination that no damage mechanisms exist and was the rationale documented?		
2	Was the DMR report for the process unit available to the PHA team?		
3	Are the known corrosion and materials damage mechanisms understood and controlled? (Note: Refer to sections II through V of this checklist)		
4	Was a DMR conducted where a damage mechanism exists on (1) new processes and/or (2) major changes that have been implemented in the process unit prior to initial startup and/or implementation of the change?		
5	Have any incident investigations for the unit identified a damage mechanism as a contributing factor of the incident?		
II. High Temperature Damage Mechanisms			
1	Is there a potential for carbon steel and/or 0.5Mo steel to exceed the Nelson curve and be exposed to high temperature hydrogen attack (HTHA)? Include consideration of piping in a normally closed bypass being overheated.		
2	Is there a potential for accelerated creep due to operating outside of the operating window (e.g., higher temperature)?		

Table 1. Process Hazards Analysis Corrosion and Materials Damage Mechanism Review (DMR) Checklist (cont'd)

Item No.	Question	Response	Recommendations
3	Is there a potential for liquid metal embrittlement?		
4	Is there a potential for accelerated sulfidic corrosion from a gradual increase in sulfur content?		
5	Is there a potential for accelerated corrosion from increased concentration of naphthenic acids in crude oil or various distillation cuts?		
6	Is there a potential for accelerated corrosion as a result of organic chloride contamination in crude oil or other feeds?		
7	Is there a potential for spontaneous ignition of materials (including titanium, zirconium, structured packing, iron sulfide, etc.) from HCl, Cl_2, O_2, and/or air?		
8	For alloys operating above 700 °F, are there any high temperature aging embrittlement phenomena that might lead to brittle behavior when equipment is pressurized at lower temperatures?		
III. Low Temperature (Aqueous) Corrosion and Stress Corrosion Cracking Damage Mechanisms			
1	Is there a potential for severe corrosion downstream of alloyed equipment due to process upsets that cause unneutralized corrosive materials or contaminants to flow downstream into systems constructed of materials not designed for such corrosive conditions?		

Table 1. Process Hazards Analysis Corrosion and Materials Damage Mechanism Review (DMR) Checklist (cont'd)

Item No.	Question	Response	Recommendations
2	Is there a potential for rapid corrosion at or downstream of injection or process mixing points due to heating/cooling, condensation/evaporation, reaction, between the injecting and mixing streams, etc.?		
3	Is there a potential for rapid corrosion at or near an injection point due to a change in flow rate, a change in flow pattern, or failure of the injection system to perform as intended?		
4	Is there a potential for rapid corrosion due to a change in flow rate or flow regime (e.g., vaporization, flashing, or other multiphase flow conditions)?		
5	Is there a potential for localized corrosion or cracking of austenitic stainless steels during shutdowns where polythionic acids may be present?		
6	Is there a potential for rapid localized corrosion at hot spots as a result of direct contact between heat tracing and process piping (e.g., lack of standoff or improper use of heat transfer material)?		
7	Is there a potential for accelerated corrosion from fluids and/or solids trapped in deadlegs?		
8	Is there a potential for process upsets to introduce moisture into moisture-free environments (e.g., anhydrous HCl or Cl_2) or to result in insufficient moisture in environments reliant on moisture for the protection of the system (e.g., anhydrous NH_3 and alcohols)?		

Table 1. Process Hazards Analysis Corrosion and Materials Damage Mechanism Review (DMR) Checklist (cont'd)

Item No.	Question	Response	Recommendations
9	Is there a potential for accelerated corrosion by the activation or deactivation of a piping segment (e.g., a bypass line or other alternate flow path) or the installation of a temporary flow path?		
10	Can changes in pump or compressor capacity (including the use of a spare pump/compressor) lead to increased corrosion rates?		
11	Can increase or decrease in pH (inadvertent or undetected) lead to increased corrosion or stress corrosion cracking?		
12	Can minor changes in feed compositions (e.g., loss of trace amounts of corrosion inhibitors, presence of organic chlorides, etc.) adversely affect corrosion rates?		
13	Can non-post weld heat treated equipment containing caustics or amines be steam cleaned/purged leading to a potential for cracking?		
14	Is there a potential for air and/or moisture ingress into otherwise oxygen-free and/or dry environments (e.g., breather valves on tanks) that could cause accelerated corrosion or cracking?		
15	Can solids be present causing increased erosion and/or corrosion (e.g., catalyst carryover, accumulation of corrosion products, etc.)?		

Table 1. Process Hazards Analysis Corrosion and Materials Damage Mechanism Review (DMR) Checklist (cont'd)

Item No.	Question	Response	Recommendations
16	Is there a potential for increased corrosion or cracking downstream of piping specification breaks due to process changes or upsets?		
17	Is there potential for increased corrosion in a pipe or tube inlet zone or at a vapor/liquid interface?		
18	Is there a potential for caustic cracking of bolted connections from small leaks of boiler feed water or other caustic solutions? Is there potential for caustic cracking of non-stress relieved equipment from heat exchanger tube leaks or caustic carryover?		
19	Is there a potential for bolting to be exposed to process environments that could lead to catastrophic fracture due to stress corrosion cracking such as caustic, amine and wet H_2S, etc.?		
20	Is there a potential for process contamination that could cause stress corrosion cracking (e.g., wet H_2S, caustic, amines, chlorides, polythionic acids) or corrosion fatigue of deaerators?		
21	Is there a potential for accelerated corrosion from the installation of incorrect materials of construction in highly hazardous systems?		
22	Is there a potential for localized corrosion or cracking of heat-affected zones in sensitized stainless steels?		
23	Is there a potential for hydrogen embrittlement or hydrogen blistering from increased concentrations of hydrogen sulfide or cyanides?		

Table 1. Process Hazards Analysis Corrosion and Materials Damage Mechanism Review (DMR) Checklist (cont'd)

Item No.	Question	Response	Recommendations
24	Is there a potential for sulfuric acid concentrations falling below critical limits due to process upsets or changes and causing accelerated corrosion?		
25	Is there a potential for temperature, pressure, or other process changes that cause shifts in dew point, and therefore, create or shift areas where corrosive compounds condense leading to accelerated corrosion?		
IV. External Damage Mechanisms			
1	Is there a potential for unmitigated and undetected external corrosion that can lead to rupture (e.g., soil-to-air interface, downstream of cooling water sprays, etc.)?		
2	Have there been changes in the monitoring or maintenance of cathodic protection systems that could lead to increased corrosion of buried piping or tank bottoms?		
3	Can changes in process conditions lead to increased corrosion under insulation, (e.g., idling of normally hot equipment, equipment in cyclic service above and below 250 °F, or exposure of stainless steel equipment to external chloride cracking)?		

Table 1. Process Hazards Analysis Corrosion and Materials Damage Mechanism Review (DMR) Checklist (cont'd)

Item No.	Question	Response	Recommendations
V. Mechanical and Other Damage Mechanisms			
1	Is there any potential for low temperatures causing freezing of water in piping deadlegs, especially in light hydrocarbon systems, resulting in piping rupture?		
2	Is there a potential for liquid slugging of piping (including flare lines) that could cause piping failure due to hydraulic shock, water hammer, and other transient overstress conditions?		
3	Is there a potential for accelerated corrosion, fouling, or plugging of inlet or outlet piping of relief devices or flare systems?		
4	Has the impact on relief capacity been reviewed for any increases in unit throughput?		
5	Can temperature changes on piping systems cause mechanical overload because of lack of expansion capability in piping design?		
6	Is there a potential for overstressing or shocking brittle materials of construction (e.g., cast iron, aged or embrittled steels)?		
7	Is there a potential for overpressuring equipment or piping when using high pressure fluids and/or positive displacement pumps to unplug a line?		
8	Is there a potential for brittle fracture of heavy wall equipment from being fully pressurized before the metal temperature has reached approximately 200 °F?		

Table 1. Process Hazards Analysis Corrosion and Materials Damage Mechanism Review (DMR) Checklist (cont'd)

Item No.	Question	Response	Recommendations
9	Is there a potential for pressurization or impact during low temperature conditions that may cause brittle fracture of equipment that is not designed for such conditions (i.e., stress applied below minimum design metal temperature)?		
10	Is there a potential for vibration (either applied or induced) that could lead to fatigue fracture of vibrating piping, over-stressed threaded connections, unsupported or over-hung weight, or exchanger tubes?		
11	Is there a potential for thermal fatigue cracking due to sudden or severe swings in temperature?		
12	Is there a potential for cavitation from rapid pressure or temperature changes?		
13	Could heat tracing be inadvertently shut off on critical equipment or relief systems and result in impairment (e.g., plugging leading to failure of a relief system to function on demand)?		
14	Are there check valves in the process that can lead to a process safety incident if they fail to function .properly?		
15	Are there any vents and drains downstream of block valves that may require additional protection from pipe plugs/caps to prevent releases?		
16	Are temporary repairs and clamps identified for scheduled removal at the next maintenance opportunity?		

Table 1. Process Hazards Analysis Corrosion and Materials Damage Mechanism Review (DMR) Checklist (cont'd)

Item No.	Question	Response	Recommendations
17	Can the failure or malfunction of internal elements such as exchanger bundles, trays, distributors, steam rings, etc., lead to overpressure, accelerated corrosion, and/or metallurgical damage to equipment and piping?		
18	Is there a potential for temperature changes that might cause brittle fracture of equipment constructed of materials with low toughness due to (1) thermal shock during steady-state operation or (2) transient conditions experienced in normal startup or shutdown of heavy walled vessels?		
19	Is there a potential for excessive temperatures or hot spots that could cause rupture from short-term overheating of furnace tubes, transfer lines, or catalyst-containing vessels?		
20	Is there a potential for presence of molten salts or liquid metals that could cause very high rates of corrosion or cracking?		
21	Is there a potential for liquid carryover into gas streams or velocity changes of mixed phase streams causing accelerated corrosion-erosion of elbows, tees, and other areas subject to downstream turbulence (e.g., downstream of control valves)?		
22	Is there a potential for changes in feed compositions where trace amounts of certain chemical compounds previously provided corrosion inhibition?		

Table 1. Process Hazards Analysis Corrosion and Materials Damage Mechanism Review (DMR) Checklist (cont'd)

Item No.	Question	Response	Recommendations
23	Is there potential for accelerating corrosion rates in sour hydrocarbon service from gradual increase of sulfur content or inadvertent increase in content of other sulfur compounds such as H_2S or mercaptan?		
24	Is there a potential for changes in water treating chemistry or procedures that could accelerate water-side corrosion?		
25	Is there any potential for ammonium hydrosulfide, ammonium chloride, or amine salt deposition and resultant corrosion rate acceleration with increased concentrations of ammonia, sulfides, or chlorides or insufficient water washing?		
26	Are there any other known corrosion mechanisms that may be of concern during normal, abnormal, or non-routine operating modes that are unique to the process unit? If so, are they currently managed by the mechanical integrity group?		
27	Is there a potential for brittle fracture (e.g., hydrogen embrittlement, temper embrittlement, 0.5Mo steel) of heavy wall hydro processing equipment from rapid heating/cooling or pressurizing when the equipment temperature is below the minimum pressurization temperature?		
28	Is there a potential for catastrophic brittle fracture (hydrogen embrittlement) of hydro processing equipment from rapid cooling?		

The Integration of Inherently Safer Design and Control

Denis Su-Feher
Mary Kay O'Connor Process Safety Center, Texas A&M University
College Station, 77843-3122, USA
denis@tamu.edu

Yogesh Koirala[1,2], Efstratios Pistikopoulos[2], & M. Sam Mannan[1,2]
[1] Mary Kay O'Connor Process Safety Center, Texas A&M University
[2] Artie McFerrin Department of Chemical Engineering, Texas A&M University
College Station, 77843-3122, USA
ykoirala@tamu.edu
stratos@tamu.edu
mannan@tamu.edu

Prepared for Presentation at
American Institute of Chemical Engineers
2017 Spring Meeting and 13th Global Congress on Process Safety
San Antonio, TX
March 26 – 29, 2017

The Integration of Inherently Safer Design and Control

Denis Su-Feher
Mary Kay O'Connor Process Safety Center, Texas A&M University
College Station, 77843-3122, USA
denis@tamu.edu

Denis Su-Feher[1,2], Yogesh Koirala[1,2], Efstratios Pistikopoulos[2], & M. Sam Mannan[1,2]
[1] Mary Kay O'Connor Process Safety Center, Texas A&M University
[2] Artie McFerrin Department of Chemical Engineering, Texas A&M University

Keywords: Process Design & Development, Safe Design, Process Automation & Control

Abstract

Current inherently safer design strategies in the conceptual design stage focus on reducing the overall hazard of a new refinery without considering the operability of the process. The process is first designed to be inherently safer with respect to a nominal, steady-state case and then, after the process is designed, layers of protection are added and operability issues are addressed. However, this sequential design approach does not account for the impact of the design itself on the operability of the refinery. A particular design may be safer with respect to its steady-state operation, but may suffer from operability issues. For example, an intensified process may contain less of a hazardous substance and thus be inherently less hazardous, but the design may restrict the controllability of the process, thus making the design have a higher risk and be less safe overall. A considerable amount of research has been done to simultaneously optimize the design and control system of refineries, but no such approach has integrated inherent safety, only seeking to produce an economically optimal design rather than a safer one.

The objective of this research is to implement a strategy to simultaneously design and control an inherently safer refining plant. Parametric Optimization and Control (PAROC), a simultaneous design and control framework will be used as a basis for the design and control of a refining process. Different inherently safer design indices will be integrated into PAROC, and the operability, safety, and economic profitability of the results will be compared.

An extensive literature review has identified indices by which controllability of a refining process can be quantitatively characterized and optimized along with the design, as well as inherently safer design indices that can be implemented into the PAROC framework. Simultaneous use of these inherently safer design and controllability indices will be compared to create a new index for the integration of inherently safer design and control.

With refinery systems and their dynamics becoming increasingly complex, consideration of operability issues in the design stage becomes even more necessary to prevent incidents. The integration of inherently safer design and control will substantially reduce operability issues that result from an uncontrollable process design and allow for greater tolerance and ease of control.

1 Introduction

The objective of an industrial refinery is to be both profitable and safe. Traditionally, the oil and gas industry has optimized the design of refineries primarily for profitability, accepting the existence of hazards in a design and managing the risk involved by implementing prevention, mitigation, and response methods after the design has been implemented. However, in 1943, Ziegler and Nichols determined that the design of a process can influence how well process control can reduce the effects of deviations and uncertainty in the process [1]. As a result of this realization, it became necessary to optimize both profitability and operability, as failing to consider operability could lead to a shorter refinery life and potentially unsafe conditions.

A number of operability characteristics were identified since then, such as controllability and flexibility [2, 3]. Controllability is the process's ability to reach and maintain a steady-state. Flexibility is the process's ability to stay at steady-state operation despite disturbances and uncertainty. Each of these characteristics deal with how well the process control system can maintain control on the design, and as such they are affected by the design itself and must be accounted for even in the initial stages of the design.

If a design has high operability, then it is likely to be a safer design. A process with high operability is capable of operating within its boundaries and uncertainties and deviations are unlikely to have an effect on a high operability design. If deviations to the process can be accounted for and an appropriate process control system is put in place, then process safety incidents are less likely to occur.

Unfortunately, high operability of the process control system does not guarantee safety. The plant safety depends on both consequence and probability of incidents that may occur. Although increasing operability will decrease the likelihood of incident occurrence, operability fails to incorporate the consequences involved with the failure. As a result, although the total number of failures are minimized, high consequence failures are not treated with the severity they require. Therefore, it is necessary to incorporate consequence measurement into the initial refinery design.

The aim of the present work is to define a process to integrate inherently safer design and control in order to create simultaneously profitable, operable, and safe processes. Several inherent safety indices are introduced to quantify consequence, and several controllability indices are introduced to quantify operability. These safety and controllability indices are used simultaneously with the Parametric Optimization and Control (PAROC) Framework as a case study to demonstrate how safety can be integrated into optimization for inherently safer design and control.

1.1 Relevant Works

1.1.1 Inherently Safer Design

Inherently Safer Design (ISD) is the idea that the level of safety in a refinery can be improved by modifying the base design of a process by reducing or removing hazards. The main principles that it defines and quantifies are shown in Table 1 [4].

Table 1. Inherent Safety Principles and Definitions [4]

Inherent Safety Principle	Definition
Intensification	Using smaller quantities of hazardous substances
Substitution	Using less hazardous materials in the process
Attenuation	Using less hazardous conditions; using the same material in a less hazardous form
Simplification	Reducing complexity in operational design to make operational errors less likely
Limitation of Effects	Reducing the severity of consequences associated with process design
Error Tolerance	Making the process resilient to disturbances
Ease of Control	Reducing the amount and complexity of control necessary to operate the process

Although there are methods such as Hazard and Operability Studies (HAZOP), Fault Tree Analysis (FTA), and Failure Modes and Effects Analysis (FMEA) that can be used to determine the consequence of incidents, all of them require detailed information that may not be available in early stages of plant design. Quantitative inherently safer design indices, on the other hand, require relatively little information and can be used to assist risk-based decision-making in the early stages of refinery design.

Furthermore, ISD is ideal for the initial design stage because the number of available inherent safety options is most abundant then, and will only decrease as the design proceeds, as shown in Figure 1 [5]. Although opportunities for prevention and mitigation appear in the middle of the design process, preventative and mitigative barriers later are often costly and not as effective at reducing the hazard inherent to the design. Much like traditional operability, inherent safety must be considered as early as possible, or costlier and/or less effective prevention and mitigation methods will have to be used instead.

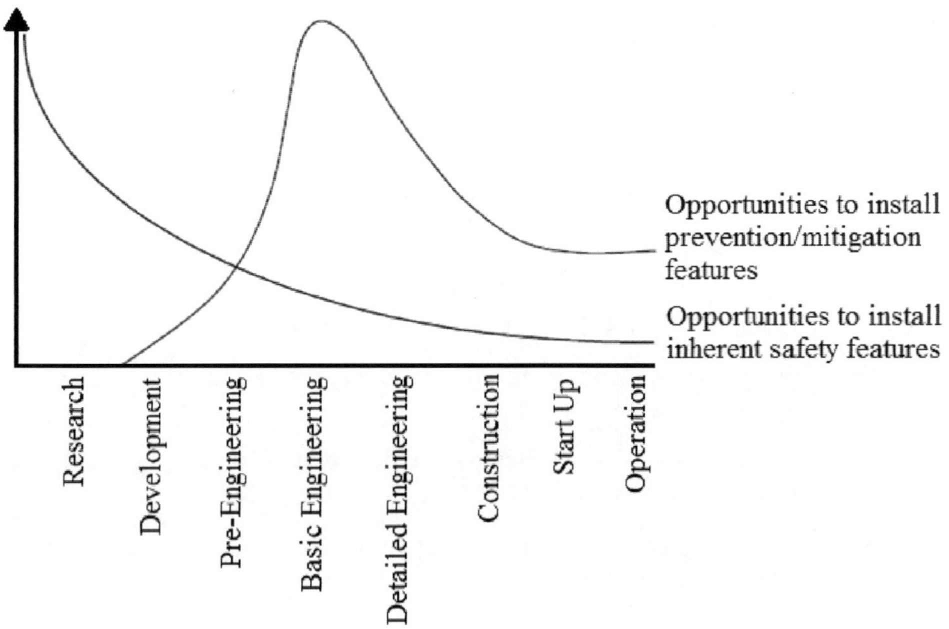

Figure 1. Inherent vs preventative safety in various stages of plant design [5].

The need for better risk assessment methods for the safer design of chemical processes has driven a substantial amount of research for comprehensive ISD indices. Ever since Trevor Kletz popularized ISD in his 1978 article entitled, "What you don't have can't leak," there have been numerous indices published that attempt to quantify different principles of inherent safety [6]. Table 2 lists these indices and summarizes some of these indices and their contributions.

Most of the current literature on inherent safety has focused on identifying the effects of intensification, substitution, attenuation, simplification, and, to an extent, limitation of effects. These principles quantify the hazard, or the potential consequences that might occur in a safety incident.

Ease of control and error tolerance, on the other hand, represent the effect of the design on operability, and the complexity of the controller needed to maintain control over the process. These principles quantify the likelihood of the hazard causing a process safety incident, and can be quantified with controllability indices.

Table 2. Inherently Safer Design Indices and their Contributions

Inherently Safer Design Indices	Contributions
Dow F&EI [7]	First safety index reported in literature, and most commonly used in industry. Evaluates material properties and process conditions.
Dow CEI [8]	Evaluates toxic effects and distance to neighbouring plants and communities to assess acute health hazard potential to them.
Mond Index [9]	Extends Dow F&EI by evaluating toxicity of chemicals.
INSET Toolkit [10]	Provides a guide to creating an inherent health and safety index.
HIRA [11]	Integrated material factor with operating conditions based on thermodynamic properties and expert opinions to find a damage radius under given overall conditions.
PIIS [12]	First inherent safety index reported to analyse the predesign of a process. Quantifies substitution by measuring the reaction route hazards.
ISI [13]	Expanded PIIS. Quantifies minimization, moderation, substitution and simplification in terms of a ranking.
I2SI [14]	Extends ISI by weighing thermodynamic data and expert opinions to determine a risk ranking in terms of damage radius.
PSI [15]	Quantifies inherent safety of process streams during simulation work in the preliminary design stage that influences explosions.
RISI [16]	Extends I2SI by evaluating frequency as well as consequence to obtain a risk ranking.

1.1.2 Controllability

Controllability is measure of a design's operability, specifically its ability to achieve and maintain a desired equilibrium value despite disturbances and uncertainties. Certain process designs can have regions of instability that an inherently more controllable design may not have [1]. For example, if there is a region of instability near a steady-state, then attempting to control the system at that steady-state might be impossible, and another, less economically feasible steady-state would have to be used instead [17]. These unstable regions would have to be addressed through either more control action, or less optimal set points, making the process system more complex, costlier, and potentially less reliable as a whole. A number of methods have been developed to measure different aspects of the controllability of both linear and nonlinear processes. They are displayed in Table 3. All of these methods take into account dynamic aspects of the process.

Table 3. Controllability Indices

Controllability Index	Summary	Linearity
RGA & BRGA [18] [19]	RGA defines a matrix that measures the effect of manipulated variables on controlled variables. BRGA extends it to block decentralized control.	Linear
NBRGA [20]	Extends BRGA to nonlinear systems.	Nonlinear
IMC [21]	Creates an upper bound of controllability by assuming that the ideal controller is the inverse of the chemical process transfer function.	Linear
RHTP zeros [22]	RHTP zeros define areas where instabilities in control response may occur. The only way to move a RHTP zero is to change the process design. A controllable process must remove as many RHTP zeros as possible.	Linear
Input-Output Controllability [23]	Defines eight rules for a controllable process.	Linear
Nonlinear Zero Dynamics [24]	Quantifies the effect of the inverse responses in nonlinear systems for limited types of systems.	Nonlinear
Relative Order [25]	Defines the pole excess of the transfer function. Quantifies how inputs directly affect outputs.	Nonlinear

It has been shown that for nonlinear systems, controllability analysis based on a linearized form of the system in many cases provides a close estimate, even if the system is strongly nonlinear [26]. However, this does not hold for all cases. At start-up, shutdown, and in batch or semi-batch processes, the controllability of highly nonlinear processes may not be accurately represented, and the accuracy of different control methods have to be validated. In this case, simulation may be needed to validate the controllability index.

The Parametric Optimization and Control (PAROC) framework, as shown in Figure 2, is a tool that provides the means for the simultaneous design and control of process systems [27]. ODE's and PDE's are implemented into optimization software with a first principle based, high-fidelity model analysis of the refinery design. However, because the design of many refineries is too complex to optimize with an economically feasible computation time, model reduction techniques are used to simplify the high-fidelity model into an approximate model. Then, the control system is implemented using multiparametric receding horizon policies. Finally, the difference between these policies and the original high-fidelity model are used to validate the process. With exception of the model reduction of the design and multi-parametric programming to determine the control structure, both of which only need to be performed once, the process is iterated until the error is minimized.

PAROC allows for rapid, repeatable simulation of realistic control structures and the creation of optimal process designs under various conditions. PAROC has been implemented in the gPROMS software, and can be used to solve the simultaneous design and control problem to obtain the optimal process design and controller.

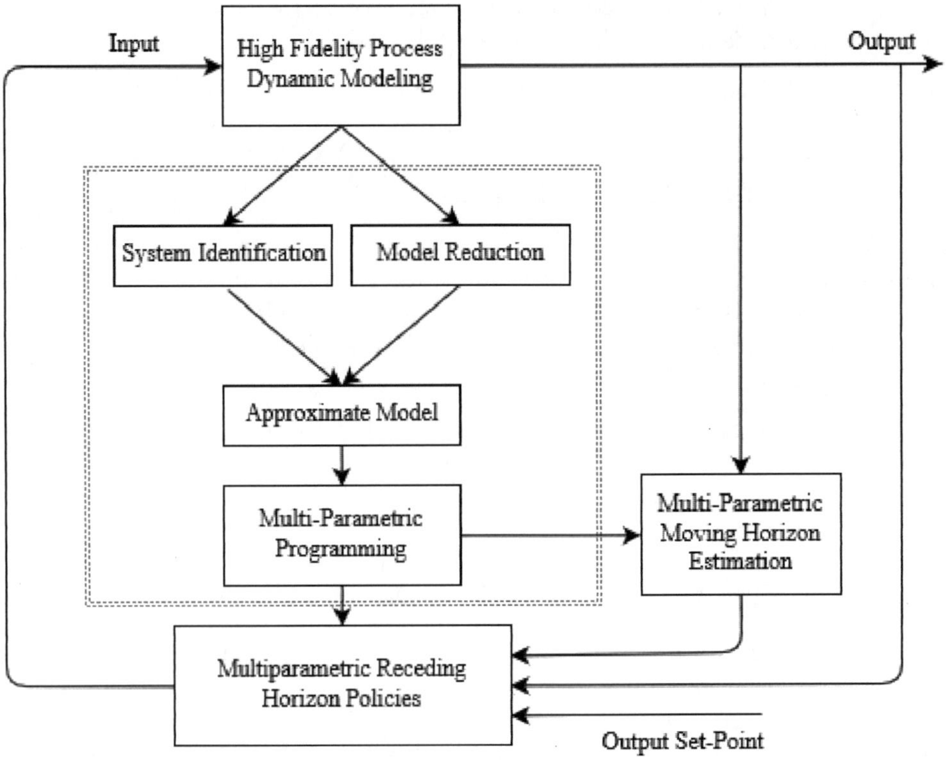

Figure 2. The PAROC Framework [27]. Actions in the dotted box happen once, offline.

2 The Integration of Inherently Safer Design and Control

The integration of ISD and control can be seen as a multi-objective optimization problem, where the objective function is given by (1), and is subject to design constraints.

$$\min\ TAC + w_C I_C + w_S I_S \tag{1}$$

TAC is the design's total annual cost and is represented by a net present value. I_C is the measured controllability based on controllability indices, and is scaled with its weighting factor, w_C. I_S is the measured safety index value, and is scaled with its weighting factor, w_S. Each of these factors respectively represent the utility cost of having a more expensive, less controllable, or higher risk design. Since the range of values for each index is different, and each designer places a different value on each different index, it is difficult to decide on the weighting factors for such a formulation.

To solve this multi-objective optimization problem, ϵ-constraint optimization can be used to reduce the multi-objective problem to a single objective function with two new constraints in addition to the original design constraints [28]. The objective becomes (2) controllability constraints become (3), and the safety constraints become (4).

$$\min\ TAC \tag{2}$$

subject to new constraints:

$$I_C \leq \epsilon_C \tag{3}$$

$$I_S \leq \epsilon_S \tag{4}$$

By varying the controllability and safety parameters, ϵ_C and ϵ_S, a pareto-optimal solution set can be generated. However, if the system poses high degrees of nonlinearity, then the controllability index will need to be validated. Furthermore, the ease of control factor must be quantified based on the complexity of the required controller. Figure 3 shows how this factor may be quantified.

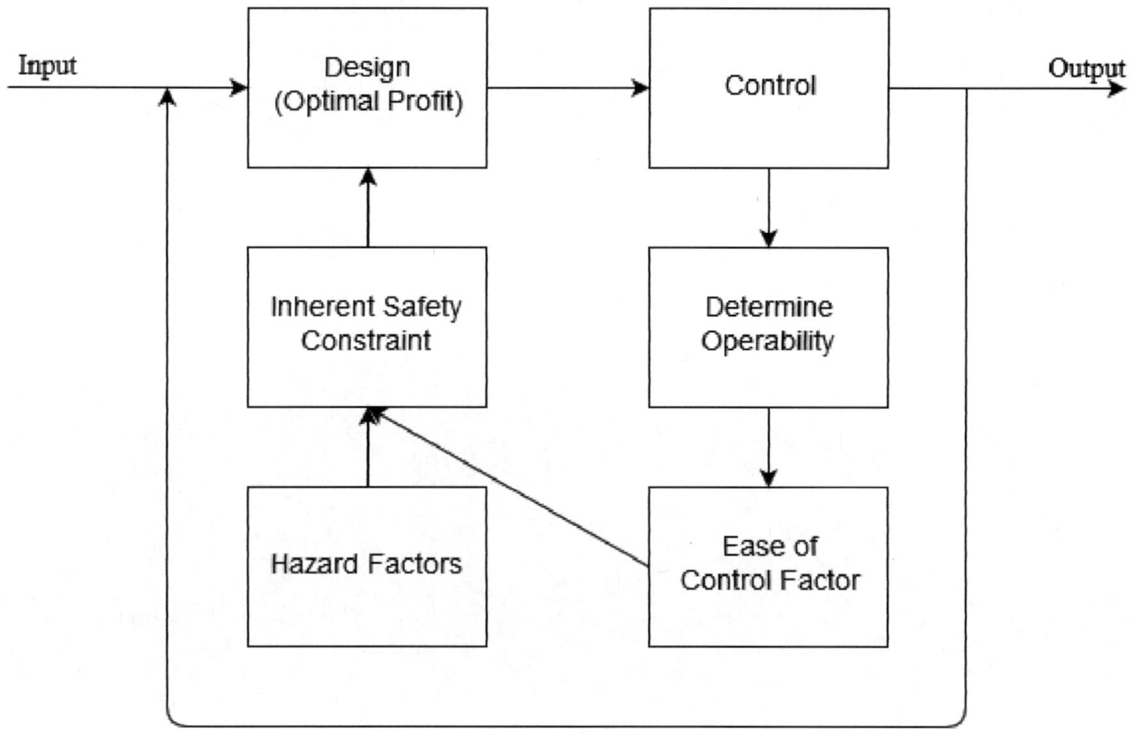

Figure 3. Process for Integrating Design, Control, and Safety

Since a nonlinear system may be too complex to be analyzed accurately with controllability indices, an estimation of the range of control must be performed with variations of the design. The effectiveness of the resulting controllers should be compared with the expected value given by the controllability index, and the controllability index should be evaluated and weighed based on its accuracy. The optimal control scheme of each design must then be analyzed for its complexity, which should be factored into the inherent safety index to quantify the ease of control of the design, a factor of the inherent safety of the design.

3 Case Study in Simultaneous Design and Control: PAROC

Simultaneous design and control guided with controllability and safety indices can be used to produce more optimal designs. The process for integrating safety into simultaneous design and control with PAROC can be seen in Figure 4. The objective function of the design remains the same (2), and controllability (3) and inherent safety (4) index variables and constraints are added to the design of the high-fidelity model of the process.

Model reduction is performed, and an optimal controller is designed from the reduced model. The controller's effectiveness is used to validate the controllability index, and its complexity is used to define the ease of control for the inherent safety index. The weighing factors of the design constraints are updated, and the process repeats until the optimal design is found.

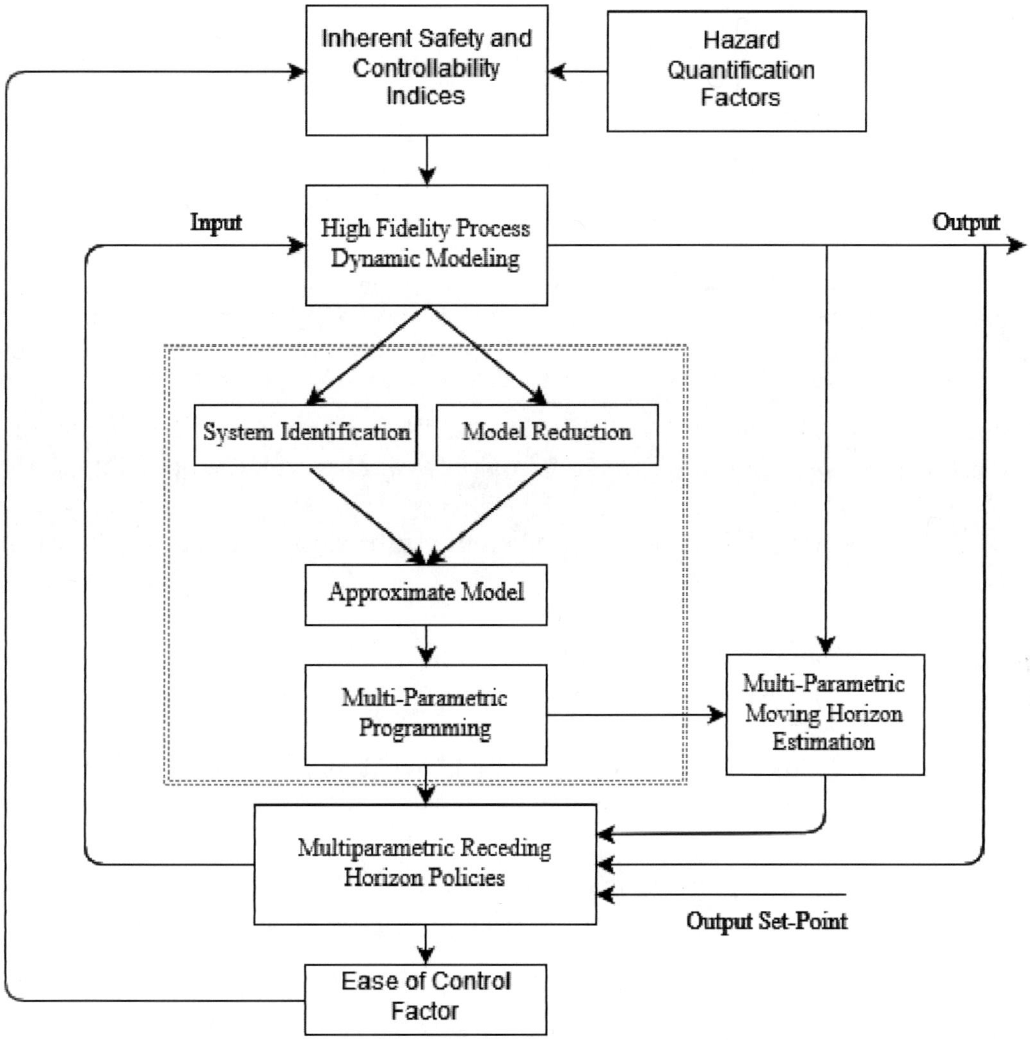

Figure 4. Integrating Safety into the PAROC Framework [27].

4 Conclusions

The background and process for integrating inherently safer design and control are discussed in this paper. A list of inherent safety indices and controllability indices are presented. A multi-objective formulation for integrating inherently safer design with controllability is presented. A case study is performed where inherent safety is integrated into the PAROC framework for simultaneous design and control.

5 References

[1] J. G. Ziegler and N. B. Nichols, "Process Lags in Automatic Control Circuits," *Trans. ASME,* vol. 65, pp. 433-444, 1943.

[2] A. M. Blanco and J. Bandoni, "Interaction between process design and process operability of chemical processes: An eigenvalue optimization approach," *Comp Chem Eng,* vol. 27, pp. 1291-1301, 2003.

[3] V. Sakizlis, J. D. Perkins and E. N. Pistikopoulos, "Recent advances in optimization-based simultaneous process and control design," *Comp Chem Eng,* vol. 28, pp. 2069-2086, 2004.

[4] T. Kletz, Process Plants: A handbook for inherently safer design, Bristol, PA: Taylor & Francis, 1998.

[5] M. Hurme and M. Rahman, "Implementing inherent safety throughout the process lifecycle," *Journal of Loss Prevention,* vol. 18, pp. 238-244, 2005.

[6] T. Kletz, "What you don't have, can't leak," *Chemistry & Industry,* vol. 9124, pp. 289-292, 1978.

[7] AIChE, Dow's Fire and Explosion Index Hazard Classification Guide, 7th ed., New York: American Institute of Chemical Engineers (AIChE), 2016.

[8] AIChE, Dow's Chemical Exposure Index Hazard Classification Guide, 1st ed., New York: American Institute of Chemical Engineers (AIChE), 1994.

[9] B. J. Tyler, A. R. Thomas, P. Doran and T. R. Greig, "The Mond fire, explosion and toxicity index - A development of the Dow Index.," Manchester, 1979.

[10] D. Mansfield, "The INSET Toolkit, INherent SHE Evaluation Tool," London, 1997.

[11] F. I. Khan and S. A. Abbasi, "Multivariate Hazard Identification and Ranking System," *Process Safety Progress,* vol. 17, no. 3, pp. 157-170, 1998.

[12] D. W. Edwards and D. Lawrence, "Assessing the inherent safety of chemical processes," *Trans IChemE,* vol. 71B, pp. 252-258, 1993.

[13] A.-M. Heikkila, "Inherent Safety in Process Plant Design, An Index-Based Approach," Helsinki University of Technology, Espoo, 1999.

[14] F. I. Khan and P. R. Amyotte, "Integrated Inherent Safety Index (I2SI): A Tool for Inherent Safety Evaluation," *Process Safety Process,* vol. 23, no. 2, pp. 136-148, 2004.

[15] A. M. Shariff, C. T. Leong and D. Zaini, "Using process stream index (PSI) to assess inherent safety level during preliminary design stage," *Safety Science,* vol. 50, pp. 1098-1103, 2012.

[16] F. K. P. A. Samith Rathnayaka, "Risk-based process plant design considering inherent safety," *Safety Science,* vol. 70, pp. 438-464, 2014.

[17] R. Aris and N. R. Undson, "An analysis of chemical reactor stability and control-I The possibility of local control, with perfect or imperfect control mechanisms," *Chemical Engineering Science,* vol. 7, no. 8, pp. 121-131, 1958.

[18] E. H. Bristol, "On a new measure of interactions for multivariable process control," *IEEE Trans. Automat. Control,* Vols. AC-11, no. 1st, pp. 133-134, 1966.

[19] V. Kariwala, J. F. Forbes and E. S. Meadows, "Block Relative Gain: Properties and Pairing Rules," *Ind. Eng. Chem. Res. ,* vol. 42, no. 20th, pp. 4564-4574, 2003.

[20] C. Manousiouthakis and M. Nikolaou, "Analysis of Decentralized Control Structures for Nonlinear Systems," *AIChE Journal,* vol. 35, no. 4th, pp. 549-558, 1989.

[21] C. E. Garcia and M. Morari, "Internal Model Control 1. A Unifying Review and Some New Results," *Ind. Eng. Chem. Process Des. Dev.,* vol. 21, no. 2nd, pp. 308-323, 1982.

[22] B. R. Holt and M. Morari, "Design of resilient processing plants—VI, The effect of deadtime on dynamic resilience," *Chemical Engineering Science,* vol. 40, no. 7th, pp. 1229-1237, 1985.

[23] E. Zafiriou, Integration of Process Design and Control, 1st ed., Baltimore: IFAC, 1994.

[24] K. Trickett, "QUANTIFICATION OF INVERSE RESPONSE FOR CONTROLLABILITY ASSESSMENT OF NONLINEAR PROCESSES," University of London, London, 1994.

[25] P. Daoutidis and C. Kravaris, "Structural Evaluation of Control Configurations for Multivariable Nonlinear Processes," *Chemical Engineering Science,* vol. 47, no. 5th, pp. 1091-1107, 1992.

[26] M. Morari, "Effect of Design on the Controllability of Chemical Plants," *Interactions Between Process Design and Process Control,* vol. 1, no. 1st, pp. 3-16, 1992.

[27] E. N. Pistikopoulos and N. A. Diangelakis, "Towards the integration of process design, control and scheduling: Are we getting closer?," *Computers and Chemical Engineering,* vol. 91, no. 1st, pp. 85-92, 2015.

[28] Y. Y. H. Vira Chankong, Multiobjective Decision Making: Theory and Methodology, Mineola: Dover Publications, 1983.

13TH GLOBAL CONGRESS ON PROCESS SAFETY

Criteria for Determining the Need for Installation of Fixed Fire and Gas Detectors in Petroleum Refineries

Laurence Siqueira
PETROBRAS
28, Henrique Valadares Av. Rio de Janeiro-RJ– 20231-030
laurencelima@petrobras.com.br

Bruno Ferreira Barsotti
PETROBRAS
bruno.barsotti@petrobras.com.br

Socrates Fofano
PETROBRAS
fofano@petrobras.com.br

Prepared for Presentation at
American Institute of Chemical Engineers
2017Spring Meeting and 13thGlobal Congress on Process Safety
San Antonio, TX
March 26 – 29, 2017

Criteria for Determining the Need for Installation of Fixed Fire and Gas Detectors in Petroleum Refineries

Laurence Siqueira
PETROBRAS
28, Henrique Valadares Av. Rio de Janeiro-RJ– 20231-030
laurencelima@petrobras.com.br

Bruno Ferreira Barsotti
PETROBRAS
bruno.barsotti@petrobras.com.br

Socrates Fofano
PETROBRAS
fofano@petrobras.com.br

Keywords: Safety, Detectors, Flame, Gas, Frequency, Probability, Ignition, Flammable, Toxic

Abstract

For determining of what accidental scenarios need the installation of fixed fire and gas detectors was necessary the preparation of a criterial and a technical guideline that could determine the probability of ignition of a released flammable mass, the probability of toxic gas plumes, the equipament failure rate and maximum tolerable frequency.

The probability of a flammable mass release due to leakage takes into account the operating conditions of the process plant such as pressure, temperature and physical state of the most significant flammable substance after the leakage. The physico-chemical data of the substance are also considered, such as molecular weight, auto ignition temperature, LEL (Lower Explosion Limit), MIE (Minimum Ignition Energy), Flash Point and Boling Point.

With the ignition probability of the leakage and the failure rate of the considered systems, it was possible to determine the probable frequency of its occurrence. For those scenarios with results above the tolerable frequency, it was considered that they were eligible for installation of fire and gas detectors.

The technical guideline has been applied in more than 200 process units of refineries and has allowed not only to reevaluate the need of installation of fire and gas systems, but also to increase its effectiveness, resulting in a reduction of consequence severity from leakage of flammable mass. Additionally it has contributed to the optimization of installation and reduction of the maintenance costs of the plants.

1. Introduction

Fixed fire and gas detection systems have been used in industrial installations where any leakage of gas or vapors (flammable or toxic), as well hydrocarbon and hydrogen fires represent significant hazards to people, assets and/or environment.

Significant variations in fire and gas systems installed in refining process units of the same type have been observed, like as some units have few, if any, detectors, while others have dozens of detectors installed. Due to this discrepancy, it was noted that different criteria have been adopted for installation of fixed fire and gas detectors.

This present paper aims to present criteria for the use of fire and gas systems focused on oil refineries. It sought to adopt criteria using a risk-based approach, observing the specific characteristics and limitations of these systems.

In this context, it was used the probability of ignition of a released flammable mass, the failure rates per type of process equipment, calculation of the risk, as well defining the tolerable frequency for the selected scenarios.

2- Criteria for Scenarios Selection

As initial tasks, qualitative risk studies was carried out, identifying the accidental risk scenarios of each process unit. In this stage, the respective hazards, causes and consequences associated with each scenario are related.

The next step was select among the identified scenarios, which should be evaluated as the need for detection. The scenarios of interest were scenarios whose danger was leakage to the environment, the cause was loss of containment, and the consequence was formation of flammable / toxic cloud, fire and/or explosion.

2.1 - Criteria for Consequences of Scenarios

Observing the mitigation (and non-prevention) characteristic of fire and gas systems, a possible approach to be adopted is to consider for evaluation only the critical scenarios, that is, scenarios with a high consequence severity (critical or Catastrophic). See risk tolerance matrix in Figure 1. Scenarios of lower severity (medium or marginal severity) will not be considered for evaluation as the need for detection.

	Consequency Severity			Frequency of Occurrence				
	People	Assets / Production	Environment	Extremely Remote	Remote	Not likely	Probable	Frequent
Catastrophic	Multiple fatalities	Catastrophic demages	Catastrophic demages	M	M	NT	NT	NT
Critical	One fatality	Severe demages	Severe demages	T	M	M	NT	NT
Medium	Severe injuries	Moderate demages	Moderate demages	T	T	M	M	NT
Marginal	Light injuries	Light demages	Light demages	T	T	T	M	M
Negligible	No injuries	Insignificant demages	Insignificant demages	T	T	T	T	M

T	Tolerable
M	Moderate
NT	Not Tolerable

Figure 1: Risk Tolerance Matrix

2.2 - Scenario Frequency Criteria

Each of the critical scenarios has associated a likely frequency of occurrence. Therefore, as an additional criterion, it is also necessary to establish a maximum tolerable frequency of occurrence, below which the use of detection is not considered necessary.

In the present work the concept of ALARP (As Low As Reasonably Practicable) developed by Health and Safety Executive, UK was used. This concept, which is well accepted in the oil and gas industry, including regulatory agencies, presents a range of risk values divided into three regions, namely:

a) "Bottom-up" region, in which risk values are clearly acceptable, and there is no need to adopt additional actions and/or measures to reduce these values. Reference: frequency of occurrence values equal to or less than 10^{-6}/year;

b) "Top-of-the-range" region, in which risk values are clearly unacceptable, and additional actions and/or measures have to be taken to reduce these values. Reference: frequency of occurrence values equal to or greater than 10^{-2}/year; and

c) "Mid-scale" region, called the ALARP region, in which the operating company has to demonstrate that it applies actions and/or measures to reduce the so-called "practicable" risk values, and the adoption of additional actions and/or measures are not "practicable". Reference: frequency values occurring in the range of 10^{-5}/year up to 10^{-3}/year.

It is up to the operating company to define the region of ALARP that it considers suitable to apply in its installations, especially the maximum frequency value of this interval. Applying the concept of ALARP in the risk tolerance matrix presented (see figure 2), the value of 10^{-4}/year can be adopted for maximum tolerable frequency (not likely frequency).

Thus, only the critical (leakage) scenarios, whose frequency of occurrence are greater than or equal to 10^{-4}/year, are then selected for evaluation as to the need for detection.

Consequency Severity				Frequency of Occurrence				
	People	Assets / Production	Environment	Extremely Remote	Remote	Not likely	Probable	Frequent
Catastrophic	Multiple fatalities	Catastrophic demages	Catastrophic demages	M	M	NT	NT	NT
Critical	One fatality	Severe demages	Severe demages	T	M	M	NT	NT
Medium	Severe injuries	Moderate demages	Moderate demages	T	T	M	M	NT
Marginal	Light injuries	Light demages	Light demages	T	T	T	M	M
Negligible	No injuries	Insignificant demages	Insignificant demages	T	T	T	T	M

T	Tolerable
M	Moderate
NT	Not Tolerable

Figure 2: Risk Tolerance Matrix

3- Evaluation of Selected Scenarios

It is considered a basic leak events tree presented in Figure 3.

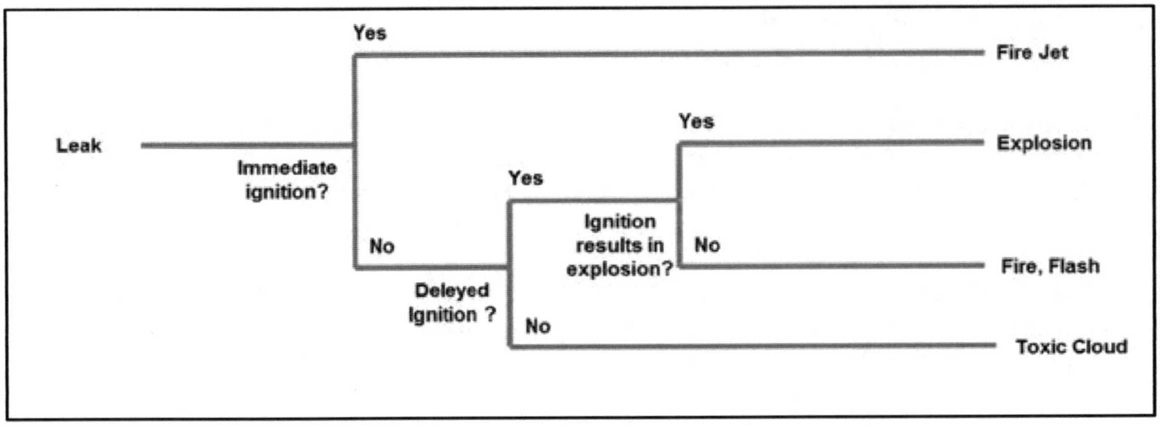

Figure 3: Leak events tree

For each scenario selected for evaluation, it is necessary estimate the respective frequency of leakage (loss of containment). For this, values of failure rate that were used are indicated in table 1, and it was necessary to correlate the type or set of equipment to the associated scenario.
The failure rates frequency adopted has considered data from the following sources: DNV - Failure Frequency Guidance, DNV - A Guide to Quantitative Risk Assessment of Offshore Installations, Energy Institute – Ignition Probability Review, Model Development and Look Up Correlations and HSE - Off Shore Ignition Probabilities

Table 1: Equipment failure rates per year adopted

Reciprocating Compressors (e.g. Make-up HDT)	**2,0 x 10^{-2}/year**
Pumps	**5,0 x 10^{-3}/year**
CentrifugalCompressors (ex: Gas, Recycling)	**2,5 x 10^{-3}/year**
Exchangers / Pipes and Valves in Preheating Batteries	**2,0 x 10^{-3}/year**
Scenarios with Other Static Equipment	**1,0 x 10^{-3}/year**
Scenarios with Pipes and Accessories only	**3,3 x 10^{-4}/year**

For all leak scenarios, the hole diameter of 10 mm can be adopted, since this value characterizes small leaks (suitable for detection by detectors), has higher frequency of occurrence, and is difficult to detect by process monitoring instruments, such as flow or pressure meters. This hole diameter adopted has used as reference the CCPS literatures: Guidelines for Determining the Probability of Ignition of a Released Flammable Mass, Continuous Monitoring for Hazardous Material Releases, Guidelines for Fire Protection in Chemical, Petrochemical, and Hydrocarbon Processing Facilities and Estimating the Flammable Mass of a Vapor Cloud

As a next step, it was necessary estimate, for each scenario selected for evaluation, the following probabilities of occurrence:

- Probability of immediate ignition;
- Probability of delayed ignition; and
- Probability of ignition with explosion.

For estimate all of these probabilities the literature that has been used was CCPS - Guidelines for Determining the Probability of Ignition of a Released Flammable Mass. Once these probabilities of occurrence (and non-occurrence) values were estimated, the selected scenarios were then evaluated as follows:

a) The values of the leakage frequencies were multiplied by the probabilities values of immediate ignition, delayed ignition and explosion;

b) The obtained results were the scenarios frequencies of occurrence, which were then compared with the maximum tolerable frequency of occurrence adopted (10^{-4}/year);

c) Scenarios which frequency of occurrence value were greater than the maximum tolerable were considered scenarios that are eligible for detection;

d) The detection type to be used in each scenario (flame, flammable gas or toxic gas) will depend on the numerical results of the calculations performed, as exemplified in Figure 4.

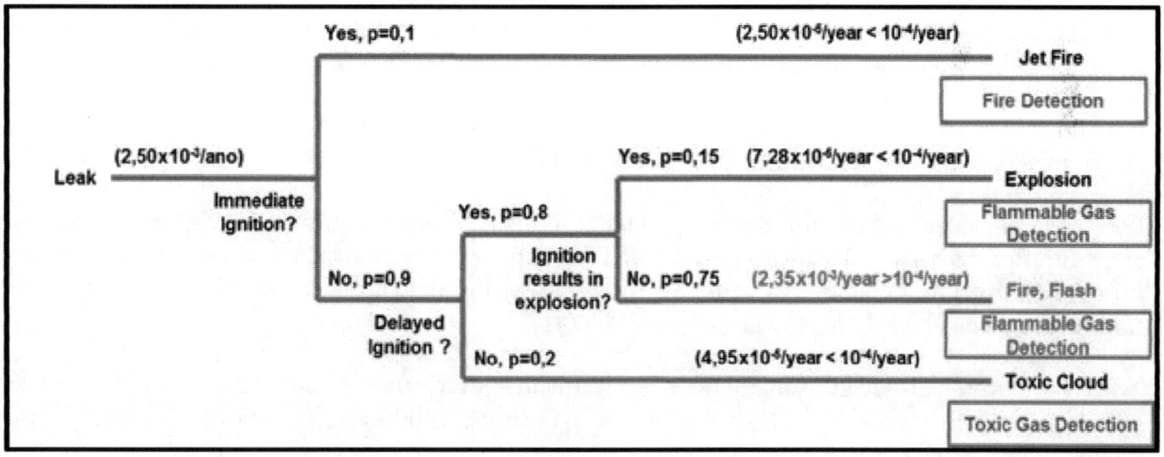

Figure 4: Leak events tree and examples of calculations performed

4- Ignition Probability Estimates

In order to estimate the ignition probabilities, it is necessary to obtain process and facility data for the scenarios selected for evaluation.

The basic process data required are pressure, temperature and flow in the operating conditions of the leaked product. However, it is also necessary to know various physicochemical properties of the leaked products, such as, physical state, molecular weight, density, Cp/Cv, vapor pressure, auto ignition temperature, flash point, boiling point, lower explosive limit and minimum ignition energy.

One way of obtaining such data is correlating the leaked products with a reference substance, considered as more representative, as exemplified in table 2.

Table 2: Examples of reference substance

Leaked Product	Reference Substance	Molecular weight	Auto ignition temperature (°C)	Flash Point (°C)	Boling Point (°C)	LEL - Lower Explosive Limit (%)	MIE (mJ)
FUEL GAS	Methane	16	600	-187,2	-162,5	5	0,28
GASOIL	n-decane	142,3	210	46	174	0,8	0,2
PETROLEUM	n-decane	142,3	210	46	174	0,8	0,2
GASOLINE	n-hexane	86	225	-23	69	1,6	0,24
DIESEL OIL	n-nonane	128	205	31	150	0,8	0,2
NAPHTHA	n-pentane	72	243	-40	36,1	1,5	0,28
LPG	Propane	44	450	-102	-42	2,6	0,25

If the critical scenario contains process equipment that runs with toxic substances, it was also necessary define the concentration of this substances in the gas phase. In this way, it was verified if the concentration is above the level of IDLH - Immediately Danger to Life and Health.

For those scenarios where the streams contains only toxic substances and the concentration is above of IDLH, then it is recommend to use toxic detectors.

For those scenarios where the streams contains both toxic and flammable substances and the concentration of toxic substance is above of IDLH, then it is recommend to use toxic detectors prior to flammable detectors, because in these cases, the detection is rated in ppm, while for flammable gas in % of LEL (lower explosive level).

The most common toxic substances considered for this study were H_2S concentration above 100 ppm, CL_2 above 10 ppm, NH_3 above 300 ppm, CO above 1200 ppm, HF above 30 ppm, Phenol above 250 ppm and ClO_2 above 5 ppm.

5 - Additional Criteria for Detection

The following are some additional criteria that can be applied to rationalize the scenarios selected for installation of fixed gas and flame detection:

5.1 - The detection of flammable gases/vapors may be dispensed for those scenarios where the leaked liquid has operating temperature lower than the Boiling Point value.

5.2 - The Hydrocarbon flame detectors can be dispensed for scenarios where there are passive protection (fireproofing) for the equipment inside the fire envelope.

5.3 - For the cases of the process units where the results of frequencies of occurrence are above the 10^{-4}/year (maximum tolerable frequency of occurrence adopted) that already has deluge systems installed and it can be interlocked with the detection system, than hydrocarbon flame detection is recommended, otherwise, it can be dispensed.

5.4 – For the cases where the frequency of occurrence results are above the 10^{-4}/year (maximum tolerable frequency of occurrence adopted) but the process unit doesn´t have any automatic or semiautomatic active protection, the decision for installation of the fixed hydrocarbon flame detectors by itself needs to be evaluated in terms of reducing the consequences of injuries to persons, assets and/or environment

5.5 – The use of fire detection can then be limited to scenarios that produces hydrogen or methanol flames, because these flames are not "visible" to the human eye. Thus, only leak scenarios with a frequency of occurrence greater than 10^{-4}/year, and with high potential of occurrence of fire involving hydrogen or methanol, were indicated for fire detection.

Examples considered:

- Compressor operating with H_2 gas;
- Equipment operating with H_2 gas and having a temperature above 500 ^oC; and
- Pump operating with methanol.

6 – Diagrams and formulas used to determine and calculate the risks

In order to maximize the understanding of the work that has been done, refer to the diagrams as follow:

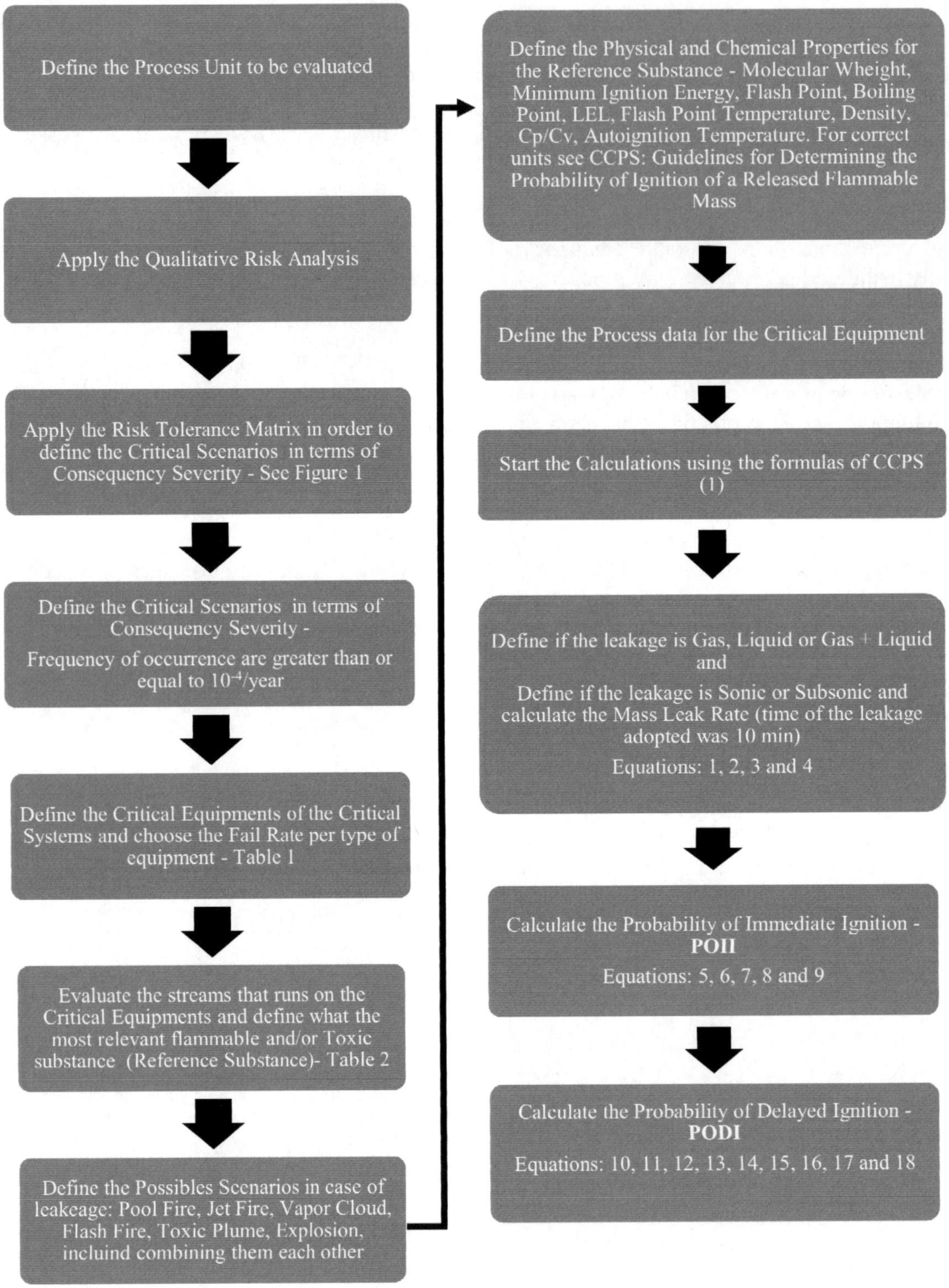

Diagram continues next page

Fire Analysis

If the POII x Failure Rate of Critical Equipment is greater or equal than Maximum Tolerable Frequency of occurrence of 10^{-4}/year, so the scenario is Elegible for Flame Detection.

Define techonology for Detection

(See criteria on items 4 and 5 of this paper)

Calculate the Probability of Explosion Given a Delayed Ignition - POEGDI

Equations: 19, 20 and 21

Explosion Analysis

If the POEGDI x Failure Rate of Critical Equipment is greater or equal than Maximum Tolerable Frequency of occurrence of

10^{-4}/year, so the scenario is Elegible for Flammable Gas Detection.

Define techonology for Detection

Toxic Plume Analysis

If the reference toxic substance is above the IDLH and the probability of toxic plume multiplied per failure rate of critical equipment is greater or equal than Maximum Tolerable Frequency of occurrence of

10^{-4}/year, so the scenario is Elegible for Toxic Gas Detection.

Define techonology for Detection

7- Equations

The criteria presented in this paper have used the following equations and the descriptions, special notes as well the meaning of each formula presented bellow can be found on literature of CCPS - *Guidelines for Determining the Probability of Ignition of a Released Flammable Mass* and *Guidelines for Fire Protection in Chemical, Petrochemical, and Hydrocarbon Processing Facilities.*

$$W_n = C_d \cdot K_{v,n} \cdot \rho_l \cdot \frac{A_n}{C_1} \sqrt{\frac{2 \cdot g_c \cdot (P_s - P_{atm})}{\rho_l}} \quad (1)$$

$$P_{trans} = P_{atm}\left(\frac{k+1}{2}\right)^{\frac{k}{k-1}} \quad (2)$$

$$W_n = \frac{C_d}{C_2}\cdot A_n \cdot P_s\sqrt{\left(\frac{k\cdot MW\cdot g_c}{R\cdot T_s}\right)\left(\frac{2}{k+1}\right)^{\frac{k+1}{k-1}}} \quad (3)$$

$$W_n = \frac{C_d}{C_2}\cdot A_n \cdot P_s\sqrt{\left(\frac{MW\cdot g_c}{R\cdot T_s}\right)\left(\frac{2\cdot k}{k-1}\right)\left(\frac{P_{atm}}{P_s}\right)^{\frac{2}{k}}\left(1-\left(\frac{P_{atm}}{P_s}\right)^{\frac{k-1}{k}}\right)} \quad (4)$$

$$MIE_v = MIE_{reported} \times (10{,}000/P_{liquid})^{0.25} \quad (5)$$

$$MIE_{adj} = MIE_v \times \exp[0.0044(60 - T)] \quad (6)$$

$$P_{AIT} = 1 - 5{,}000e^{-9.5(T/AIT)} \quad (7)$$

$$POII_{static} = 0.003 \times P^{1/3} \times MIE_{adj}^{-0.6} \quad (8)$$

$$POII_{Level\ 2} = P_{ai} + (1 - P_{ai}) \times POII_{static} \quad (9)$$

$$PODI_{S/D} = 1 - [(1 - S^2) \times e^{-St}] \quad (10)$$

$$M_{MAG\_Amount\ Released\ (liquid)} = (\text{Amount Released}/5{,}000)^{0.3}$$

$$M_{MAG\_Amount\ Released\ (vapor)} = (\text{Amount Released}/1{,}000)^{0.5}$$

$$M_{MAG\_Hole\ Diameter\ (liquid)} = (\text{Hole Diameter})^{0.6}$$

$$M_{MAG\_Hole\ Diameter\ (vapor)} = (\text{Hole Diameter})$$

(11,12, 13 and 14)

$$M_{MAT} = 0.5 - 1.7 \log(MIE) \quad (15)$$

$$M_T = 0.4 + (T - 1.3 \times FP)/230 \quad (16)$$

$$PODI_{Level\ 2} = PODI_{S/D} \times M_{MAG} \times M_{MAT} \times M_T \times M_{IN/OUT} \quad (17)$$

$$PODI_{final} = PODI_{calc}\,(1 - POII) \quad (18)$$

$$POEGDI_{Level\ 2} = 0.3 \times M_{CHEM} \times M_{MAGE} \times M_{IN/OUT} \quad (19)$$

$$M_{MAGE} = (PODI\ M_{MAG})^{0.5} \quad (20)$$

$$POEGDI_{FINAL} = PODI_{FINAL}\ X\ POEGDI \quad (21)$$

8- Results

The criteria presented for selection and evaluation of scenarios were applied in a set of 80 different process units, covering a total of 10 refineries. The results obtained, in terms of average amounts of scenarios, are presented in table 2, grouped by type of refining process.

Table 2: Scenarios by type of process units

PROCESS UNIT	Scenarios A	Scenarios B	Scenarios C	Scenarios D
Catalytic Cracking	119	88	57	35
Delayed Coking	74	61	46	25
Atmosp. and Vacuum Distillation	67	53	35	13
Diesel Hydrotreating	52	40	23	15
Catalytic Reform	51	37	34	22
Sulfur Recovery	33	21	20	11
Hydrogen Generation	25	15	11	5

a) Scenarios A - Total amount of risk scenarios studied, with preliminary risk analysis technique.

b) Scenarios B - Subset of Scenarios A. Those are scenarios of leaks (to the environment) that have the potential to be evaluated for the need for flame or gas detection.

c) Scenarios C - Subset of Scenarios B. Those are scenarios B which the severity of the consequence is critical or catastrophic (see Figure 1).

d) Scenarios D - Subset of Scenarios C. Those are scenarios C which frequency of leakage, combined with the probability of ignition (see Figure 5), is greater than the tolerable frequency of occurrence (see Figure 2). The additional criteria (see item 5) are also considered.

9 - Conclusions

The use of the presented criteria for selection and evaluation of scenarios, especially the use of the probability of ignition, leads to an expressive reduction of scenarios indicated for detection. Considering a hypothetical refinery, that has the process units listed in Figure 7, the total amount of scenarios indicated for detection is 126 (Figure 7, scenarios D), against a total of 226 (Figure 7, scenarios C). In this case, the use of the presented criteria represents a reduction of 100 (or 44%) in the indicated scenarios for detection.

Considering an average value of 1.4 detectors per scenario, and the cost of US$ 8,000 for a detector installation, and the cost of US$ 1,000.00 per year for a detector maintenance (checks, calibrations and replacements), we have for the hypothetical refinery:

- Reduction of investment value (CAPEX) in fixed fire and gas detection system: US$ 1,120,000.00;
- Reduction of the maintenance costs value (OPEX) in fixed fire and gas detection system: US$ 140,000.00/year.

10 - Bibliography

(1) CCPS – *Guidelines for Determining the Probability of Ignition of a Released Flammable Mass*, Center for Chemical Process Safety, American Institute of Chemical Engineers, New York, NY, 2014.

(2) CCPS – *Consequence Analysis of Chemical Releases.* Center for Chemical Process Safety, American Institute of Chemical Engineers, New York, NY, 1999.

(3) CCPS - *Guidelines for Fire Protection in Chemical, Petrochemical, and Hydrocarbon Processing Facilities.* Center for Chemical Process Safety, American Institute of Chemical Engineers, New York, NY, 2003.

(4) CCPS – *Guidelines for Evaluating the Characteristics of Vapor Cloud Explosions, Flash Fires and BLEVEs.* Center for Chemical Process Safety, American Institute of Chemical Engineers, New York, NY, 1994.

(5) CCPS – *Estimating the Flammable Mass of a Vapor Cloud.* Center for Chemical Process Safety, American Institute of Chemical Engineers, New York, NY, 1998.
(6) CCPS - *Guidelines for Chemical Process Quantitative Risk Analysis – Second Edition.* Center for Chemical Process Safety, American Institute of Chemical Engineers, New York, NY, 2000.
(7) CCPS - *Guidelines for Process Equipment Reliability Data – Second Edition.* Center for Chemical Process Safety, American Institute of Chemical Engineers, New York, NY.
(8) Lee's Loss Prevention in the Process Industries – Hazard Identification, Assessment and Control Volume 1 – Fourth edition, Texas, USA, 2012
(9) Electrical Installations in Hazardous Areas – Eur Ing McMillan C Eng. FIEE FInstMC, Oxford OX2 8DP, Woburn, MA, 1998
(10) DNV - Failure Frequency Guidance, Process Equipment Leak Frequency Data for Use in QRA
(11) DNV - A Guide to Quantitative Risk Assessment of Offshore Installations
(12) Handbook Failure Frequencies – Safety Report, Flemish Government, LNE Department, Environment, Nature and Energy Policy Unit – Safety Reporting Division. 2009
(13) Energy Institute – Ignition Probability Review, Model Development and Look Up Correlations
(14) HSE – Health and Safety Executive, London, UK. Off Shore Ignition Probabilities, Report number HSL/2005/50, Harpur Hill, Buxton, Derbyshire SK17 9JN
(15) HSE – Health and Safety Executive, Guidance on Risk Assessment for Offshore Installations, Sheet N.3/2006
(16) IOGP – International Association of Oil & Gas Producers Risk Assessment Data Directory – Report N.434-6.1 – Ignition Probabilities, 2010
(17) IOGP - International Association of Oil & Gas Producers Risk Assessment. Data Directory – Report N.434-6.1 – Safety Performance Indicators Process Safety Events, 2014 Data
(18) IOGP - International Association of Oil & Gas Producers - Risk Assessment Data Directory – Report N.434-6.1 – Safety Performance Indicators Process Safety Events, 2015 Data
(19) IOGP – International Association of Oil & Gas Producers – Risk Assessment Data Directory, Storage Incident Frequencies. Report N.434-3, March 2010.
(20) ANSI/ISA - 5.1 Instrumentation Symbols and Identification
(21) ANSI/ISA - 92.00.02 - 2013 Installation, Operation and Maintenance of Toxic Gas-Detection Instruments;
(22) API 580 – Risk Based Inspection
(23) API 581 – Part 3, Consequence Analysis in an API RBI Assessment;
(24) API 1996 - Final Report 1996 American Petroleum Institute/National Petroleum Refiners Association Survey of Refining Operations and Product Quality First Edition
(25) API 2021- Fire Protection in Refineries
(26) API 2218 – Fireproofing Practices in Petroleum and Petrochemical Processing Plants
(27) API 752 – Management of Hazards
(28) NFPA 497 – Recommended Practice for the Classification of Flammable Liquids, Gases or Vapors and of Hazardous (classified) Locations for Electrical Installations in Chemical Process Areas

480396 Exploring the Catalytic Options in Low-Mid Pressure Diesel Hydrotreating

Abstract:

Regularly revised, worldwide diesel fuel specifications continue to press impurity levels lower and product quality higher. Such fuel specification revisions naturally encourage the semi-continuous development of superior hydrotreating catalysts, to satisfy the new requirements and improve the economics of hydrotreacting where applicable. The goal of this talk is to draw rational boundaries where the most active examples of supported NiMo and supported CoMo catalysts ought to be applied, given the inhibiting influences of organic nitrogen species and aromatics species, for example. Included in the scope shall also be application of both types of catalysts simultaneously, where a bed of one of the catalysts is loaded above the other. In consideration of the pressing need for improved hydrotreating catalysts, Honeywell UOP has recently commercialized an entirely new portfolio of supported hydrotreating catalysts under the Unity(TM) trademark. Several of these catalysts shall be featured in this presentation, since they were developed for diesel hydrotreating in the low-to-mid pressure range.

480176 Next-Generation Utilities Optimisation for Refineries and Large-Scale Chemical Production Sites

Abstract:

Refineries and chemical production sites are major consumers of energy in the form of electricity, steam and hydrocarbon feedstocks. Given that tariffs, costs and demands are constantly changing, there is much scope for optimising on-site production, conversion and distribution of energy to minimise cost and emissions, by managing the options available in the most cost effective way while meeting all constraints of the system. However many current 'optimisation' applications do not provide full capabilities for maximising economic value or minimising emissions, particularly in situations where demands from major consumers and the prices of the various components can change on an hourly basis.

This paper describes an advanced optimisation platform for managing and optimising utility operation that not only helps planners rapidly optimise equipment selection and load allocation to improve overall efficiency and reduce emissions and operating costs, but also presents operators with a ranked list of possible actions from which they can choose the best course of action given the situation 'on the ground'.

The approach uses medium-fidelity models of the utilities system and major devices that are coupled with plant operating data via data validation and reconciliation facilities. An advanced optimisation system capable of both continuous and integer decisions determines the economically optimal operating point taking into account equipment and operational constraints, including availability. The resulting mathematical problem is solved within an equation-oriented framework, providing robustness and speed of execution well in advance of most current systems.

Because of the speed of solution, a key feature is the ability to run multiple optimisations and provide operators with a ranked list of potential combinations of changes and their corresponding benefits, within a dashboard tailored for the site. This allows operators clearly to evaluate and discuss which changes are the best to apply when, resulting in advice that is practical and easy to implement and verify. It addresses one of the biggest obstacles to practical realisation of the benefit of optimisation systems, which is gaining operator buy-in to proposed changes.

The paper is illustrated with an examples of implementation on a major European chemical manufacturing site.

478714 Study on Regional Air Quality Impact from Olefin Plant Shutdown Operations

Abstract:

Ground-level ozone is a pervasive air pollutant, which can be potentially aggravated by flaring emissions from olefin plant shutdown operations. Although, flaring is crucial to chemical plant safety, the excessive flaring, for instance the intensive flaring from plant shutdown operations, emits huge amounts of ozone pollution precursors (VOCs and NOx). This will cause tremendous industrial material loss and environmental damage; meanwhile result in potential adverse impact to the regional air quality. Thus, the plant shutdown flare emissions should be minimized at any possibility. In this paper, plant-wide dynamic simulations for flare emissions with regional air-quality modeling are coupled together to quantify the air-quality impact and investigate potential opportunities to mitigate the adverse air-quality impact during plant shutdown operations. It proposed a systematic methodology on air-quality conscious study for shutdown operations of an olefin plant. Through studies of original and optimal cases, it shows that the original plant shutdown operation procedure has a serious and negative effect on the regional air quality; however, the optimal shutdown operation procedure would significantly reduce such impact together with the saving of tremendous emission sources. This study can provide valuable and quantitative supports for decision makings involving multiple stakeholders, including environmental agencies, regional plants, and local communities.

2017 AIChE Spring Meeting & 13th GCPS Global Conference on Process Safety

Up-to-date Engineering Solutions to Estimate and Minimize Thermal Breathing Losses from Storage Tanks

Dr. D. Moncalvo & Dr. M. Davies & Dr. T. Heidermann

Some examples of Tank Vacuum Collapse

Maybe you won't hear about it in the news but it surely costs lots of money!

PROTEGO
for safety and environment

Tank Breathing in API 2000 7th Ed. and ISO 28300

Tank breathing in API 2000 7th Ed. and ISO 28300
is distinguished into **Outbreathing** and **Inbreathing**

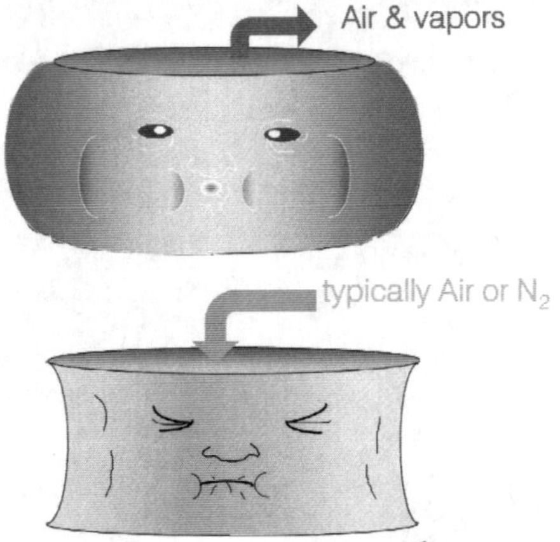

These standards distinguish between two scenarios:

- Normal in- and out-breathing
  - Liquid pumping
  - Weather changes (sun exposure, rain)
  - Internal failure of heat transfer devices
  - Control valve failures
  - Inert purge and utility failures
  - Vaporization or condensation
  - Steam collapse
  - Other causes
- Emergency out-breathing (fire exposure)

PROTECTO
for safety and environment

Tank Thermal Breathing in API 2000 7th Ed. and ISO 28300

$$\dot{V}_{thermal\ out} = Y \cdot V_{tk}^{0.9} \cdot R_i$$

$$\dot{V}_{thermal\ in} = C \cdot V_{tk}^{0.7} \cdot R_i$$

Insulation factor R_i

$$R_i = 1/[1 + h \cdot l_{in}/\lambda_{in}]$$

Not considering condensation or evaporation of product

Latitude		Y Factor	C Factor			
			Vapor pressure			
			Hexane or similar		Higher than hexane or unknown	
			Average storage temperature			
			< 77°F	≥ 77°F	< 77°F	≥ 77°F
0 - 42°	Houston	0.32	4	6.5	6.5	6.5
42° - 58°	Rotterdam	0.25	3	5	5	5
58° - 90°	Norway	0.20	2.5	4	4	4

PROTEGO
for safety and environment

Tank Thermal Breathing in API 2000 7th Ed. Annex A (*API 2000 5th Edition*)

Tank capacity	Thermal breathing (*SCFH*)	Max. heat transfer rate
Between 20 000 und 180 000 bbl.	2 SCFH each ft² of roof and shall area	Between 50 and 100°F/h
Below 20000 bbl.	1 SCFH each bbl. of tank volume	100°F/h

Annex A method is applicable, if following ***Limitations*** are satisfied:

Tank volume is less than 180 000 bbl.
Max. operating temperature in vapor space is 120°F
Tank is uninsulated
Temperature of stored products and feed is below boiling temperature at max. oper. pressure

Not considering condensation or evaporation of product

PROTEGO
for safety and environment

Tank thermal breathing in the scientific literature

Model	Year	Thermal modeling	Evaporation modeling	Condensation modeling	Sensitivity analysis	Measurements
Sigel et al.	1983	X		X		X
Förster et al. (*PTB model*)	1984	X				
Fullarton et al.	1987	X	X	X		X
Holtkötter et al.	1997	X		X		X
Salatino et al.	1999	X				
Brooks (DIERS pres)	2012				X	
Davies and Moncalvo (1)	2016	X		X	X	
Davies and Moncalvo (2)	2016	X	X	X		X

(1)	Journal of Loss Prevention in the Process Industry , volume 43, doi: 10.1016/j.jlp.2016.06.006
(2)	12th Global Congress on Process Safety (GCPS), Houston, April 2016, ISBN: 978-0-8169-1096-0

PROTEGO
for safety and environment

Davies and Moncalvo (2016)

- Balance of energy within the tank

$$\left(M_{vap}c_{vap} + M_{liq}c_{liq} + M_{air}c_{air}\right)\frac{dT_B}{dt} = -h_{int}A(T_B - T_E) - \dot{M}_{air}c_{air}(T_B - T_{amb})$$

- Balance of energy at the tank walls

$$M_E c_E \frac{dT_E}{dt} = h_{int}A(T_B - T_E) - h_{rain}A(T_E - T_{rain}) + \rho_B \dot{V}_{cond}\Delta h_{cond}$$

- Balance of mass within the tank

$$-\left(V_{air} + V_{vap}\right)/T_B \cdot dT_B/dt + \dot{V}_{cond} = \dot{V}_{air}$$

- Condensation model

$$\dot{V}_{cond} = \beta \cdot 1/p_B \cdot dp_{sat}/dT \cdot [T_B - T_E]A$$

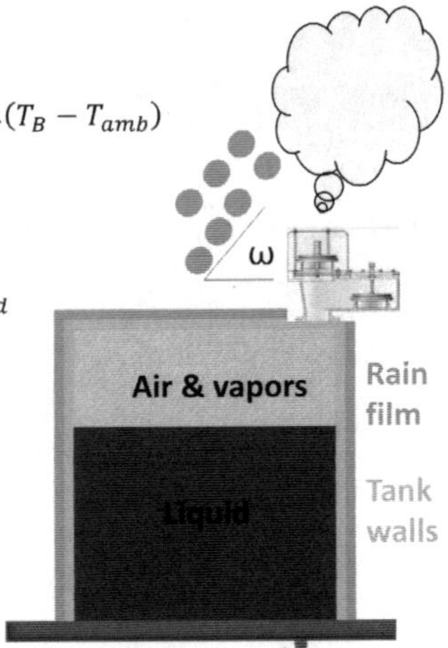

Source	Journal of Loss Prevention in the Process Industry, volume 43, doi: 10.1016/j.jlp.2016.06.006

PROTEGO for safety and environment

Some realistic examples of thermal inbreathing

	Tank 1	Tank 2
Tank height	49.2 ft.	
Tank diameter	49.2 ft.	82.0 ft.
Tank volume (V_{tank})	16,650 bbl.	46,250 bbl.
Initial temperature (T_0)	120°F	
Ambient temperature	60°F	
Average storage temperature	68°F	
Rain intensity	46.1 lb/h ft²	
Rain inclination	68°	
Wall material and thickness	Stainless Steel; $^1/_5$ in	
Insulation	Mineral wool	
Stored product	Dry Air Saturated Humid Air Saturated Heptane / Air Saturated Methanol / Air	

PROTEGO for safety and environment

Results for ***uninsulated*** *tank 1 (D = 49.2 ft.)* and *tank 2 (D = 82.0 ft.)*

Required Inbreathing [SCFH]	API 2000 Main body (max)	API 2000 Main body (min)	API 2000 Annex A	PTB (numerical)	Davies Moncalvo *without* condensation	Davies Moncalvo *with* condensation
Dry air	37,179	23,237	16,659	30,931	31,258	31,258
Humid air				30,730	30,682	48,138
Sat. Heptane / Air				19,576	19,608	31,640
Sat. Ethanol / Air				24,761	24,794	52,825

Required Inbreathing [SCFH]	API 2000 Main body (max)	API 2000 Main body (min)	API 2000 Annex A	PTB (numerical)	Davies Moncalvo *without* condensation	Davies Moncalvo *with* condensation
Dry air	76,013	47,508	35,910	68,807	69,576	69,576
Humid air				67,978	68,149	102,569
Sat. Heptane / Air				41,455	41,680	65,512
Sat. Ethanol / Air				53,485	53,807	108,608

PROTEGO for safety and environment

Worst realistic cases for rain intensity on *Tank 1*

Required Inbreathing [SCFH]		Rain intensity [lb/h ft²]		API 2000 Main body (max)	Davies Moncalvo *without* condensation
Dry air Tank 1	Scenario 1	46.1	API 2000	37,179	31,258
	Scenario 2	62.4	US historical highest (Holt, MO)		34,020
	Scenario 3	85.9	Worldwide hist. highest (Cherrapunji, IND)		36,910

Lesson learned:

API 2000 7th Ed. Main Body is safe for non-condensing products even under extreme weather

PROTEGO for safety and environment

Reducing Tank breathing: Insulation Layer and Tank Wall Thickness on *Tank 1*

Required Inbreathing [SCFH]	Insulation Thickness [in]	API 2000 Main body (max)	Davies Moncalvo without condensation
Dry air Tank 1	uninsulated	37,179	31,258
	1	8,905	1,364
	2	5,058	949
	4	2,714	551

Required Inbreathing [SCFH]	Wall Thickness [in]	API 2000 Main body (max)	Davies Moncalvo without condensation
Dry air Tank 1	0.2	37,179	31,258
	0.25		28,989
	0.3		27,221
	0.4		24,404

Lessons learned:

Increasing tank wall thickness may not suffice to reduce inbreathing requirements

Perform an engineering and economical study to find out the best insulation strategy for your tank

PROTEGO
for safety and environment

Tests on the influence of overpressure on tank leakage losses

- Leakage, tested with PROTEGO® portable test rig, may begin at 75% of set pressure as per API 2000 5th Edition Appendix C
- The valves were supposed to be fully open at 1 oz./in² (4.3 mbar) gauge.
- The first test was performed at 100 % overpressure
- The second test was performed at 10% overpressure

Relieving pressure [oz/in²g]	Overpressure (%$_{overpr}$)	Set pressure [oz./in² a]	Leakage Pressure [oz./in² g]
1.0	100% (1)	0.5	0.375
1.0	10% (0.1)	0.9	0.675

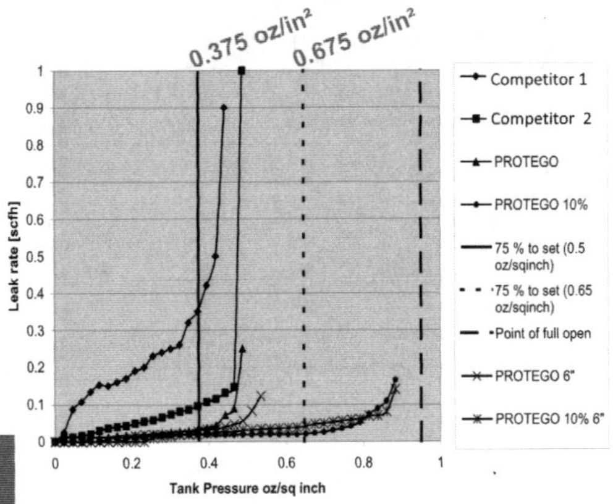

$$p_{rel\,g} = p_{set\,g}(1 + \%_{overpr})$$

$$p_{leak\,g} = p_{set\,g} \cdot \%_{leak}$$

PROTEGO
for safety and environment

Impact of test results in improvement of API 2000 standard

- Two main results from this research test:
  - Avoid choice of valves requiring large overpressures
  - Leakage pressure at 75% set pressure underestimates leakage losses

- Improvement of API 2000 Standard between 5th (1998) and 7th Edition (2014)

Requirements	API 2000 5th Edition	API 2000 7th Edition
Leakage pressure	75 %	90 %
Blowdown	0 %	Dependent on design

- Additional drawbacks to be considered when selecting a PRV with large overpressures:
  - Leakage pressure of PRV may be higher than set pressure of blanketing devices
  - PRV with low set pressures tend to cycle more (additional cost of spare parts)

PROTEGO
for safety and environment

Conclusions

API 2000 7th Ed. Main Body is safe for tanks with non-condensing gases even under extreme weather

Do not underestimate condensation in atmospheric tanks and perform an engineering study, especially if your tank contains volatile products, like (m)ethanol or gasoline

Increasing tank wall thickness may not suffice to reduce inbreathing requirements

Perform an engineering and economical study to find out the best insulation strategy for your tank (trade-off between costs of insulation purchase and maintenance against savings on breathing losses)

An effective minimization of product losses suggests the selection of PRV with minimum overpressure: the authors support the 10% overpressure as standard also for PVRV

PVRV with low overpressures usually guarantee also a chatter-free opening and avoid conflict with blanketing valves and ERV.

PROTEGO
for safety and environment

Braunschweiger
Flammenfilter
Thank you for your kind attention!
Reproduction, in part or in full, subject to prior written approval of Braunschweiger Flammenfilter GmbH.

478031 The Effect of Nitrogen and Sulfur Containing Molecules on Standard Methods for Accurate Determination of Oxygenates in Bio-Oils

Abstract:

Several thermochemical routes exist for deconstructing solid cellulosic biomass to produce bio-oil, including thermal pyrolysis and hydrothermal liquefaction. Historically, due to the complex nature of bio-oil, there have been large gaps in available analytical information. Bio-oil analysis, for raw oils and upgraded oils, requires time consuming and difficult analytical methods for quantitative identification of species present in the oil. These measurements are necessary for evaluating chemical differences between pyrolysis oils, evaluating the activity and selectivity of catalysts used to upgrade oil, and they are vital for evaluating the fuel quality (and subsequent blending limitations) of the oil. To address our knowledge of chemical active sites in the oils we extensively tested standardized methods to accurately describe oxygenates present in pyrolysis oils. In recent work conducted we have investigated the applicability of these methods for oils rich in sulfur and nitrogen such as HTL oil based on algae. We found that the presence of some sulfur and nitrogen functional groups in bio-oil can affect the values generated by methods that were developed to measure the oxygen content in bio-oils. A discussion of the study will be presented.

480325 Biolubricants Development, Formulation and Friction Test

Abstract:

Biolubricants have exhibited a lot of advantages over the conventional lubricants: excellent lubricity, minimizing of corrosion of metal surfaces, non-toxicity, and biodegradability. Biolubricants provide energy saving which makes them suitable to be used in all fields of industry. In this study, three biolubricants were developed and the friction force study was conducted and compared with mineral oil. The results indicated that the friction force of two biolubricants was significantly lower than that of mineral oil during the 36 minutes of testing time.

Pentanol: A Promising Fuel and Petrochemical Building Block

Kevin M. Van Geem[1], Cato A.R. Pappijn[1], Ruben Van de Vijver[1] and Judit Zador[2], (1) Department of Chemical Technology, Ghent University, Ghent, Belgium, (2) Sandia National Laboratories, California, USA

Introduction

The increasing worldwide energy demand and the major environmental constraints on greenhouse gas emissions are strongly driving the search for alternative fuel and base chemical resources. Many studies have reported industrial, experimental and theoretical results on feedstocks derived from non-fossil processes, among which urban waste and biological feedstocks are the most promising. Biofuels produced from biomass show strong potential for reducing harmful emissions and alleviating petroleum shortages, making research in this area very attractive [1-4]. Primary alcohols have already been considered for decades as sustainable fuel. Ethanol, belonging to the first generation biofuels, is widely spread as substitute for conventional diesel. Other alcohols such as propanol and n- and iso-butanol gained significant interest in the past for their higher energy density, higher cetane numbers and lower hygroscopic properties. To increase the energy content even more, the second generation biofuel n-pentanol is now extensively studied. *n*-Pentanol is a promising candidate for diesel blends not only due to its higher heating value, but also due to its low autoignition temperature and higher reactivity [4-6]. Furthermore, it is much easier to blend *n*-pentanol into conventional hydrocarbon fuels due to its longer carbon chain. Its performance in engines has been studied to minimize the CO, hydrocarbon, NOx, and smoke emission,[7,8] particulate emissions,[9] mixing characteristics,[10] and optimize the cetane number[11] of a pentanol/diesel blend. Experimental and modeling work on the combustion properties such as laminar flame speeds[12,13] and ignition characteristics[14] have been reported. In this work, the pyrolysis and combustion chemistry of n-pentanol is studied in two-fold. First, an experimental dataset was obtained on a Bench Scale set-up. Second, a kinetic model was constructed using the automatic kinetic model generation tool Genesys enabling a comparison between experimental and simulation results.

Experimental and simulation method

Experiment

The pyrolysis of n-pentanol has been studied experimentally on a Bench Scale set-up, which has been described previously by Djokic et al.[15] and Harper et al.[16]. The reactor is an Incoloy 800HT tube of 1.475 m long with an internal diameter of 6 mm, placed in a vertical furnace consisting of four heating sections controlled by four thermocouples inside the reactor tube. The pressure is measured at the in- and outlet of the reactor, with a set point of 0.17 MPa. The pressure drop along the reactor can be neglected. Two sets of experiments were done, one with at a pentanol dilution of 1:4 in nitrogen gas, another one with a dilution of 1:1. For the first set, the pentanol inlet flow amounts up to 48 g h^{-1}, while for the second one it is equal to 120 g h^{-1}. The temperature inside the reactor was varied between 913 and 1073 K with increments of 20 K, in order to cover a wide conversion range. The reactor effluent was analyzed using two single GC's and one GC×GC, which allows on-line qualification and quantification of the complete product spectrum, using a time of flight mass spectrometer and a flame ionization detector respectively.

Kinetic model

A fundamental part of the computer simulations for chemical processes is the availability of a chemical kinetic model, i.e. a set of rate expressions of the occurring reactions. For the pyrolysis of n-pentanol, a kinetic model was constructed using Genesys[17], an automatic kinetic model generator recently developed at the Laboratory for Chemical Technology (LCT). Genesys is integrated with advanced open-source chemo-informatics libraries, allowing the generation of kinetic models independent of the reactants, elements, or chemistry. Furthermore, Genesys makes use of a rule-based termination criterion. Reaction families relevant for hydrocarbon and oxygenates pyrolysis have been constructed, consisting of hydrogen abstraction reactions, β-scission reactions and the reverse radical addition reactions, homolytic bond scissions and the reverse recombination reactions, and CO formation reactions. Each reaction family is constrained to only allow relevant reactions to be generated. These constrains are defined based on the immediate surroundings of reactive centers, e.g. valency or hybridization, as well as on the molecule in its entirety, such as molecule size, presence of rings or aromaticity.

Kinetic models for thermal decomposition, steam cracking, and combustion processes can contain up to thousands of reactions, for which most of the rate coefficients have never been measured experimentally. Hence, estimation methods are necessary in order to bridge the data gap. In the kinetic model for the pyrolysis of n-pentanol, constructed with Genesys, most kinetic and thermodynamic parameters are estimated using group contribution methods. Furthermore, the n-pentanol pyrolysis sub-model generated by Genesys is merged with the Aramcomech[18] base mechanism, which describes the pyrolysis and oxidation chemistry of small compounds, with up to 4 carbon atoms. The kinetics for the homolytic scission reactions of C-H, C-C, C-O and O-H bonds are taken from the literature, as well as the H_2O elimination in n-pentanol.[19]

Although estimation methods are the main fallback in the case of data shortage, their accuracy beyond the training set with which they were developed is questionable. Therefore, with the growing computational power and availability of high-performance computing solutions, many rate coefficients are nowadays calculated ab initio. In this work, the kinetics of the reactions for n-pentanol radicals have been determined from in-house ab initio calculations. During the pyrolysis of n-pentanol the first radicals appear when the C–C, C–O, C–H, or O–H bonds break. While unimolecular dissociation of the fuel molecule continues to be an important process as long as the fuel is still present in significant concentrations, abstraction reactions mostly by X = H, CH_3, OH radicals are driving the secondary chemistry. These abstraction reactions yield six different pentanol radicals, which will be denoted in the following as α-R, β-R, γ-R, δ-R, ε-R, and o-R, alluding to the location of the radical site.

$$
\begin{aligned}
CH_3CH_2CH_2CH_2CH_2OH + X &\rightarrow CH_3CH_2CH_2CH_2C{\bullet}HOH + XH \\
&\rightarrow CH_3CH_2CH_2C{\bullet}HCH_2OH + XH \\
&\rightarrow CH_3CH_2C{\bullet}HCH_2CH_2OH + XH \\
&\rightarrow CH_3C{\bullet}HCH_2CH_2CH_2OH + XH \\
&\rightarrow C{\bullet}H_2CH_2CH_2CH_2CH_2OH + XH \\
&\rightarrow CH_3CH_2CH_2CH_2CH_2O{\bullet} + XH
\end{aligned}
$$

The kinetic model generated with Genesys has been used in a reactor model to simulate the experimental data. The Bench Scale set-up can be modelled as an ideal plug-flow reactor[16]. Chemkin[20] is used as kinetic model simulation tool. The reactor conditions and inlet flow rates are taken from the experimental measurements. The temperature profile and pressure are imposed on the reactor.

Results and Discussion

The experimental and simulation results are given in Figure 1 and Figure 2 for the major and minor compounds respectively. Simulations have also been done with three models found in the literature by Heufer et al.[19], Togbé et al.[21] and Wang et al.[22]. Compared to the other models, the current work gives the closest agreement of the conversion with the experimental data. The predictions of the three main products, ethene, methane and CO, as a function of conversion is comparable to the experiments for the four models. Propene and ethane both exhibit a maximum in their mass fraction profile at a conversion of 90%. The simulations are able to capture the maximum, but it lies at a higher conversion. The model of Heufer et al. leads to the best propene simulations, although the differences are small between the models. For ethane, Togbé et al. shows a better agreement with the experimental data, as the models of Heufer et al. and Wang et al. are not able to capture the trend of the selectivity towards ethane as a function of the conversion. The water yields are similar for the four models, and the current work gives the best butadiene predictions, although secondary reaction pathways of butadiene are probably missing, which explains the overprediction only at high temperatures. Although the relative deviations of the simulations on these compounds are higher compared to the major products, the absolute deviations are of a similar order of magnitude.

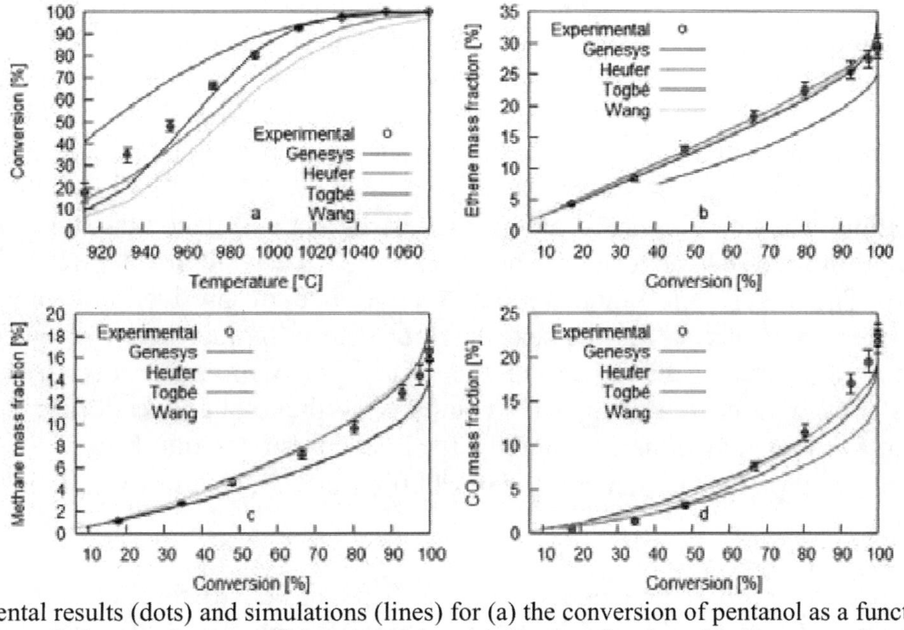

Figure 1. Experimental results (dots) and simulations (lines) for (a) the conversion of pentanol as a function of temperature, (b) the mass fraction of ethene as function of the conversion, (c) the mass fraction of methane as a function of conversion and (c) the mass fraction of CO as a function of conversion. The model generated by Genesys is compared to the models of Heufer et al.[19], Togbé et al.[21], and Wang et al.[22].

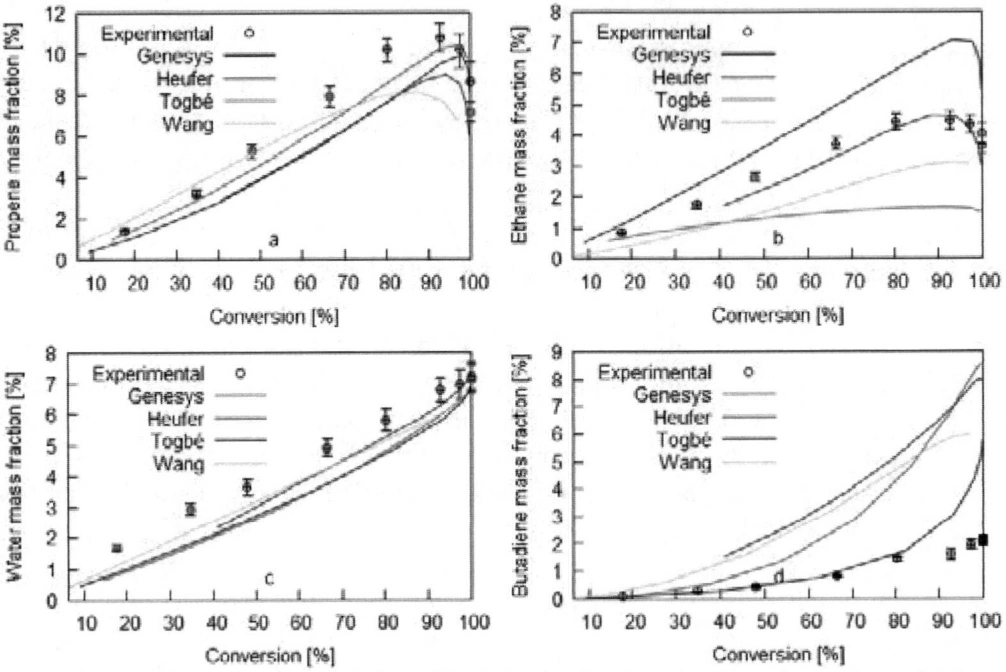

Figure 2. Experimental results (dots) and simulations (lines) for (a) the mass fraction of propylene, (b) the mass fraction of ethane, (c) the mass fraction of water and (c) the mass fraction of butadiene as a function of conversion. The model generated by Genesys is compared to the models of Heufer et al.[19], Togbé et al.[21], and Wang et al.[22].

From the reactor simulations, the main decomposition pathways can be identified and their relative importance can be quantified. The initial decomposition of the n-pentanol radical happens through either a C-C bond cleavage or the direct elimination of a water molecule, forming pentene. Homolytic scission of C-O, C-H and O-H bonds do not play an important role. The weakest bond is, in agreement to studies of pentanol[19,23] or other alcohols[24,25], the bond in β-position of the hydroxyl group, followed by the terminal C-C bond. Once a radical pool is created, these scission reactions decrease very rapidly in importance, and the main decomposition route is hydrogen abstractions, as shown in Figure 3.

The reaction scheme is recorded at the highest n-pentanol decomposition rate, which is at a distance of 37 cm into the reactor for the simulation at 973 K. At 37 cm, the set point temperature is already reached. The percentages indicate the relative rate of production compared to the total decomposition rate of n-pentanol. For the sake of clarity, no intramolecular hydrogen abstractions and well-skipping reactions have been added to the figure. Two intramolecular hydrogen abstraction reactions have a contribution of more than 1% compared to the total conversion rate of n-pentanol: the net conversion of δ-R to o-R amounts 2%, and the total net conversion of ε-R to α-R is around 1%. Also, one well-skipping reaction has a noteworthy contribution: o-R reacts to propene and the CH_2•CH_2OH with a production rate of 3%.

Around half of the o-R first undergoes a C-C-β-scission to formaldehyde and n-butyl. The remainder of the o-R decomposes through the well-skipping reaction mentioned above, while the α-R decomposes to n-propyl and vinyl alcohol. The β-R reacts via a C-O-β-scission, forming 1-pentene and a hydroxyl radical. The C-C-β-scission is also somewhat important, leading to ethyl and 1-buten-3-ol. Two C-C-β-scissions are possible in γ-R, forming methyl and 1-buten-4-ol, and 1-butene and C•H_2OH, respectively. As already mentioned, δ-R is consumed by an intramolecular hydrogen abstraction reaction to o-R. Furthermore, it decomposes through two pathways: the formation of propene and C•H_2CH_2OH and a well-skipping reaction to formaldehyde and n-butyl, accountable for almost 10% of the total decomposition of n-propanol.

76453

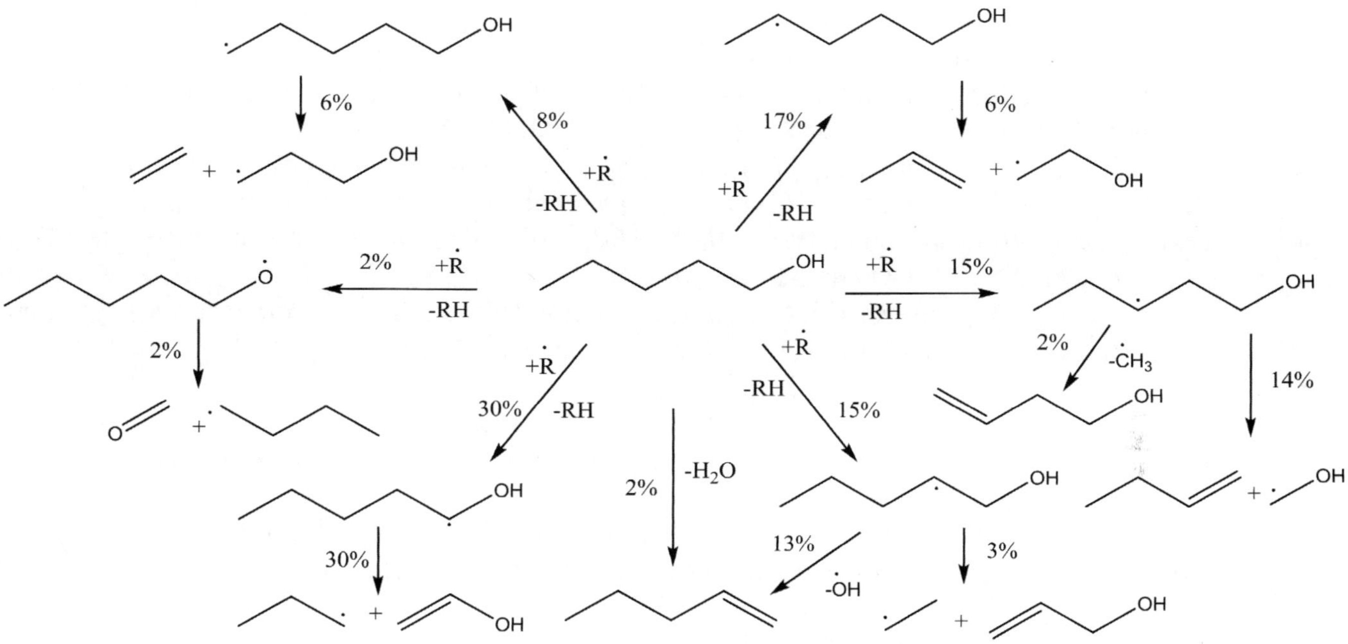

Figure 3. Initial decomposition of n-pentanol at a position of 37 cm in the Bench Scale set-up at 973 K. The percentages are the relative rates to the total rate of consumption of n-pentanol.

Conclusions

In view of the increasing worldwide energy demand and the major environmental constraints on greenhouse gas emissions, n-pentanol can be considered as a promising alternative fuel candidate. In this work, the pyrolysis and combustion chemistry of n-pentanol was studied in two-fold. First, the decomposition of n-pentanol has been performed in a flow reactor at various temperatures covering a wide conversion range. Second, in order to simulate the experimental data, a kinetic model was developed making use of the automatic kinetic model builder Genesys. For assignment of thermodynamic and kinetic parameters, data from literature and values estimated by group contribution methods were combined with newly calculated rate coefficients for reactions of the n-pentanol radical.

The product fractions of simulations and experiments are compared and a satisfactory agreement is achieved. The kinetic model developed by Genesys succeeds in describing both the conversion and selectivities of the main products. Rate-of-production analysis allows to identify the important reaction pathways and intermediates under pyrolysis conditions. The initiation reactions are homolytic scissions of C-C and the C-O bond, forming an initial pool of radicals. The C-C bond in β-position of the alcohol group is the weakest and has the highest contribution to the initial decomposition. Once radicals are created, chain reactions mainly consisting of hydrogen abstraction reactions and β-scission reactions follow each other to form small olefins and oxygenates.

References

1. Agarwal AK. Biofuels (alcohols and biodiesel) applications as fuels for internal combustion engines. *Progress in Energy and Combustion Science.* 2007;33(3):233-271.
2. Al-Hasan A. Effect of ethanol-unleaded gasoline blends on engine performance and exhaust emission. *Energy Conversion and Management.* 2003;44(9):1547-1561.
3. Black G, Curran HJ, Pichon S, Simmie JM, Zhukov V. Bio-butanol: Combustion properties and detailed chemical kinetic model. *Combust. Flame.* Feb 2010;157(2):363-373.
4. Gautam M, Martin DW. Combustion characteristics of higher-alcohol/gasoline blends. *P I Mech Eng a-J Pow.* 2000;214(A5):497-511.
5. Campos-Fernandez J, Arnal JM, Gomez J, Lacalle N, Dorado MP. Performance tests of a diesel engine fueled with pentanol/diesel fuel blends. *Fuel.* May 2013;107:866-872.
6. Yacoub Y, Bata R, Gautam M. The performance and emission characteristics of C1-C5 alcohol-gasoline blends with matched oxygen content in a single-cylinder spark ignition engine. *Proceedings of the Institution of Mechanical Engineers. Part A. Journal of power and energy.* 1988;212(5):363-379.
7. Rajesh Kumar B, Saravanan S, Rana D, Nagendran A. Combined effect of injection timing and exhaust gas recirculation (EGR) on performance and emissions of a DI diesel engine fuelled with next-generation advanced biofuel – diesel blends using response surface methodology. *Energy Conversion and Management.* 9/1/ 2016;123:470-486.
8. Wei L, Cheung CS, Huang Z. Effect of n-pentanol addition on the combustion, performance and emission characteristics of a direct-injection diesel engine. *Energy.* 6/1/ 2014;70:172-180.
9. Zhang Z-H, Chua S-M, Balasubramanian R. Comparative evaluation of the effect of butanol–diesel and pentanol–diesel blends on carbonaceous particulate composition and particle number emissions from a diesel engine. *Fuel.* 7/15/ 2016;176:40-47.
10. Li L, Wang J, Wang Z, Xiao J. Combustion and emission characteristics of diesel engine fueled with diesel/biodiesel/pentanol fuel blends. *Fuel.* 9/15/ 2015;156:211-218.
11. Imdadul HK, Masjuki HH, Kalam MA, et al. Influences of ignition improver additive on ternary (diesel-biodiesel-higher alcohol) blends thermal stability and diesel engine performance. *Energy Conversion and Management.* 9/1/ 2016;123:252-264.
12. Nativel D, Pelucchi M, Frassoldati A, et al. Laminar flame speeds of pentanol isomers: An experimental and modeling study. *Combustion and Flame.* 4// 2016;166:1-18.
13. Köhler M, Kathrotia T, Oßwald P, Fischer-Tammer ML, Moshammer K, Riedel U. 1-, 2- and 3-Pentanol combustion in laminar hydrogen flames – A comparative experimental and modeling study. *Combustion and Flame.* 9// 2015;162(9):3197-3209.
14. Heufer KA, Bugler J, Curran HJ. A comparison of longer alkane and alcohol ignition including new experimental results for n-pentanol and n-hexanol. *Proceedings of the Combustion Institute.* // 2013;34(1):511-518.
15. Djokic M, Carstensen HH, Van Geem KM, Marin GB. The thermal decomposition of 2,5-dimethylfuran. *Proceedings of the Combustion Institute.* 2013;34:251-258.

16. Harper MR, Van Geem KM, Pyl SP, Marin GB, Green WH. Comprehensive reaction mechanism for n-butanol pyrolysis and combustion. *Combustion and Flame.* 2011;158(1):16-41.

17. Vandewiele NM, Van Geem KM, Reyniers MF, Marin GB. Genesys: Kinetic model construction using chemo-informatics. *Chemical Engineering Journal.* 2012;207:526-538.

18. Metcalfe WK, Burke SM, Ahmed SS, Curran HJ. A Hierarchical and Comparative Kinetic Modeling Study of C-1 - C-2 Hydrocarbon and Oxygenated Fuels. *International Journal of Chemical Kinetics.* 2013;45(10):638-675.

19. Heufer KA, Sarathy SM, Curran HJ, Davis AC, Westbrook CK, Pitz WJ. Detailed Kinetic Modeling Study of n-Pentanol Oxidation. *Energy & Fuels.* 2012;26(11):6678-6685.

20. *CHEMKIN Release 4.1.1* [computer program]. San Diego, CA, USA: Reaction Design, Inc.; 2007.

21. Togbé C, Halter F, Foucher F, Mounaim-Rousselle C, Dagaut P. Experimental and detailed kinetic modeling study of 1-pentanol oxidation in a JSR and combustion in a bomb. *Proceedings of the Combustion Institute.* // 2011;33(1):367-374.

22. Wang G, Yuan W, Li Y, Zhao L, Qi F. Experimental and kinetic modeling study of n-pentanol pyrolysis and combustion. *Combustion and Flame.* 9// 2015;162(9):3277-3287.

23. Zhao L, Ye L, Zhang F, Zhang L. Thermal Decomposition of 1-Pentanol and Its Isomers: A Theoretical Study. *The Journal of Physical Chemistry A.* 2012/09/20 2012;116(37):9238-9244.

24. Cai J, Zhang L, Zhang F, et al. Experimental and Kinetic Modeling Study of n-Butanol Pyrolysis and Combustion. *Energy & Fuels.* 2012/09/20 2012;26(9):5550-5568.

25. Frassoldati A, Cuoci A, Faravelli T, et al. An experimental and kinetic modeling study of n-propanol and iso-propanol combustion. *Combustion and Flame.* 2010;157(1):2-16.

480098 Advanced Process Control in a Mobile Pilot Plant for the Evaluation of Long-Term Dynamic Performance of a CO_2 Capture Process Under Industrial Conditions

Abstract:

The removal of carbon dioxide (CO_2) from industrial gases is of great interest and is an important technology for the reduction of greenhouse gas emissions. One common method is the application of reactive solutions in absorption/desorption processes. This technology is well-established and has commercially been implemented in industry [1]. However, most processes suffer from unstable operation conditions, such as fluctuations in feed gas compositions or changes in the required load [2]. In addition, non-optimal operation conditions for regeneration lead to increased costs for the whole process. The flexible operation of plants for carbon capture is of great interest and is essential for future applications. The investigation of the dynamic performance of CO_2capture processes by dynamic test campaigns provides information to compensate changes in the operation conditions. For this purpose two scenarios are considered. Firstly, fluctuations in the gas feed, such as changes in the flow rate or gas compositions, are observed. Due to the coupling of the process to power plants, daily changes of the load in a power plant require the flexible operation of the carbon capture process.

Therefore, a modular and transportable pilot plant is designed and constructed by the *Process Dynamics and Operations Group at Technische Universität Berlin*[3]. Regarding varying compositions of the gases and the operation of the pilot plant a flexible and reliable design of the process is required. In addition to that, the plant is operated under real conditions at a steel mill in Duisburg, which highlights the importance of a flexible implementation.

These aspects need to be considered for the conceptual design, the construction of the pilot plant, and the process control design. For this reason, the concept of the pilot plant contains two absorption columns with a total height of 3 m of structured packing and a diameter of 0.11 m each. The desorption column for the regeneration contains 2.5 m packing with the same diameter. Preceding pretreatment columns are installed to maintain stable operation conditions at the inlet of the absorber and the removal of acidic components. Following the absorption and desorption columns posterior treatment columns are implemented to ensure sustainable process conditions at the gas outlet and to reduce the loss of the amine solution. The pretreatment and posterior treatment columns contain 1 m of structured packing each. The pilot plant is automated using ABB´s Freelance 700F process control system. To ensure reliable operation conditions and to provide the required information for the process control more than 30 temperature, 13 pressure, 8 level, and 13 flow indicators are installed.

For the studies of the dynamic responses of the carbon capture process advanced process control loops are applied by the implementation of soft sensors. Thus, the CO_2 removal is controlled by the feed flow rate of the amine solution to the absorption. Another issue is the change in the gas load. Due to the coupling of the process to power plants, the operation of the CO_2capture process is adjusted to the load of the power plant. For this reason, a switch in operation modes is considered, which compensates partial gas loads and leads to reduced heating duty for the desorption.

The main purpose of this contribution is the systematic evaluation of dynamic responses of the pilot plant towards unstable operation conditions, such as fluctuations in the gas flow or changings gas loads. Therefore, advanced process control loops are introduced by the implementation of soft sensors. These aspects are considered and implemented in a pilot plant.

[1] S. Ziaii, T. F: Edgas, G. Rochelle „Dynamic Modeling to Minimize Energy Use for CO_2Capture in Power Plants by Aqueous Monoethanolamine", Ind. Eng. Chem. Res. 2009, Vol. 48, 6105-6111

[2] P. Tait, B. Buschle, I. Ausner, P. Valluri, M. Wehrli, M. Lucquiaud "A pilot-scale study of dynamic response scenarios for the flexible operation of post-combustion CO_2capture", International Journal of Greenhouse Gas Control, Vol. 48, 2016, 216-233

[3] M. Müller, G. Wozny, H. Thielert „Innovative Produkt- und Prozessentwicklung mittels mobiler und modularer Mini-plant Technik", Jahrestreffen der Fachgemeinschaft Prozess-, Apparate- und Anlagentechnik, Fulda 2011

Optimal Design of Separations for Pilot Plants and Laboratory Bench Units

Joseph B. Powell, PhD, Shell Chief Scientist – Chemical Engineering, Houston, TX

Abstract

Separations for pilot plants and laboratory bench units typically will not scale directly to commercial. Attempts to reproduce the actual extent of separation targeted for commercial operation may lead to unnecessarily large and expensive pilot or demonstration units. It is important to design research and development units which can provide target recycle compositions to test integrated process chemistry, and to provide separation data for use in design of commercial units. This seminar will present some examples of pilot and bench units used in design of commercial separations via distillation, liquid extraction, adsorption and ion exchange, and crystallization, with emphasis on use of shortcut methods augmented by numerical modeling.

Introduction

Pilot plant and demonstration projects are expensive relative to bench-scale experiments and modeling (Figure 1). Time required to build, safeguard and operate larger units can also be detrimental to project economics, where R&D expenditure prior to commercial production results in negative cash flows which offset project value. In some cases, cost and timing for demonstration plants completely offset the future value of the project, and become a "show stopper" for the technology development. Optimization of process development and costs requires careful consideration of what needs to be piloted and why?

Technical Risk Registry

Assessment of technical risks involved in process development for a new concept can help prioritize R&D activities, and assess scales needed for de-risking. Often new feed stocks, or catalysts or reactor concepts comprise the principal challenges and risks to address. Primary separations may have critical new technology development needs for some concepts (reactive distillation, selective crystallization), but may be noncritical for others especially where they can be calculated from physical property data using process simulators. Often however, it is the trace component or secondary separation problem that impacts process viability. "Trace chemistry" often is not apparent until one runs an "integrated pilot plant" to discern problems in separation, which may appear as onset of foaming, or failure to achieve target purity.

Proper sizing of pilot plants and possible demonstration units can thus depend on:

1. Scale required to de-risk hydrodynamics or residence time distribution

2. Requirements for running an integrated process to determine impacts of trace chemistry and build-up to steady state compositions.
3. Requirements for producing product for product or market development.

For product development (#3), larger scale pilots or demonstration plants may be needed. For testing and scale-up of process steps (#1 and #2), however, it may be necessary to test a specific unit operation at scale (fluidized bed reactor), but the primary separation steps may be calculable from first principles and not require piloting in the same type of separation as proposed for commercial, or at the scale needed to match the rest of the pilot design at steay state.

Distillation Optimization

Distillation is the most common method for separations in the process industries, and performance can generally be calculated *a priori* if components are known, using process simulators with known physical properties. Trace chemistry remains a challenge for process development. If bulk compositions are known, then process simulations can be used to predict temperature profiles for small pilot plant columns. One can then conduct pilot studies in a relatively small column (e.g. 25 mm diameter) using a calculated temperature profile imposed by direct wall heating. If the compositions are not known, then a nearly adiabatic vacuum-jacketed glass column can be considered at larger diameter (75-mm) to avoid significant heat losses. This increases the required throughput nearly 10-fold, or requires operation at higher reflux ratios.

For one process development program, a 7-meter tall x 75-mm distillation column was constructed for a new product development. Use of protruded metal packing provided an estimated 80 theoretical stages for separation. A smaller once-through reactor / extractor pilot was operated to produce intermediate product for purification by distillation. Accumulation of this product from the smaller pilot allowed inventory of the larger-sized distillation column for operation in batch reflux mode (infinite recycle ratio). Addition of known markers and use of shortcut distillation theory [1] allowed fitting of thermodynamic property parameters for the unknown components, and could establish whether desired purity product could be produced as a side draw.

For this case, on-spec product was not produced. A smaller batch distillation unit (500-ml, 250 mm column height) was streamed, and a batch distillation sequence was conducted. It was found that the concentration of a volatile component was increasing with each successive distillation cut, and not separated from the heavier key component to be purified (Figure 2). This indicated that undesired reactive distillation was occurring, whereby heavy components were reacting in the column to produce light byproducts. An inverted distillation sequences was thereby designed and tested in the batch still, where the heavy bottoms were first topped to produce a majority distillate, and the tops distillate then redistilled to produce product where purity now increased with increasing extent of distillation.

Given that the batch column afforded only about 10 theoretical stages, the pilot column was used to distill the larger accumulated inventory of material, first in topping mode, with distillate then recycled to conduct a second distillation free of the initially removed heavy ends. On spec product was now demonstrated, using the information learned in the small batch distillation sequence.

In this manner, the separations studies needed for process development were conducted using a reaction system front end that was only 1/10th the size of what would have been required if a fully continuous, nearly adiabatic distillation train were required to obtain integrated performance data. A larger distillation column was constructed, but this did not drive the size of the continuous reactor pilot unit, as it was not directly coupled to the reactor / liquid extraction front end. Rather, integrated or accumulated product from the continuous reactor pilot provided batch feed to the distillation pilot, stand-alone unit. Two distillation columns were found to be needed for the continuous process to conduct the inverted distillation sequence needed to overcome undesired reactive distillation. By running distillation in semi-batch mode, construction of only one column was needed, with collection of topped intermediate product from an initial distillation to send through the same column to demonstrate performance of the second, finishing column. Process development costs and timing requirements were both substantially reduced.

Entrainment, Skin Temperature, and Purge rates

Process equipment often does not scale up because of inability to keep linear velocities and residence times constant during scale up. Thus, a process separator for disengaging gas-from-liquid or two liquids following extraction with a rule-of-thumb 10-minutes residence time, will give a much smaller linear velocity for a pilot unit with overall equipment sizes well less than a meter, vs. commercial units with a characteristic dimension of several meters. While coalescence and separation is designed on the basis of residence time, phase entrainment or onset of foaming is a function If linear velocity. Thus, commercial equipment will often foam or entrain, where pilot or lab bench equipment did not, due to higher linear velocities required to achieve the same residence times.

One can calculate this offset however and test for entrainment or foaming using semi-continuous testing with representative intermediate product, where recycles are at least partially closed. Thus, the pilot or lab equipment can be operated for short periods of time at high linear velocities to test for onset of foaming or entrainment, without having this feature dictate the scale of the overall pilot program.

Similar issues hold true for skin temperatures for heat exchangers or reboilers. Lab and pilot equipment readily provide high surface area per unit volume for heat transfer. Larger scale commercial equipment does not provide similar area, on the basis of a surface A to volume V ratio of $[A \propto d^2] / [V \propto d^3] = 1/d$. One can however conduct specific experiments e.g. in stirred autoclaves with fixed wall temperature to examine the impact of calculated skin temperatures for commercial designs, on decomposition kinetics. Kinetic measurements are thus used to calibrate a model for prediction of impact on the specific commercial design assumption for heat transfer area.

Samples taken from pilot plants or bench reactors can represent huge purge rates for impurities relative to commercially practiced processes. One can examine the buildup of impurities and compare with calculated concentrations basis purge rate differences upon scale up, and then either concentrate or dope the corresponding impurities to consider impacts on foaming or emulsions, or ability to separate in purification. Commercial performance can again be calibrated and modeled, despite the implausibility of operating the pilot unit at low commercial purge rates.

Using known performance equipment for pilot or bench studies

If for example a packed liquid extractor or distillation column is used for separation under pilot or lab conditions, the number of theoretical stages may not be known, unless calibrated via known markers as shown for the previous distillation example. However, it may not be necessary to pilot the exact type of separator to be deployed commercially, if commercial performance can be calculated once appropriate property or separation factor data is available. Thus, use of one or more mixer-settlers to rigorously define the number of stages and measure liquid-liquid extraction equilibria may be more valuable at the pilot stage, than using a packed or sieve-tray extractor that would attempt to match the proposed commercial design type, but for which the number of stages and hence performance is more difficult to calibrate.

For at least one development, the mixer-settler approach was used to obtain design data, and a small pilot sieve tray extractor was deployed to demonstrate lack of fouling in continuous operation. Tray design for the 50 mm extractor was too crude to enable a detailed estimate of tray efficiency, but this was not a problem for design of the commercial unit [2, 3].

Summary

In summary, separations do not have to drive the scale of continuous pilot plants, unless the separation itself is the main issue to be de-risked in the development program. The separation devices chosen should be capable of generating reliable design data where needed, and may be operated in semi-continuous mode to examine separation factors, or diagnose potential for foaming, entrainment, or skin-temperature derived degradations. In this manner, modeling plus experiment in separations can work in tandem to reduce the require scope, cost, and timing for process development.

References

1. S. Sundaram and L. B. Evans, *Shortcut procedure for simulating batch distillation operations*, **Ind. Eng. Chem. Res., 1993, 32** (3), 511-518.
2. R. E. Treybal, **Mass Transfer Operations**, McGraw-Hill, NY (1980).
3. R. E. Treybal, **Liquid Extraction**, McGraw-Hill, NY (1963).

Figure 1: Relative cost of research, development, and demonstration phases

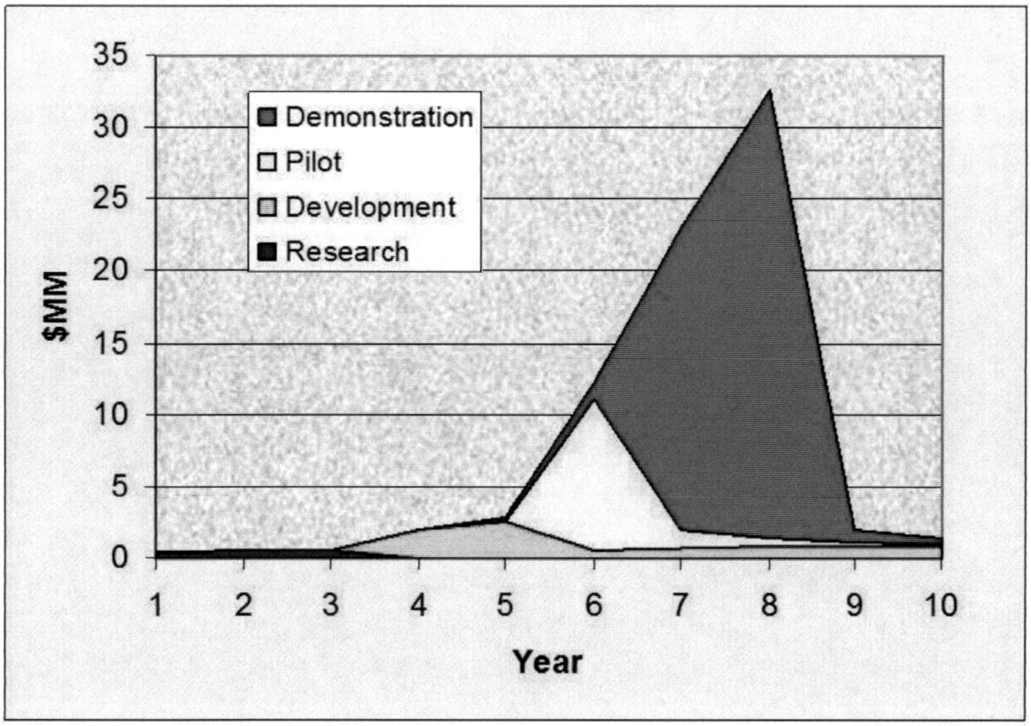

Figure 2: Batch distillation revealing reactive formation of volatile component from heavy ends

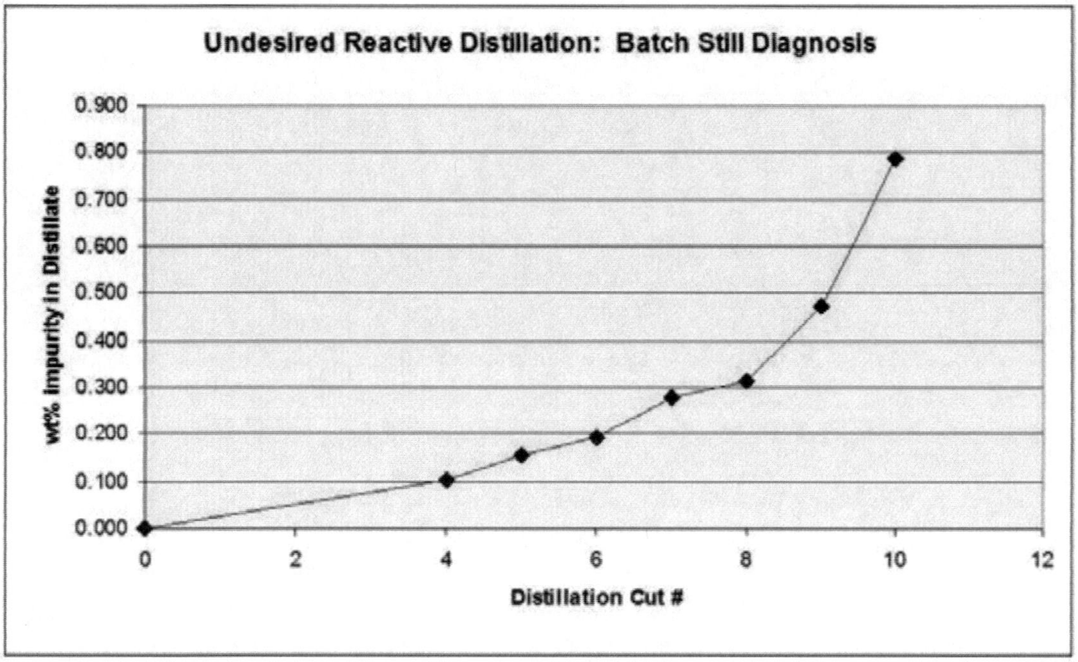

Figure 3: Inverted Distillation Sequence to Resolve Undesired Reactive Distillation

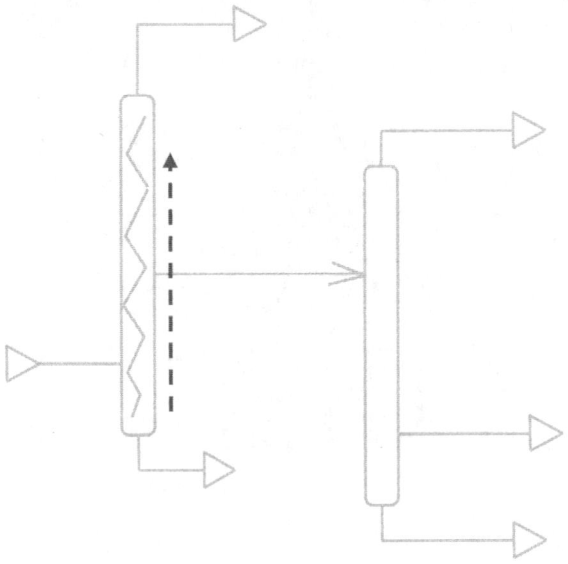

479574 Distillation: An Under-Plowed Field

Abstract:

We all know and agree that the heart of a chemical process resides in the reactor section of a chemical plant, where product forms. However, the majority of the invested capital goes to the unit operation separating desired product from by-product and solvent. The same unit operation also comprises the majority of the operating costs for a given chemical process.

Distillation is the most common separation unit operation in the chemical processing industry. It was studied intensively from the 1890's through the 1980's. However, since the early 1990's, distillation has become rather passe with regard to academic investigators. But, that response does not mean distillation is devoid of interesting problems to investigate.

Advances in computer power and user-friendly software opens new opportunities for investigating the fundamental features of distillation. In this presentation, we present a dynamic model for describing the separation that occurs on each tray of a laboratory-scale distillation unit. We then show how this model can be used to scale a distillation operation from laboratory-sized equipment to commercial-sized equipment.

PERFORMING OPTIMAL PILOT PLANT TESTS FOR DESIGN OF LIQUID-LIQUED EXTRACTION COLUMNS

Authors:

Donald J. Glatz
Brendan Cross
Thomas Lightfoot

Koch Modular Process Systems, LLC

Prepared for Presentation at

AIChE Spring Meeting
San Antonio, TX
March 2017

INTRODUCTION

The best and most reliable method for the design of a liquid-liquid extraction (LLE) column is by pilot plant testing with actual feed and then relying on known scale up correlations for the specific type of extraction column[1]. In general, the development procedure consists of the following steps; (1) selecting the solvent and generating liquid-liquid equilibrium data, (2) using this information and the project requirements to select the column type to use, (3) performing a pilot plant test to generate the data for accurate scale up to the production column, and (4) designing and subsequent fabrication of the column. This development procedure has been well documented in a previous paper[2].

The purpose of this paper is to focus on the pilot plant test method. During process development work, the ability to pilot plant test is often limited by a number of factors. Some of the key factors are:

- Availability of the feed materials. Often very little if any actual feed material is available.
- Known purity of the solvent. Since the solvent is a recycle stream, the solvent will generally not be pure during real commercial operation.
- Time restraints. Often testing must be done quickly to meet deadlines.
- Disposal of waste materials. Generally it is desirable to minimize the amount of waste material generated during pilot plant testing.

For all of these reasons it is important to have an optimized test method that can be completed in a short time frame using a minimal amount of feed and solvent. For basic R&D work, experimental design is a well proven and accurate method for process optimization. However, for pilot plant testing, this method requires a significant amount of feed and solvent and can take a long period of time. As presented in this paper, a Step-by-Step optimization method is outlined which saves both time and material. This method does require some experience but when it is applied correctly it can insure that an extraction column is designed correctly and performs as required.

LABORATORY WORK PRIOR TO PILOT TESTING

As mentioned in the Introduction, the first step for an extraction process is selecting the solvent and generating equilibrium data (distribution coefficient vs solute concentration). There are a number of methods for doing this. One such method was outlined in a previous paper[3]. In addition to the raw equilibrium data, observation of the phase behavior is just as important, especially for selecting the right type of column for scale up and developing a test plan for that column. The following observations are recommended:

- Mixing intensity required to generate good phase dispersion.
- Mass transfer rate/the time required to reach equilibrium. Generally we have learned that equilibrium can be reached in 2 minutes of high speed mixing, except if one or both phases are viscous (> 100 cps).
- After mixing, the time for the phases to separate and achieve a clear interface.

- The interface behavior – Is there a rag layer (solids) or an emulsion band that forms at the interface?
- Entrainment – Is some entrainment observed in one or both phases?

A typical data sheet for generating liquid-liquid equilibrium data is shown in Figure 1. A review of this sheet shows that for each mix-decant it will provide the material balance, the solute concentration in both phases, the cross solubility of the phases, and the observation of the phase behavior. With some experience, this data will help the engineer select both the best extraction column for the process and the starting point for pilot plant testing which will be discussed below.

PILOT TEST METHOD – STEP-BY-STEP OPTIMIZATION

A summary of the proposed method for pilot plant testing an agitated extraction column is provided here. This method requires fairly quick analysis of the samples generated in the pilot column, with the solute concentration in the raffinate usually being the key marker for the performance. It is also highly desirable to perform the test in a glass-shell extraction column so that the operators can observe the phase behavior and "tweak" the column operating conditions accordingly. The procedure is outlined as follows:

1. Select the continuous and dispersed phase for the process. There are a number of factors which can affect which phase is continuous and which phase is dispersed, but this is not the topic for this paper.

2. Select the column for scale-up and then set up the pilot column with multiple inlets. If possible it is best to set up a column that is expected to be taller than required. A manifold system can be included to vary the inlet location for the dispersed phase. The continuous phase must be fixed at the end where it enters, whereas the manifold system can be used to vary the inlet location for the dispersed phase.

3. Select the initial operating conditions for the test, i.e. the height of the column, the capacity, the solvent to feed ratio, the temperature and the initial agitation speed. Suffice to say that this is difficult to do without some previous experience with similar extraction processes. However, with some experience an engineer or scientist can usually determine acceptable starting conditions for a given type of extraction column.

4. Once the initial operating conditions are set, the column should be started up in the usual manner(4), and the agitation set at a low speed. This speed should be significantly below the flooding point for the process. Observation of the phase behavior in the glass shell column will determine this speed. The turnover time for the process should be calculated. Turnover time is the volume of the column divided by the feed + solvent rates. During the first two column turnovers, the agitation speed can be tweaked up slightly if warranted by observation of the phase behavior. However, after two turnovers, the agitation speed should remain fixed for the duration of the run which is a total of five (5) turnovers from the initial start-up. After five turnovers, raffinate and extraction samples are taken and analyzed for solute

concentration.

5. Next the agitation speed is increased slightly and then fixed for an additional three (3) turnovers before sampling and analyzing. This procedure is repeated, progressively increasing the agitation speed and sampling after three turnovers each time until flooding is observed in the column.

   NOTE: For most extraction process, we have learned that five column turnovers upon start up followed by three turnovers after making slight system modifications are sufficient to reach steady state. However, it is possible that for some systems, more turnovers may be required.

6. An evaluation of the results from this first series of tests will determine the next direction for the pilot plant testing. Usually the next direction is optimization of the solvent to feed ratio (S/F) vs the number of stages (height of the column). If the results from the first series of testing are good (the required solute extraction is achieved), then the next test runs should be at progressively lower solvent to feed ratios and/or lower height of the column. However, if the initial tests do not achieve the desired solute extraction, then the next test runs should be at progressively higher solvent to feed ratios and/or more agitated stages. Which way to go is generally determined by the expected economics for the process. Usually the more economical solution is to use more stages and thereby minimize the solvent usage.

7. After optimizing the S/F vs column height, the next variable to test is capacity, pushing the capacity to progressively higher and higher capacities to determine the optimal capacity for the process. Sometimes it is also desirable to test at lower capacity, especially if the production column is scheduled to begin operating at a capacity that will be significantly lower than the design capacity.

8. Finally, if necessary, additional variables such as temperature, pH and variable mixing intensity throughout the column (due to variable hydraulic behavior between the two phases from one end of the column to the other) can be tested.

If applied correctly, this test procedure can usually be completed in three operating days. Day 1 is finding the optimal agitation speed, Day 2 is optimizing S/F ratio vs column height, and Day 3 is optimizing the capacity A fourth day may be required if any other variables mentioned in Step 8 are tested, and/or multiple feeds are required. Two actual examples are now presented which demonstrate the Step-by-Step optimization procedure.

Example 1 – Carboxylic Acid Extraction

The application required the extraction of acetic acid from a whole fermentation broth using ethyl acetate as the solvent. The goal was to achieve >95% extraction of the acetic acid at the minimum solvent to feed ratio. As expected for fermentation broths, this extraction system was extremely prone to emulsification. As such the efficiency of an extraction column was expected to be low (high HETP) and only gentle agitation could be applied. The distribution coefficient for this process is approximately 0.9. Using the Kremser equation[5], the number of theoretical stages vs S/F are calculated as follows:

S/F = 2.0, number of theoretical stages = 3.9
S/F = 1.7, number of theoretical stages = 4.9
S/F = 1.5, number of theoretical stages = 6.1

Based upon these calculations plus observation of the slow phase separation and high tendency towards emulsification a 1" diameter x 12' plate stack height, glass-shell KARR® Column[6] was selected for the testing. The aqueous phase was selected as the continuous phase. The equipment set-up is shown in Figure 2. The results from three days of pilot plant testing are shown in Table 1. The progression of the pilot plant test was as follows:

- Runs 1 – 6 were performed on Day 1 at a broth feed rate of 83 cc/min, S/F = 1.7 weight basis, and room temperature. The starting point for Run 1 was 30 strokes per minute (SPM). This was increased to 40 SPM for Run 2 and 50 SPM for Run 3 where the column flooded. The desired acid extraction was achieved. Runs 4 and 5 with 8' plate stack height failed to achieve the desire acid extraction whereas 10' plate stack height in Run 6 did achieve the desired extraction. It was decided to conservatively use 12' plates stack height for all additional runs.

- Runs 7-9 were performed on Day 2 at the lower S/F ratio of 1.5. Initially the feed rate and solvent rates were adjusted to maintain the same combined flow rate as the previous runs, i.e. the new feed rate was 90 cc/min. The desired acid recovery was not achieved in Run 7. In Runs 8 and 9 the feed rate was adjust down to 83 and 65 cc/min respectively, while maintaining the S/F ratio of 1.5 and the desired recovery was achieved for both runs. It was decided to lock in the capacity at 83 cc/min and S/F of 1.5.

- During Day 3, the temperature was increased to 45 °C in the column. This proved to be an excellent operating temperature for this process. The operators reported that the phase behavior was more stable at this temperature. As shown in Run 11, successful operation at 50 SPM resulted in almost 99% acid extraction. This was determined to be the scale up run for the process.

Based upon these results, a 22" diameter KARR® Column was designed, built and has operated successfully for this process. The total amount of fermentation broth feed used for the testing was approximately 70 gallons.

Example 2 – Methanol Extraction

The application required the extraction of methanol from an organic stream using water as the solvent. The target was to reduce the methanol concentration in the organic phase from 2.5% to less than 0.1% using a specified amount of water (S/F = 0.32, weight basis). The liquid-liquid equilibrium data showed that 5-6 theoretical stages were required. A pilot KARR® Column was set up as shown in Figure 3. Due to the low S/F ratio (low water rate relative to the organic feed rate) the organic phase (light phase) was selected as the continuous phase and thus the interface was controlled at the bottom. Results from two days of pilot plant testing are shown in Table 2.

The test conditions and results are summarized as follows:

- All tests runs were made using a plate stack height of 6', i.e. the upper water inlet was never used. Also all runs were room temperature (20-25 °C). The first three runs were made using standard 2" plate spacing throughout the agitated zone, a feed rate of 150 cc/min and progressively higher agitation speed (SPM). The result show that the column flooded at 110 SPM, but had not achieved the desired alcohol concentration in the raffinate stream.

- Observation of the column during the first three runs showed that the dispersed phase water droplets would drop down through the column fairly rapidly, with little hold-up until they reached a point approximately 2' above the organic phase inlet. This was the point where flooding initiated during Run #3. Thus, the column was shut down and drained. The plate stack was removed and the spacing between the perforated plates was modified as follows; from the bottom of the plates stack 1' of 6" spacing, followed by 1' of 4' spacing and then the standard 2" spacing for the remainder of the plate stack. Thus, for any given agitation speed, significantly less power would be imparted into the bottom 2' of the agitated zone.

- The difference in the physical operation of the column was dramatic. As shown by the data, much higher agitation speeds were possible (without flooding) and the dispersed phase hold-up was observed to be significantly higher throughout the column. After Runs 4 and 5, the operators determined that higher capacity could easily be achieved with the modified plate spacing. As such for Runs 6-9, the feed rate was increased to 180 cc/min. The desired raffinate concentration was achieved at 200 SPM (Run 9). One additional run was made at a feed rate of 210 cc/min and the desired raffinate concentration was again achieved at 175 SPM.

Based upon these results, a 45" diameter KARR® Column was designed, built and has operated successfully for this process. The total amount of organic feed used for the testing was approximately 55 gallons.

ADDITIONAL OPTIMIZATION FROM PILOT TESTING

During pilot plant testing of liquid-liquid extraction processes, there are other process conditions that can be observed and optimized. In fact, pilot plant testing may be the only way some of these can be recognized and the correct design modification implemented. Some of the most common are listed here:

- Using the proper material of construction for the internals due to the wetting characteristics of the system. It is important for the continuous phase to preferentially "wet" the surface of the internals so that the dispersed phase does not coalesce on the internals. This is easily seen when using a glass shell extraction column.
- Minimization of entrainment – The amount of entrainment out each end of the column can usually be measured by immediate centrifuge of samples leaving the column. If the process

will not tolerate a certain level of entrainment, then modifications can include (1) expanding the disengaging chamber (2) installing a coalescing element in the disengaging chamber, or (3) building an external separator.

- Interface behavior – Does a solids "rag" layer or an emulsion band form at the interface. If so, how is this controlled?
- Co-extraction of impurities besides the desired solute – If this occurs, how can it be handled? This may require a "backwash" section in the column or using a fractional extraction technique[7].
- When flooding occurs, how is this handled in order to return the column to smooth operation? It is best to learn this during pilot plant testing so that it can be applied commercially in the future.

CONCLUSION

For most liquid-liquid extraction processes it is possible to develop and perform a pilot test plan that will save time and money but still generate the data for accurate design of a production column. The generalize procedure was outlined in this paper and supported by two real examples where this method was applied.

REFERENCES

1. Cusack, R., Karr, A., "A Fresh Look at Liquid-Liquid Extraction, Part 3, Extractor Design and Specification", Chemical Engineering, April 1991

2. Cusack, R., Glatz, D., Apply Liquid-Liquid Extraction to Today's Problems", Chemical Engineering, July 1996

3. Glatz, D., Parker, W., "Extraction Enhancement, One Step at a Time", Chemical Engineering, November 2004

4. Glatz, D., Lightfoot, T., "Using Extraction Columns for Recovery of Valuable Products from Biomass", AIChE National Meeting, November 2012

5. Cusack, R., Fremeaux, P., and Glatz, D., "A Fresh Look at Liquid-Liquid Extraction, Part 1, Extraction Systems", Chemical Engineering, February 1991

6. Cusack, R., Fremeaux, P., "A Fresh Look at Liquid-Liquid Extraction, Part 2, Inside the Extractor", Chemical Engineering, March 1991

7. Glatz, D., Lightfoot, T., "Designing Columns for Fractional Liquid-Liquid Extraction", AIChE National Meeting, November 2010

Figure 1 - Shake Test Data Sheet

Client: **Application:**

Description: **Feed:**

Solvent:

	Description		Weight (grams)				Separation		Solute Analysis			
No.	Feed	Solvent	Feed	Solvent	Raff.	Ext.	Time	Observation	Feed	Solvent	Raff.	Ext.

Figure 2 – Pilot Plant KARR Column Set-Up
Acetic Acid Extraction from Broth using Ethyl Acetate

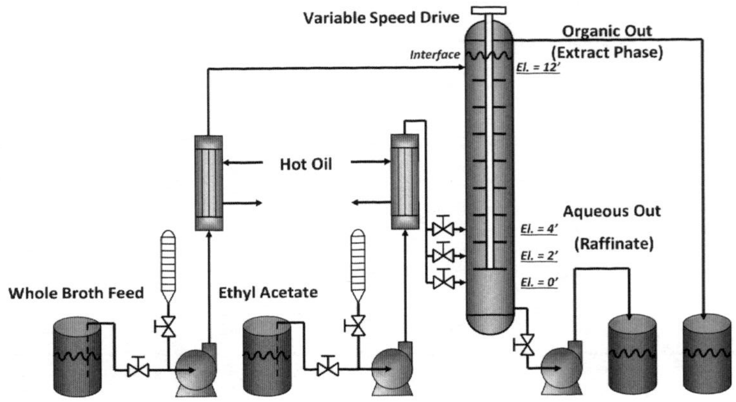

Figure 3 – Pilot Plant KARR Column Set-Up
Extraction of Alcohol from Acrylate Using Water

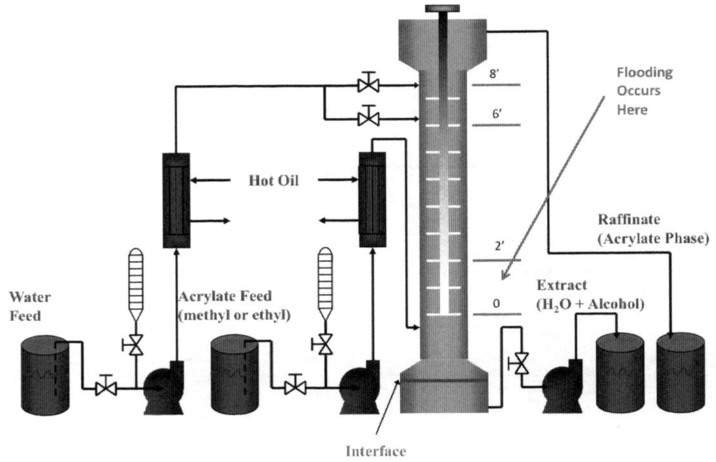

Table 1 - Pilot Plant Test Results
Acetic Acid Extraction from Broth using Ethyl Acetate

Run #	Agitated Height (ft)	Feed Rate (cc/min)	Temp. C	S/F Ratio	Agitation SPM	% Acid Recovery
1	12	83	23	1.7	30	95.0
2	12	83	23	1.7	40	97.5
3	12	83	23	1.7	50	Flooded
4	8	83	23	1.7	30	93.3
5	8	83	23	1.7	40	Flooded
6	10	83	23	1.7	30	96.3
7	12	90	23	1.5	30	94.3
8	12	83	23	1.5	30	96.5
9	12	65	23	1.5	40	97.6
10	12	83	45	1.5	30	98.3
11	12	83	45	1.5	50	98.7
12	12	83	45	1.5	60	Flooded

Data generated in 25mm diameter KARR Column
Total amount of fermentation broth used = approximately 70 gallons

Table 2 - Pilot Plant Test Results
Extraction of Methanol from Acrylic Feed Stream Using Water as the Solvent

Run #	Plate Stack	Feed Rate (cc/min)	Water Feed Rate (cc/min)	Agitation Speed (SPM)	Raff. Conc. % Alcohol Target <0.1%
1	1	150	45	75	0.165
2	1	150	45	100	0.124
3	1	150	45	110	**FLOODED**
4	2	150	45	110	0.169
5	2	150	45	140	0.112
6	2	180	54	100	0.203
7	2	180	54	125	0.146
8	2	180	54	150	0.118
9	2	180	54	200	**0.078**
10	2	210	63	175	**0.084**

Notes: KARR Column with 1" diameter x 6' plate stack height
Plate Stack #1 has constant 2" plate spacing throughout
Plate Stack #2 from bottom: 1' of 6", 1' of 4", and 4' of 2" spacing between plates
Total Amount of organic feed used = 55 gallons

Stratification During Vapor or Gas Freeing of Storage Tanks

Michael L. Hanks, Ph.D., P.E.
ESI
146 2nd St. N, #205
Saint Petersburg, FL 33701
mlhanks@engsys.com

Amy E. Gray, Ph.D.
ESI
3310 Green Park Circle
Charlotte, NC 28217
aegray@engsys.com

Ronald Pape, Ph.D., P.E.
ESI
4215 Campus Drive
Aurora, IL 60504
rpape@engsys.com

Prepared for Presentation at
American Institute of Chemical Engineers
2017 Spring Meeting and 13th Global Congress on Process Safety
San Antonio, TX
March 26 – 29, 2017

Stratification During Vapor or Gas Freeing of Storage Tanks

Michael L. Hanks, Ph.D., P.E.
ESI
146 2nd St. N, #205
Saint Petersburg, FL 33701
mlhanks@engsys.com

Amy E. Gray, Ph.D.
ESI
3310 Green Park Circle
Charlotte, NC 28217
aegray@engsys.com

Ronald Pape, Ph.D., P.E.
ESI
4215 Campus Drive
Aurora, IL 60504
rpape@engsys.com

Keywords: Vapor Freeing, API RP 2016, Stratification.

Abstract

API RP 2016 provides guidelines for vapor freeing and degassing of tanks. The instructions include warnings and methods to recognize and avoid stratification or channeling during vapor freeing, and to control and prevent ignition sources from occurring in the tank during vapor freeing. A recent incident occurred where vapor in a tank ignited shortly after vapor freeing. The investigation revealed that the vapor freeing method used most likely resulted in stratification of the vapor in the tank, and the atmosphere testing method used to test the vapor level in the tank was not capable of detecting stratification. The low vapor reading resulted in shutting down the vapor freeing operation after only 5 hours when the planned duration was 48 hours. By following API RP 2016, this event would likely not have occurred. This paper discusses methods for estimating when stratification is likely to occur during vapor freeing of tanks based on jet mixing correlations, and uses a CFD model to examine the likelihood of stratification in several cases. These methods can be used to develop criteria for vapor freeing of tanks and for investigating incidents involving stratification of flammable vapors in an enclosed space.

1 Introduction

ANSI/API Standard 2015 Requirements for Safe Entry and Cleaning of Petroleum Storage Tanks applies to atmospheric and low-pressure tanks used in all sectors of the petrochemical industry [1]. It provides requirements for safe entry for tank cleaning operations from decommissioning through recommissioning. API RP 2016 Guidelines and Procedures for Entering and Cleaning Petroleum

Storage Tanks is a recommended practice that supplements ANSI/API 2015 [2]. Section 5 of API RP 2016 provides guidance for vapor and gas freeing of storage tanks. It defines vapor freeing as:

> The removal of flammable or toxic vapors and gases from a tank by displacement or the reduction of the percentage of vapors and gases in the tank's atmosphere to a safe level by dilution with fresh air.

According to API RP 2016, "mechanically introducing fresh air into a tank is the preferred method of removing vapors or gas from a storage tank."

API RP 2016 recommends two methods for vapor freeing using fresh air: dilution and displacement. The dilution method assumes the vapor and air mix thoroughly during vapor freeing, and the guide recommends a minimum air flow rate of 5 tank volumes of air per hour to accomplish this. The displacement method assumes the air and vapor do not mix, and relies on buoyancy of the vapor compared to air (either positive or negative buoyancy) to stratify the vapor and air. This way, essentially pure vapor is discharged from the tank as air flows into the tank. The guide recommends that the air be introduced at a low flow rate to ensure the vapor in the tank stays stably stratified.

The flow rate recommended in API RP 2016 for the dilution method would be expected to mix the air and vapor in the tank once sufficient vapor has been removed that there is insufficient buoyancy remaining to maintain stratified layers. However, the air flow rate into a tank may be restricted by equipment limitations, restrictions on discharges to atmosphere by environmental boards, or by other considerations. For example, the lead author is aware of a case where the local air quality board restricted the flow rate through a thermal oxidizer used for vapor freeing to less than 2500 CFM. This means there is a maximum tank volume for which the thermal oxidizer could be run at the recommended flow rate for vapor freeing.

For cases where the flow rate through the tank must be less than the recommended amount, there is no guidance in API RP 2016 for estimating whether the air and vapor will be well mixed during the vapor freeing operation or whether the tank contents will be stratified. Also, the API RP 2016 guide does not provide information on how low of a flow rate is required to ensure the contents do remain stratified for a displacement operation other than using natural ventilation to displace the vapor.

The dilution method of vapor freeing can generally be modeled using the Perfect Mixing Model (PMM) if the jet is sufficiently strong to overcome stratification. This model is a mass balance on flows in and out of the tank and assumes the tank contents are instantaneously perfectly mixed. The model constants relate to known parameters and provide estimates of the time required to complete the vapor freeing operation. The displacement method cannot be modeled as easily, but the time required to free a tank using the displacement method can be bounded.

Air streams used for vapor freeing are generally induced in the tank by fans or eductors placed in openings on the side or top of the tank. The resulting air stream into the tank can be reasonably well modeled as a buoyant free jet, at least until the airstream encounters the tank wall, roof or base. Chow (2009) modelled vapor freeing of tanker compartments and found that published research on jets mixing in enclosed volumes was useful for comparing to vapor freeing operations [3].

This paper uses correlations developed for jet mixing in tanks to examine conditions for which the air and vapor will mix completely, and conditions for which they will remain stratified. The correlations used correspond most closely to vapor freeing of cone roof tanks that do not have significant flow obstructions inside the tank (e.g. baffles, partitions, etc.) The concepts can be applied in general to vapor freeing of a floating roof tank, but the specific correlations may not be valid for these tanks because of the extreme aspect ratio of the headspace in a floating roof tank.

2 Jet Mixing in Tanks

When fans are used to introduce fresh air into a tank during vapor freeing, the airstream for the fan can be treated as a free jet at least until the jet impacts the tank walls or other significant internal obstructions. The jet can follow three different flow regimes depending on the density of the vapor or gas in the tank and the initial momentum of the air jet: a plume, a pure jet, or a fountain. For a plume, buoyancy dominates the initial momentum of the jet and the jet momentum acts generally in the same direction as the buoyant force. A pure jet occurs when the density difference between the air and the vapor or gas is sufficiently small that the initial jet momentum drives mixing. A fountain is a buoyant jet in which the initial momentum acts in the direction opposite to the buoyant force.

The density of the vapor or gas in the tank can be less than air (e.g. methane); comparable to air (e.g. ethane, nitrogen); or greater than air (e.g. propane or other heavier volatile molecules.) The manways or ports where air is blown into or drawn from the tank are generally located in the tank sidewall near the ground, or on the top of the tank. This means there are 8 general cases for how an air stream will flow in the tank headspace during vapor freeing as shown in Table 1.

Table 1. General cases governing airstream flow in the tank during vapor freeing.

Vapor or Gas Density	Side Air Inlet	Top Air Inlet
Less than air	(A) Horizontal momentum jet	(B) Vertical momentum jet or plume
Comparable to air	(C) Horizontal pure jet	(D) Vertical pure jet
Greater than air	(E) Horizontal momentum jet possibly transitioning to a vertical plume	(F) Fountain
Combination (e.g. high density layer at the bottom of the tank, low density layer at the top)	(G) Horizontal momentum jet, possibly transitioning to a plume and/or a fountain.	(H) Vertical momentum jet or plume, possibly transitioning to a fountain.

Whether the jet will mix with all the vapor in the tank or whether a stratified layer will form depends on the above factors and on the jet flow rate, the tank geometry, and the location of the discharge from the tank.

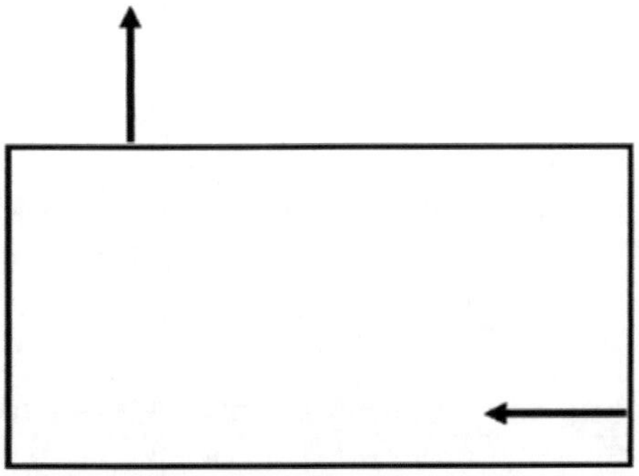 (1) Discharge: Eductor on the roof with suction to the tank. Inlet: One or more side manways. (API RP 2016-2001 5.3.4.1-5.4.3.2)	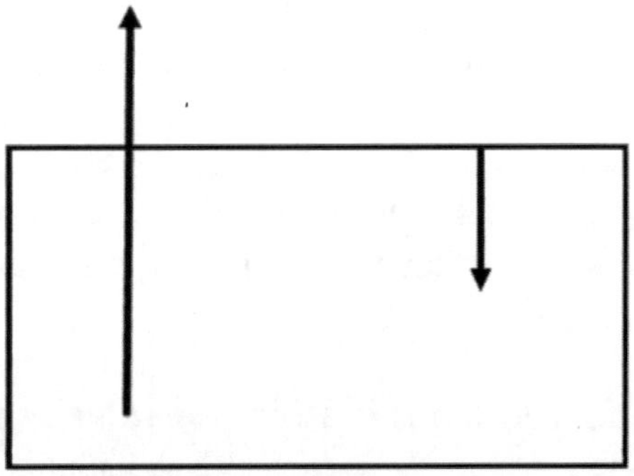 (2) Discharge: Eductor on the roof with suction to the tank. Attach a duct to the educator so it pulls air from the bottom of the tank. Inlet: A second roof manway. (API RP 2016-2001 5.3.4.3)
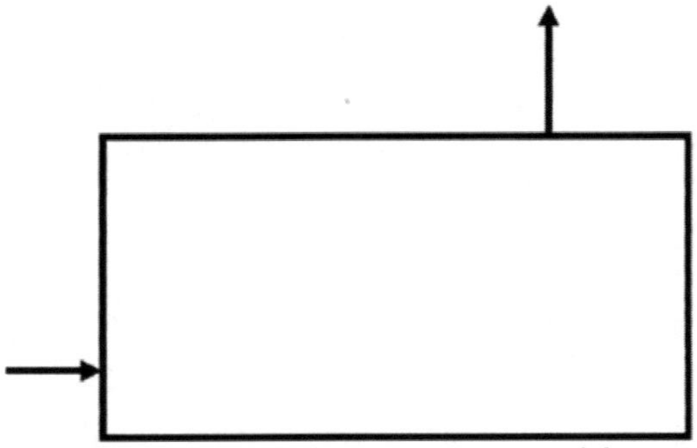 (3) Inlet: Fan blower on a side manway. Discharge: A roof manway. (API RP 2016-2001 5.3.4.4)	 (4) Discharge: Eductor on a side manway. Inlet: A roof manway (and possibly a side manway). (API RP 2016-2001 5.4.3.6)
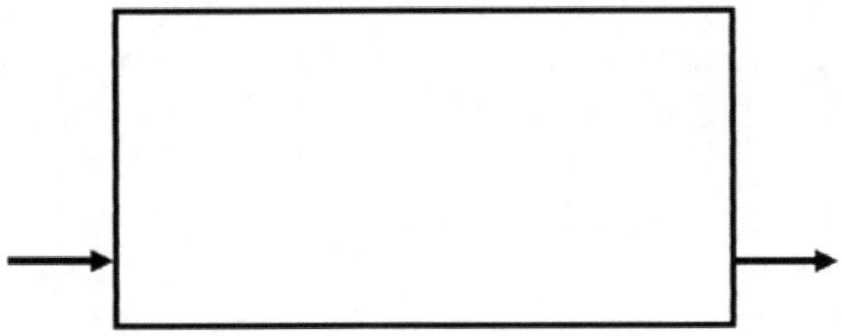 (5) Inlet: Fan blower on a side manway. Discharge: A second side manway. (API RP 2016-2001 5.3.4.5)	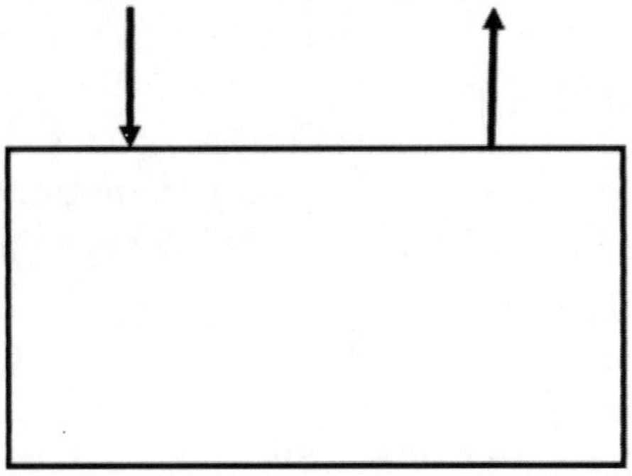 (6) Inlet: Fan blower on the roof. Discharge: A roof manway. (Note: this combination is not included in API RP 2016.)

Figure 1. API RP 2016 recommended options for locating inlet and outlet fans or eductors.

API RP 2016 recommends the combinations for positioning the air inlet and vapor discharge during vapor freeing of a cone roof tank shown in Figure 1.

Methods 1 and 3 are similar to each other in terms of air flow in the tank. Methods 2 and 4 are also similar to each other. Method 6 is not included in API RP 2016 but is the other obvious possible combination of inlet and outlet locations.

Comparing the possible vapor density vs. air density conditions from Table 1 to the recommended combinations of inlet and discharge locations in Figure 1, it is clear there are preferred combinations for dilution vapor freeing and for displacement vapor freeing. The most obvious combinations are shown in Table 2.

Table 2. Preferred combinations of inlet and outlet locations for various combinations of vapor density.

Inlet and Outlet Location Method	Dilution (High air flow rate)	Displacement (Low air flow rate)
(1) and (3)	(A), (C), (G) (E) possible	(A)
(2) and (4)	(D), (F) (B), (H) possible	(F)
(5)	(E)	(E)
(6)	(B)	(B)

Combinations that take advantage of both the momentum of the air jet and of the buoyancy of the vapor relative to air will be the most efficient and effective for vapor freeing. For example, when freeing a low-density vapor from a tank, Methods (1) or (3) will be better than would (2), (4), or (6). Therefore, (A) would be preferred over (B) for a low-density vapor and (F) would be preferred over (E) for a high-density vapor. For freeing buoyant vapors, the natural combinations with inlet and outlet location are (1 or 3)(A) and (2 or 4)(F). For cases (C) and (D) where the vapor or gas in the tank is neutrally buoyant with respect to air, the location of the air inlet and vapor discharge are less important and displacement is not possible. For cases (H) and (G) either (1 or 3) or (2 or 4) can be effective.

Several combinations would not be recommended because they could result in short circuiting of the air. For example, Method (5) with a low-density vapor and a low air flow rate could result in air short-circuiting to the discharge without freeing vapor from the upper part of the tank headspace. A similar situation could happen for Method (6) with a heavy vapor and low air flow rate.

Correlations for jet mixing can provide guidance for whether the air flow rate into the tank will result in displacement or dilution for combinations (1)(A), (2)(F), and (4)(H); and whether the air jet has sufficient momentum to adequately mix with the vapor for cases (C) and (D).

The jet mixing correlations can be divided into two parts. The first part examines whether the vapor and air will be stratified during the vapor freeing operation or whether the air jet will break the stratification. The second part examines the time to mix the tank contents for cases or conditions where the air and vapor will not be stratified.

Vapor freeing by displacement occurs when the air flow rate is low enough that stratification occurs or existing stratification remains intact. For an ideal displacement operation where there is no mixing between the air and vapor, the time to free vapor from the tank is the time required to flow one tank volume of air into the tank. The actual time required will be longer than this due to partial mixing.

Vapor freeing by dilution occurs when the air flow rate is sufficiently high to break up existing stratification, prevent stratification from occurring during the vapor freeing operation, and to mix the air and vapor fast enough to avoid short-circuiting of air through the tank. This later condition can be evaluated by comparing the mixing time to the residence time in the tank.

2.1 Stratification

Stratification can occur either because the vapor density is sufficiently different than air that introducing air results in stratification, or because there are multiple vapors in the tank that result in the tank contents being stratified before vapor freeing begins. Stratification is unlikely to occur if the relative density difference between the two fluids is:

$$\left(\frac{\rho_2-\rho_1}{\rho_2}\right) \leq 0.05. \qquad (1)$$

for $\rho_1 < \rho_2$.

2.1.1 Side-Entry Jet

For a side-entry jet mixing in a tank, when the relative density difference exceeds 0.05, there is a critical jet velocity, u_c, above which the jet may be able to mix the fluids so no stratification occurs [3]. This is estimated by:

$$u_c = \left[\frac{2gGH\left(\frac{\rho_2-\rho_1}{\rho_2}\right)}{\sin^2\theta}\right]^{0.50} \qquad (2)$$

where $\theta = (\beta+5)^{\circ}$, β is the angle of the jet from horizontal, and G is a stratification constant that is a function of the relative density ratio and the ratio of the tank height to the jet discharge diameter, H/d_j. The value of G can be estimated by:[1]

$$G = 0.8\left(\frac{\rho_2-\rho_1}{\rho_2}\right)^{-0.286} \tag{3}$$

Equation (3) is for cases where $H/d_j > 80$. When $H/d_j < 80$, the value of G is lower for a given relative density ratio. This means the calculated value of u_c tends to be higher than is actually required to mix the two fluids.

This configuration naturally corresponds with the combination of (1 or 3)(A): a side jet inlet, a roof outlet and a vapor in the tank that is less dense than the jet.

Figure 2 shows the estimated critical velocity (the velocity necessary for breaking up the stratified layers) is a strong function of the jet angle. Pointing the jet upward into the tank significantly reduces the air jet velocity needed to break up stratified layers.

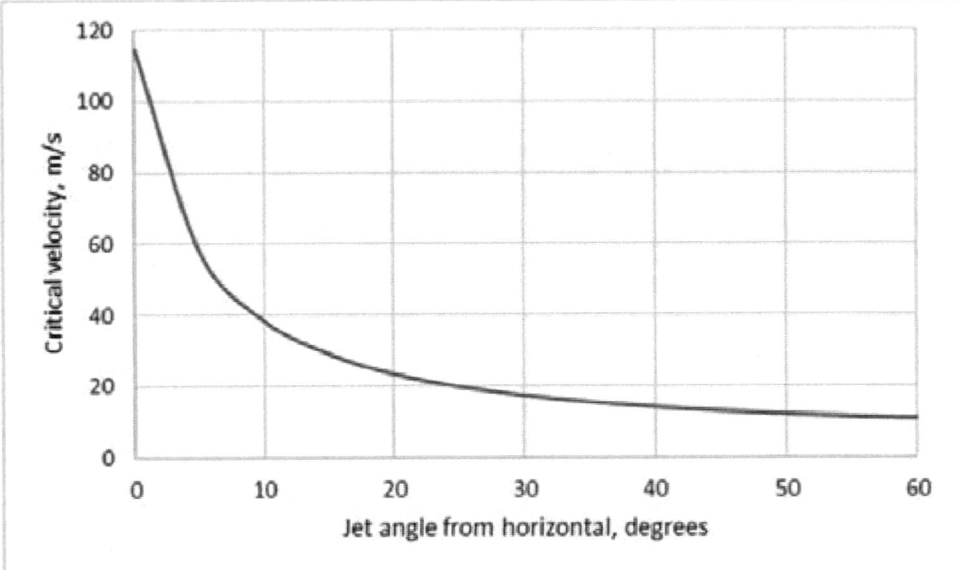

Figure 2. Estimated critical velocity from equation (2) for a side-jet of air and the tank initially filled with methane. Tank dimensions are those in API RP 2016-2001 example 5.1: tank height = 40 ft; Tank diameter = 125 ft.

2.1.2 Vertical Entry Jet or Plume

Hunt, et al. (2001) studied a filling box - an enclosed, rectangular tank with a vertical, buoyant jet but with no other inlets or outlets [4]. For the purposes of this paper, the interesting results from the filling box studies are information on the stability of a stratified layer as a function of the jet momentum and the geometry of the volume. The vapor freeing geometry corresponding to this

[1] Linear least squares fit to values scaled from [3] Figure 9.6, p. 173.

correlation is a turbulent air jet entering vertically downward from the roof into a less-dense vapor, which is the combination (6)(B).

For a single vertical jet, they report a correlation between critical jet length (the distance over which momentum of the jet dominates the flow) and the volume aspect ratio (height over width or length):

$$\left(\frac{L_j}{H}\right)_{crit} = \frac{1}{K}\left[\frac{1}{3\alpha^{0.5}}\left(\frac{H}{R}\right)^{-1} - 1\right] = 0.863\left[1.157\left(\frac{H}{R}\right)^{-1} - 1\right] \quad (4)$$

where $\frac{L_j}{H}$ is the dimensionless jet length, H is the tank height, R is the characteristic length of the base (diameter or length) and $K = \frac{1.057}{2^{1.25}\alpha^{0.5}\pi^{0.25}} = 1.159$, for $\alpha = 0.083$. For values of $\frac{L_j}{H} > \left(\frac{L_j}{H}\right)_{crit}$ the momentum of the jet is sufficient to overturn the entire volume so no stratification occurs, while for $\frac{L_j}{H} < \left(\frac{L_j}{H}\right)_{crit}$ a stable stratified layer forms.

The dimensionless jet length is a function of the jet momentum and buoyancy. The relevant factors are:

Dimensionless jet length: $$\frac{L_j}{H} = \frac{M_0^{0.75}}{HB_0^{0.5}} \quad (5)$$

Jet momentum flux: $$M_0 = \frac{Q_0^2}{A_j} \quad (6)$$

Jet buoyancy flux: $$B_0 = g_0' Q_0 = \frac{g|\Delta\rho|}{\rho_v} Q_0 \quad (7)$$

where Q_0 is the initial jet volume flow rate, A_j is the cross-sectional area of the jet at the discharge, g is the gravitational constant, $|\Delta\rho|$ is the absolute value of the density difference between the jet fluid and the vapor, and ρ_v is the density of the vapor.

For the same tank dimensions given above and using an air jet directed vertically down from the roof into a tank filled with methane, with an air jet nozzle area of 1 m^2, the correlation predicts an air flow rate of approximately 18 ACH (air changes per hour) would be required to prevent stratification during vapor freeing. It is unlikely that this high of an air flow rate is feasible. This means the vapor freeing should be expected to start as displacement and transition to dilution once most of the methane has been removed.

2.1.3 Fountain Jet

The fountain jet case is likely a common occurrence in vapor freeing of storage tanks because there will generally be a heel of liquid in the bottom of the tank emitting a vapor with heavier components that can form a layer near the bottom of the tank. Lighter components or the protective gas pad can form a layer in the top portion of the tank. In this situation, a downward vertical air jet introduced through the tank roof would flow through the light layer due to negative buoyancy and initial momentum, and then form a fountain when it encountered the heavy layer. Similarly, a horizontal side-entry jet at a low flow rate could become buoyant in the heavy layer, turn vertically upward, and become a fountain when it encountered a lighter vapor layer.

Correlations for fountain rise height can be used to estimate whether the air jet is sufficiently strong to break down the stratification, or whether the air jet will erode the layer by mixing at the interface between the heavy and light layers. It is unlikely that the location of the interface will be known, so some arbitrary rise height must be chosen for this estimate. The most conservative value is to use the tank height.

Burridge & Hunt (2015) developed correlations for rise height of fountains [5]. For turbulent air jets ($Re_0 > 2000$, where $Re_0 = \frac{4\rho_j Q_0}{\mu_j \pi d_j}$) the rise height of a fountain can be estimated by:

$$Z = \begin{cases} 2.46 \cdot L_M \; for \; Fr_0 \geq 4.0 \\ 2.8 \cdot L_M - 2.1 \cdot L_Q \; for \; 2.0 \leq Fr_0 \leq 4.0 \\ 0.82 \cdot r_0 Fr_0^{2/3} \; for \; 1.0 \leq Fr_0 \leq 2.0 \\ 0.81 \cdot L_{Hyd} \; for \; 0.3 \leq Fr_0 \leq 1.0 \end{cases} \tag{8}$$

where $L_M = M_0^{3/4}/B_0^{1/2} = r_0 Fr_0$, $L_Q = Q_0/M_0^{1/2} = r_0$, $L_{Hyd} = r_0 Fr_0^{2/3}$, and $Fr_0 = \frac{M_0^{5/4}}{Q_0 B_0^{1/2}}$. They make the constraint that $\frac{\Delta\rho}{\rho_v} \ll 1$. This means that the correlation may not be accurate for freeing very low density or very high density vapors with air. Their data were measured using liquids and cover a range of *Re* and *Fr* that do not necessarily correspond to those for gases and vapors, so the correlations may not be accurate for vapor freeing. However, the correlation does at least give a general idea of how far a fountain will penetrate the stratified layer.

It is unlikely that the location of the interface of a heavy vapor would be known. The interface layer location and the initial jet flow rate can be varied to estimate the air flow rate required to penetrate through the entire tank regardless of where the interface is located. If the jet penetrates through the entire layer, it is likely the jet will have sufficient momentum to break down the stratified layer and mix the contents of the entire tank.

If the jet is unable to break up the stratified layers, then a complex situation can result where the heavy and light layers remain intact, and air mixed with one or both vapors can form a third layer between them. The natural approach to remove both layers would be to have two outlets – one at the side to remove the heavy layer and one at the roof to remove the light layer. However, this approach may not be feasible and should only be used after a hazard review.

After sufficient time, the air jet can erode each layer and eventually remove both the heavy and light vapors as long as the air flow rate is sufficient to mix the tank headspace when buoyancy is no longer a significant factor.

2.1.4 Summary of Stratification Effects

If the air jet will not break up the stratified layers, then the vapor freeing operation will initially be by displacement. The locations of the inlets and outlets should be selected to take best advantage of the jet momentum and the vapor density. Initially, the outlet will be essentially pure vapor. As the vapor freeing proceeds, the concentration of vapor in the discharge will drop and the vapor remaining in the tank will be mixed with air. As a consequence, the relative density difference between the air jet and the vapor/air mixture in the tank will decrease.

At some point the density difference will become small enough that the air jet will break up any remaining stratification and buoyancy will no longer significantly affect the flow regime in the tank. Once this condition is reached, the vapor freeing will naturally transition to the dilution method.

2.2 Dilution Method

When stratification is not an issue or is no longer an issue, then the consideration of importance is whether the air jet will adequately mix the tank vapor space. Two factors are of importance in making this determination. First, the distance the jet travels in the tank before hitting the tank walls must be less than ~400 nozzle diameters: $L_j < 400 \cdot d_j$ [4]. Second, the tank contents must be mixed rapidly enough to prevent significant concentration fluctuations in the outlet. This can be evaluated using correlations for characteristic mixing time. If the mixing time is much less than the characteristic residence time in the tank, then the jet will likely be capable of thoroughly mixing the vapor space.

The characteristic residence time can be estimated as $t_r = \frac{V}{Q_0}$, where V is the volume of the vapor space. In the case where there are multiple inlets, Q_0 is the total flow rate of all inlets.

2.2.1 *Mixing Time Correlations*

Correlations for mixing time exist for several configurations relevant to vapor freeing [3]. Fox (1956) developed a correlation for mixing time for a side entry jet discharging into a cylindrical tank. This correlation is for fluids with the same density and viscosity [6]:

$$t_m = \frac{ADH}{u_0 d_j}\left(\frac{d_j}{H}\right)^{0.50} \tag{9}$$

where

$$A = \begin{cases} \dfrac{7.8x10^5 Fr_j^{0.17}}{Re_0^{1.33}} & for\ 250 < Re_0 < 2000 \\ \dfrac{120 Fr_j^{0.17}}{Re_0^{0.17}} & for\ 2000 < Re_0 < 160{,}000 \end{cases} \tag{10}$$

where $Fr_j = \frac{u_0^2}{gd_j}$. For a jet discharging upward from the bottom center of a cylindrical tank, mixing time can be approximated using equation (9) with the following values for *A* [7]:

$$A = \begin{cases} \frac{7.08x10^5 Fr_j^{0.17}}{Re_0^{1.33}} \; for\ 100 < Re_0 < 2000 \\ \frac{145 Fr_j^{0.17}}{Re_0^{0.17}} \; for\ 2000 < Re_0 < 100{,}000 \\ 22.8 Fr_j^{0.17} \; for\ 100{,}000 < Re_0 < 140{,}000 \end{cases} \tag{11}$$

Jets pointing down have mixing times about 90% longer than upward pointing jets, so for an air inlet on the roof, multiply *A* by 2.

The following design guidelines provide limits for using jets to mix tank contents [3]:

- Roof mounted jets (axial jet):
  - A single axial jet is acceptable if the tank height to diameter aspect ratio is in the range $0.75 \leq H/D \leq 3.0$.
  - Otherwise, use more than one jet.
- Side entry jet:
  - A single side-entry jet is acceptable for $0.25 \leq H/D \leq 1.5$
  - Use multiple side-entry jets if $H/D \leq 0.25$ or $H/D \geq 3.0$

3 Perfect Mixing Model

If the residence time is long compared to the mixing time (e.g. $t_r >> t_m$) then the vapor concentration in the outlet will not fluctuate significantly, and the perfect mixing model (PMM) may estimate the component concentration quite well. The PMM can be used to estimate the vapor concentration versus time, and can be used to estimate the time required to reach a specific concentration target.

If the inlet flow rate is fast compared to the rate at which blending occurs (e.g. $t_m >> t_r$) then the component concentration in the outlet will fluctuate. At the extreme, a component in an inlet stream could essentially short circuit the enclosure and flow to the outlet without mixing significantly. In this case, the perfect mixing model will not be a good estimate of the component concentration in the enclosure [8].

The PMM can be written as:

$$x(t) = x(0) \cdot e^{-t/t_r} \tag{12}$$

where $x(t)$ is the volume fraction of vapor in the tank, and $x(0)$ is the initial volume fraction of vapor. In the case where vapor freeing starts with the displacement method, the value of $x(0)$ is the vapor fraction at the time stratified layers break up and the freeing transitions to the dilution method, and the value of *t* is set to 0.

Solving equation (12) for t allows an estimate of the time required to reach a specified volume fraction of the component, $x(t)_s$, for $0 < x(t)_s < x(0)$. This is given by:

$$t = -t_r \ln\left(\frac{x(t)_s}{x(0)}\right) \tag{13}$$

A common final vapor concentration target is to be less than 25% of LEL. Assuming LEL = 1% by volume and the initial volume fraction of vapor is 1.0, the time required is:

$$t_{0.25 \cdot LEL} = -t_r \ln\left(\frac{0.25 \cdot 0.01}{1.0}\right) \approx 6 \cdot t_r = 6 \cdot \frac{V}{Q_0} \tag{14}$$

With appropriate units, V/Q_0 is the number of air changes per hour (ACH) in the tank vapor space. Therefore, equation (14) can be used to determine the number of air changes required to reach the target vapor concentration. In the example above, 6 air changes are required.

If the vapor freeing operation begins with stratified layers and uses the displacement method, the total time to free the tank headspace will be approximately the time to flow one tank volume of air plus the time calculated to get the remainder of the way to the target vapor concentration, using equation (13) with x(0) set to the vapor concentration after one air change.

If the jet is initially capable of breaking up any stratification and if $t_r \gg t_m$, then the time needed for vapor freeing can be estimated by equation (13) with x(0) = 1.

4 Examples

There is little published data on vapor freeing of tanks, so computer simulations were used to compare the mixing correlations given above to vapor freeing. The Fire Dynamics Simulator (FDS) software package was used to model velocity of flows and the concentration of vapor in the tank with time. FDS was developed to model compartment fires, and so is well suited to model buoyant flows and the concentration of components in the compartment. It is fundamentally a computational fluid dynamics software package. It has the added advantage that it is freely distributed software so it can be used by anyone with the requisite background and training. For the specific application to vapor freeing, FDS has weaknesses that may reduce the accuracy of the simulation. For example, FDS uses rectangular boxes for the mesh so the walls of a cylindrical tank are not smoothly modelled. Similarly, a round fan or educator on a manway must be modelled as a square vent.

Therefore, the FDS simulation results provide a qualitative analysis of the specific cases modeled, and should not be viewed as providing a quantitatively reliable prediction without comparison to relevant experimental data.

The examples given below compare FDS simulation results to the correlations given above for the side-entry jet and the vertical jet. The tank dimensions used are from Example 5.1 in API RP 2016: 125 ft diameter, 40 ft tall fixed (cone) roof tank. The roof was modeled as a flat surface rather than as a cone. The side vents were modeled as 1-meter squares. The FDS mesh used 0.5-

meter cubes. The bottoms of side vents were located 1 meter above the base of the thank, and roof vents were located 2 meters in from the edge of the tank. The geometry for one case is shown in Figure 3.

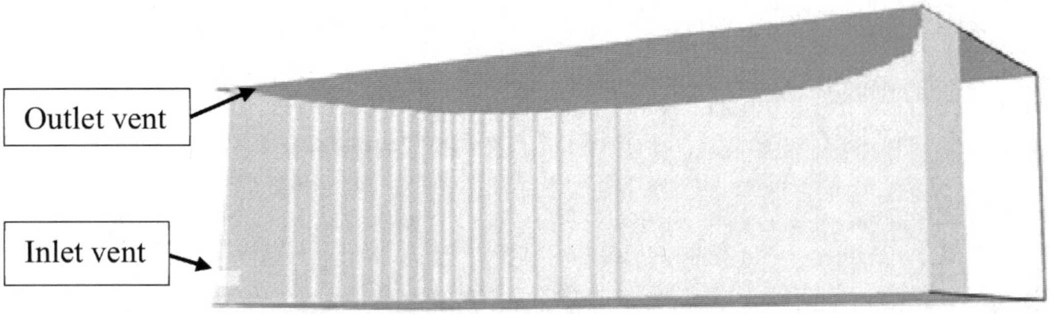

Figure 3. Cutaway view of the tank geometry used for FDS simulations. This view shows the half of the tank divided through a side-inlet vent.

4.1 Side-Entry Jet with Low-Density Vapor

The condition modeled for this case is the Combination (1)(A): an eductor on the roof drawing air from the tank with a side-entry vent for air into the tank, and the tank initially filled with methane. Simulations were run for three flow rates, as shown in Table 3.

Table 3. Combination (1)(A) Simulations (Methane).

Case Identification	Flow Rate (m^3/s)	Flow Rate (CFM)	ACH	Purpose
(1)(A) Low	2.0	4,230	0.52	Develop and maintain a stratified layer.
(1)(A) Mid	19.3	40,800	5.0	Match API RP 2016 recommended flow rate.
(1)(A) High	114	241,300	29.5	Match the critical velocity from equation (2).

Figure 4 compares the methane concentration from the FDS simulation at the tank vertical axis in 0.5 m increments to the concentration calculated by the PMM.

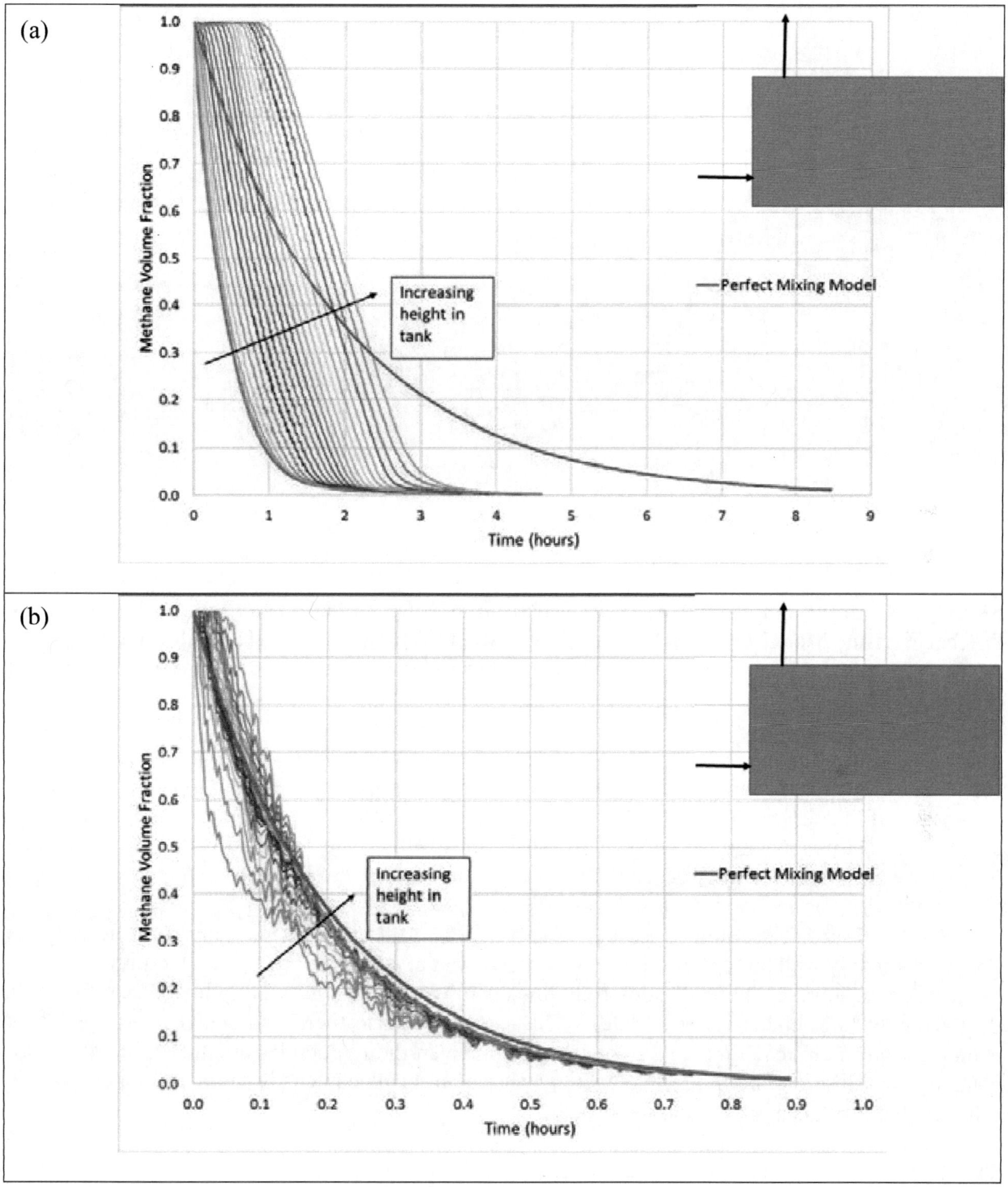
(a)
Methane Volume Fraction
Time (hours)
Increasing height in tank
Perfect Mixing Model
(b)
Methane Volume Fraction
Time (hours)
Increasing height in tank
Perfect Mixing Model

(c)

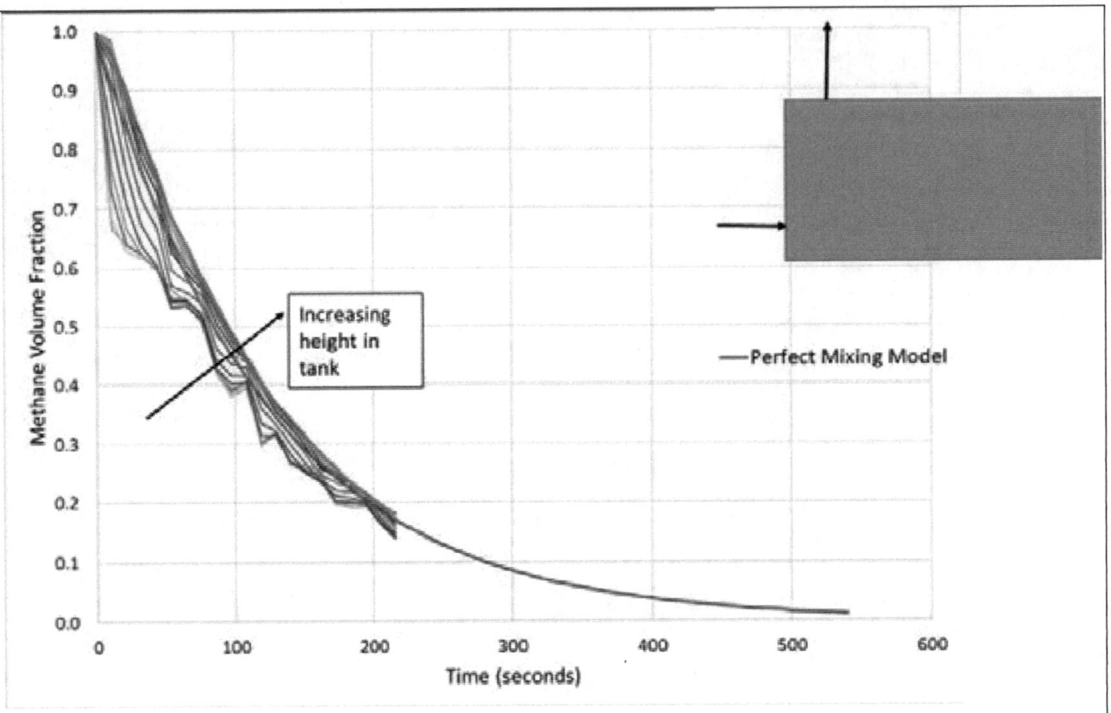

Figure 4. Comparison of methane concentration in the tank from the FDS simulation to the Perfect Mixing Model (PMM) for Combination (1)(A): (a) 0.5 ACH; (b) 5 ACH; (c) 29.5 ACH. The arrows on the blue rectangle indicate the location of the inlet and outlet vents on the tank.

The FDS results for the 0.5 ACH case of Combination (1)(A) show that the air flow rate is sufficiently low that a stratified layer forms during the vapor freeing process. This can be seen by the spread of the curves from the PMM line. These results also show that if the air flow rate that can be used is this low, and the vapor density is significantly different than that of air, then using displacement to remove vapor is faster than using dilution (e.g. the PMM line).[2]

The 5 ACH flow rate recommended by API RP 2016 results in the tank contents following the PMM reasonably well throughout the freeing process. It also shows that the time required to reach the target concentration using dilution (e.g. low air flow rate) is significantly lower than for using displacement (e.g. higher air flow rate). This is true because there is a specific number of air changes required to reach the target concentration. Even though displacement requires fewer air changes than dilution, the time to do each air change in displacement is so much longer than for dilution that displacement is a slower process.

The 29.5 ACH rate results show that the critical velocity for breaking up a stratified layer does result in much less variation of the methane concentration vertically through the tank. However, the flow rate is so high it is unlikely that any gas freeing process would come close to this flow

[2] In order to use dilution in this case, some mechanism for mixing the tank contents would need to be added. For example, placing a fan inside the tank as a mixer.

rate. The results for the 5 ACH case show that a flow rate within ~20% of the critical velocity predicted by the correlation is sufficient to allow use of the PMM.

4.2 Vertical Entry Jet With Low-Density Vapor

The condition modeled for this case is the Combination (6)(B): an eductor on the roof drawing air from the tank with a second roof vent on the opposite side of the tank for the air inlet, and the tank initially filled with methane. The results are shown for simulations with three flow rates, as shown in Table 4.

Table 4. Combination (6)(B) Simulations (Methane).

Case Identification	Flow Rate (m^3/s)	Flow Rate (CFM)	ACH	Purpose
(6)(B) Low	2.0	4,230	0.52	Develop and maintain a stratified layer.
(6)(B) Mid	19.3	40,800	5.0	Match API RP 2016 recommended flow rate.
(6)(B) High	70.3	148,900	18.2	Match the critical jet length from equation (4).

Figure 5 compares the FDS simulation results to the PMM for the three cases. Again, the PMM does not model the 0.5 ACH case well. As with the case shown in Figure (2a), displacement results in slightly faster vapor freeing than does dilution for the same air flow rate. Case (6)(B) does take best advantage of initial jet momentum and buoyancy in that the initial jet momentum acts opposite to buoyancy. This results in increased mixing during the displacement part of the vapor freeing process. The result is that it takes longer to reach the vapor fraction target for Case (6)(B) Low than it does for Case (1)(A) Low.

The PMM does model Case (6)(B) Mid reasonably well, indicating once again that the API RP 2016 recommended flow rate of 5 ACH is likely to break up stratification sufficiently that the vapor freeing will be by dilution for most of the process.

Figure (5c) for Case (6)(B) High shows that the flow rate required to achieve the critical jet length predicted by equation (4) results in little variability of the vapor fraction from the PMM line, indicating no significant stratification occurs during this process. As in Case (1)(A) High, the air flow rate is so high that it may not be feasible for any real vapor freeing process. A flow rate ~25% of the predicted critical value appears to be sufficient to allow use of the PMM to model the vapor freeing process.

(a)

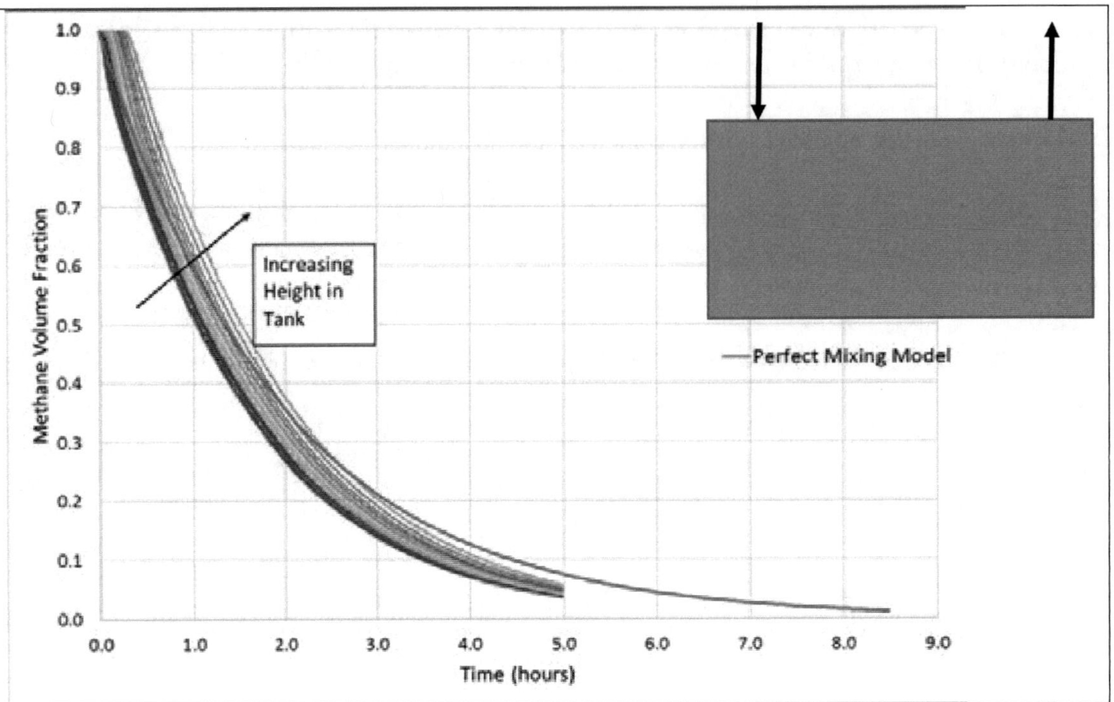
Increasing Height in Tank
Perfect Mixing Model
Methane Volume Fraction
Time (hours)

(b)

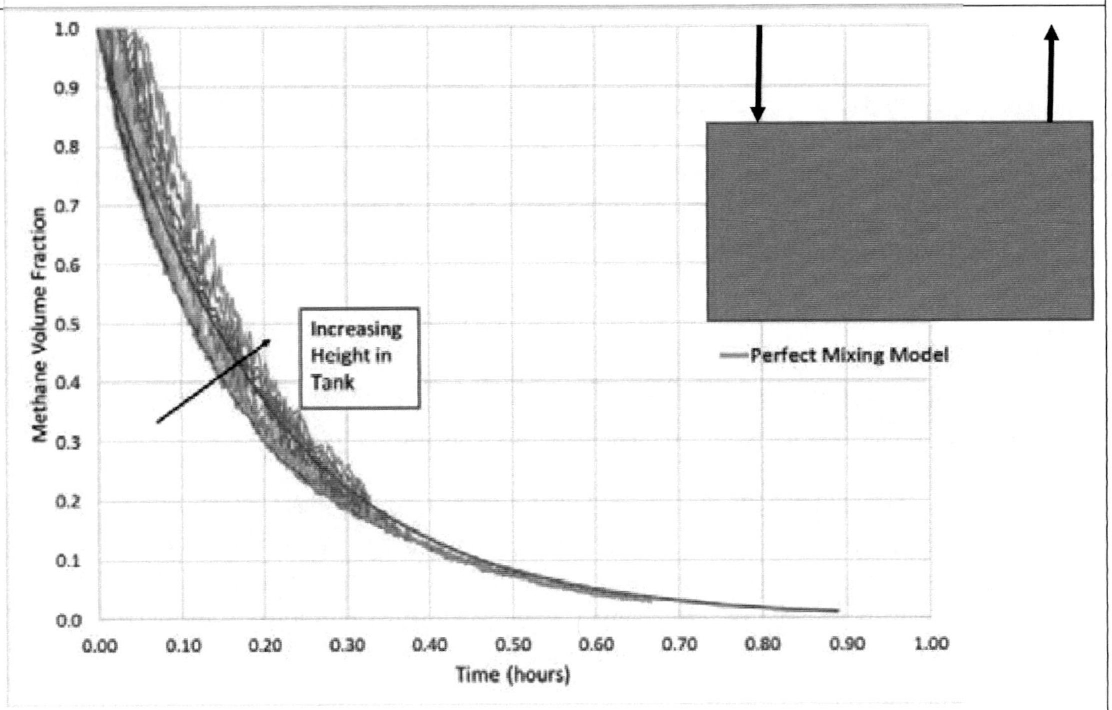
Increasing Height in Tank
Perfect Mixing Model
Methane Volume Fraction
Time (hours)

(c)

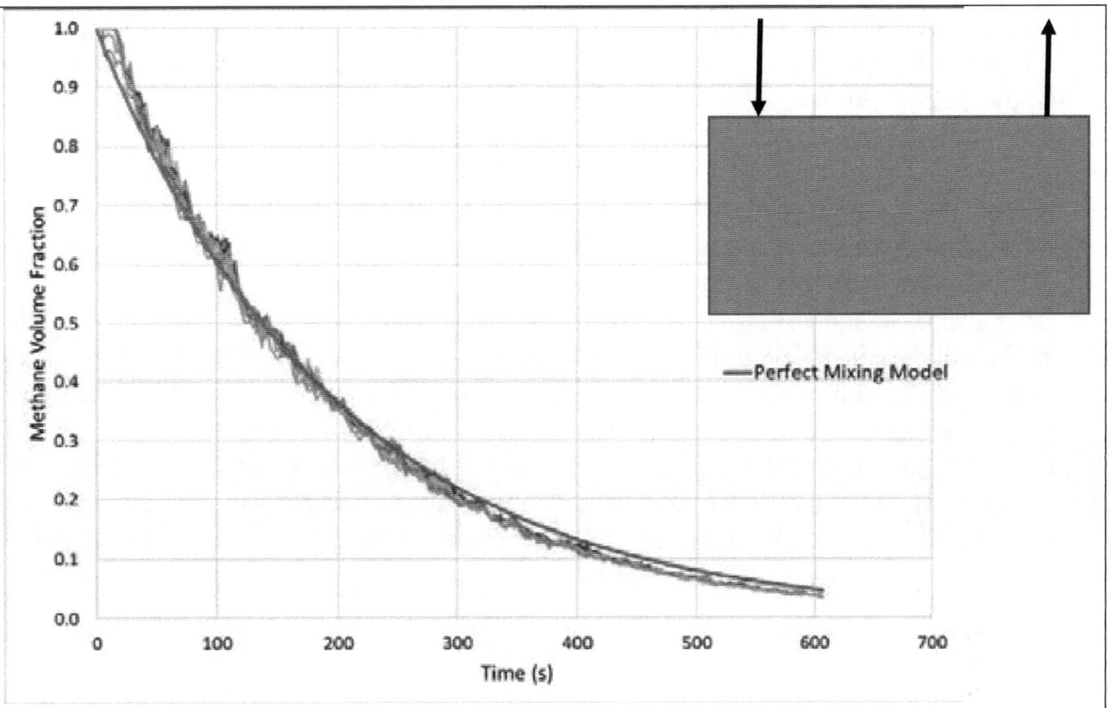

Figure 5. Comparison of methane concentration in the tank from the FDS simulation to the Perfect Mixing Model (PMM) for Combination (6)(B): (a) 0.5 ACH; (b) 5 ACH; (c) 29.5 ACH. The arrows on the blue rectangle indicate the location of the inlet and outlet vents on the tank.

4.3 Fountain Flow for Side-Entry Jet into Two Layers With Different Densities

The condition modeled for this case is the Combination (1)(G): an eductor on the roof drawing air from the tank with a side-entry vent directly below the outlet. The tank is initially filled half with methane and half with propane. The air inlet is in the propane layer. The results are shown for simulations with two flow rates, as shown in Table 5.

Table 5. Combination (1)(G) Simulations (Half Methane/Half Propane).

Case Identification	Flow Rate (m^3/s)	Flow Rate (CFM)	ACH	Purpose
(1)(G) Low	2.0	4,230	0.52	Maintain the stratified layers.
(1)(G) Mid	19.3	40,800	5.0	Match API RP 2016 recommended flow rate.

The results of the FDS simulations are shown in Figure 6 compared to the PMM model. In Figure (6), the buoyancy of the air jet in the two gases results in the total gas volume fraction remaining relatively high near the roof and floor of the tank, while the gas volume fraction near the middle of the tank decreases. This means that there is no clear trend with height in the tank for the profiles shown in Figure (6).

For the low flow rate case in Figure (6a), the stratified layers remain intact for a significant fraction of the vapor freeing process. The location of the air inlet and outlet does not take advantage of the buoyancy of air in propane. The effect of this can be seen in Figure (6a) in that the total gas volume fraction remains above the PMM line for the entire vapor freeing process. This occurs because the buoyancy of the air jet results in the air jet deflecting upward into the methane layer before it mixes well with the propane layer. Therefore, the air jet erodes the propane layer rather than mixing with and diluting the propane layer. The higher flow rate case shown in Figure (6b) is sufficient to overcome this effect.

Figure (6c) shows the FDS simulation results for a case with an air flow rate of 0.5 ACH and with two outlets in the tank: one in the roof opposite the air inlet, and one in the side opposite the air inlet. The total gas volume fraction profiles show significant variability from the PMM indicating that much of the vapor freeing process is occurring by displacement. Adding a second outlet on the side takes advantage of buoyancy for both the methane and propane. The result is that the total vapor volume fraction reaches the target sooner than for the case with no side outlet (Figure (6a)).

(a)

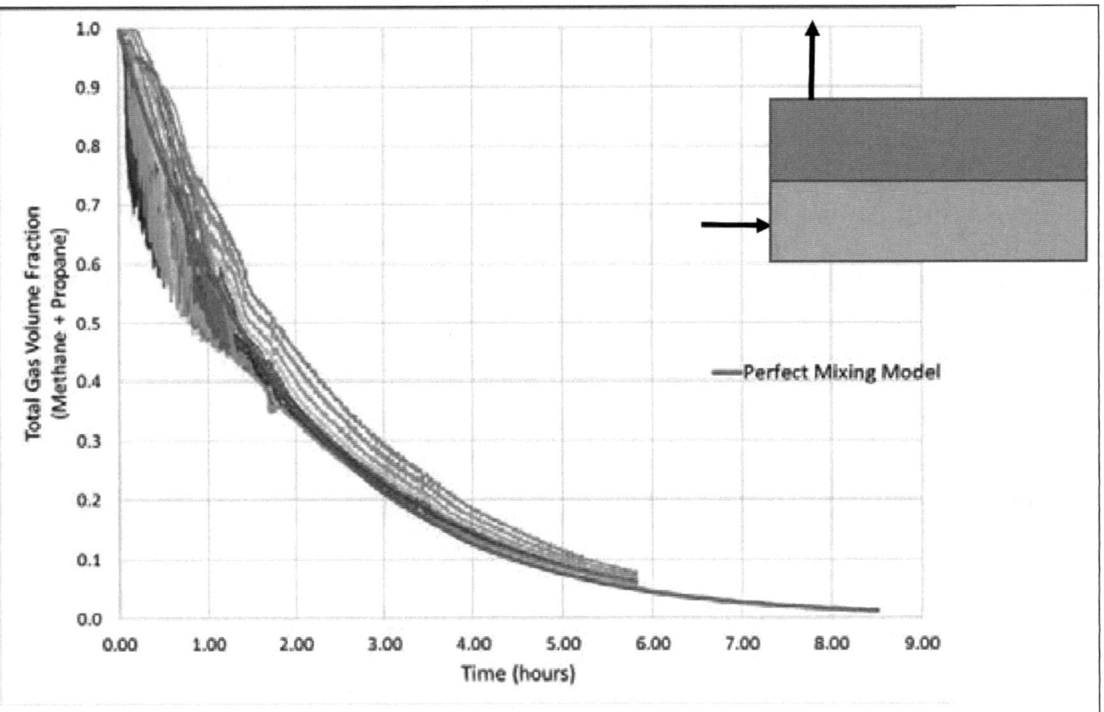

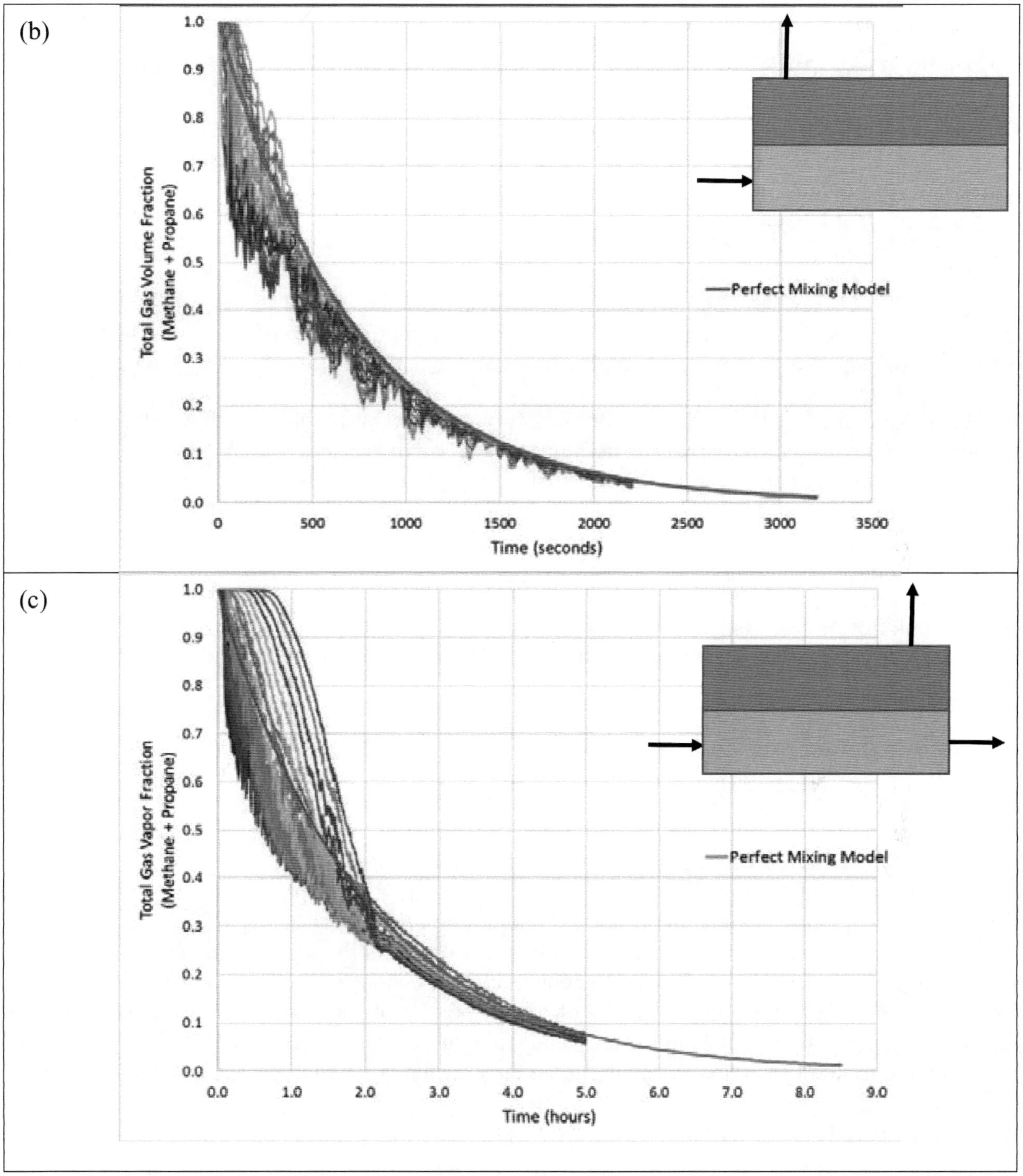

Figure 6. Comparison of total gas volume fraction in the tank from the FDS simulation to the Perfect Mixing Model (PMM) for Combination (1)(G): (a) 0.5 ACH; (b) 5 ACH ; (c) 0.5 ACH with two outlets. The rectangle with arrows indicates the location of the inlet and outlet vents on the tank. The blue area in the rectangle indicates methane, the green area indicates propane.

5 Conclusion

API RP 2016 provides recommended methods for vapor freeing storage tanks. Following these recommendations should result in safe and effective freeing of tank headspaces of toxic or flammable vapors. When using air to remove vapors from a tank headspace, it may not be feasible to achieve the 5 ACH air flow rate recommended in the guide. In some situations, air at the recommended flow rate may not break up stratified layers in the tank. In such cases, it is important to orient the air inlet and vapor outlet to take advantage of both the buoyancy of the vapors and the jet momentum.

The correlations given above provide a way to estimate whether the air jet will break up stratified layers, and provide some guidelines on when a single air inlet may be sufficient or when multiple inlets may be required to provide adequate mixing in the tank.

The FDS simulations indicate that the correlations for critical velocity of a side-entry jet and for a vertical-entry jet are conservative for the purposes of vapor freeing. It appears that a flow rate greater than approximately 20% of the critical values calculated using the correlations will be sufficient to break up stratification, so that the vapor freeing process will be by dilution. For flow rates below this, one should anticipate that at least the first air change in the tank will be mostly by displacement.

For low air flow rates, consider angling the air jet so that it traverses the maximum distance possible through the tank before encountering a wall. This will increase mixing efficiency when buoyancy effects are reduced or are nonexistent.

Safety issues posed by flammable mixtures in the tank during vapor freeing are beyond the scope of this paper. However, the Perfect Mixing Model and the FDS results do provide qualitative information on where in the tank and at what times flammable mixtures may be present. For vapor freeing by dilution (e.g. when the air flow rate is sufficiently high that the PMM can reasonably be applied) the PMM can be used to estimate the time range during which the entire tank headspace is likely to contain a flammable mixture. For vapor freeing by displacement (e.g. when the air flow rate is sufficiently low that the vapor volume fraction profiles show significant variation from the PMM model), a conservative approach is to consider that a flammable mixture exists in some part of the tank headspace for virtually the entire vapor freeing process. The region where the air jet enters the tank and begins to mix with the vapor should also be considered to contain a flammable mixture for virtually the entire vapor freeing process. Care should be taken to consider any potential mixture variations throughout the tank during any vapor freeing operation.

6 Nomenclature

B_0	Initial jet buoyancy flux, m^4/s^3
D	Tank diameter, m
d_j	Jet nozzle diameter, m
Fr_0	Froude number for jet based on initial jet properties
G	Stratification constant
g	Gravitational constant, 9.8 m/s^2
g'_0	Reduced gravity, g

H	Tank height, m
K	Constant = 1.159
L_{hyd}	Hydraulic length scale, m
L_j	Jet length scale, m
M_0	Initial jet momentum flux, m^4/s^2
Q_0	Initial jet volume flow rate, m^3/s
R	Characteristic tank base length scale, m
Re_0	Jet Reynolds number based on the nozzle diameter
r_0	Jet nozzle radius, m
t	Time, s
t_m	Characteristic mixing time, s
t_r	Characteristic residence time, s
u_0	Jet initial velocity, m/s
u_c	Critical jet velocity, m/s
V	Tank volume, m^3
x(t)	Vapor volume fraction in the tank
Z	Fountain jet penetration distance. m

Greek Symbols:

α	Constant = 0.083
β	Angle of a side-entry jet from horizontal, degrees
θ	Defined as $\beta+5^o$
π	Constant = 3.14
ρ	Density, kg/m^3

7 References

[1] ANSI/API Standard 2015, *Requirements for Safe Entry and Cleaning of Petroleum Storage Tanks,* Washington, D.C.: API, 2014.

[2] API RP 2016, *Guidelines and Procedures for Entering and Cleaning Petroleum Storage Tanks,* Washington D.C.: API, 2016.

[3] K. Chow, *Simulation and analysis of gas freeing of oil tanks,* Coventry University: Ph.D. Thesis, 2009.

[4] B. K. Revill, "Jet Mixing," in *Mixing in the Process Industries, 2nd ed.*, M. F. E. A. W. N. N. Harnby, Ed., Oxford, Butterworth Heinemann, 1992, pp. 159-183.

[5] G. R. Hunt, P. Cooper and P. F. Linden, "Thermal stratification produced by plumes and jets in enclosed spaces," *Building and Environment,* vol. 36, pp. 871-882, 2001.

[6] H. C. Burridge, A. Mistry and G. R. Hunt, "The effect of source Reynolds number on the rise height of a fountain," *Physics of Fluids,* vol. 27, p. 047101, 2015.

[7] E. A. Fox and V. E. Gex, "Single-phase blending of liquids," *AIChE Journal,* vol. 2, no. 4, pp. 539-544, 1956.

[8] A. G. C. Lane and P. Rice, "Comparative assessment of the performance of the three designs for liquid jet mixing," *Industrial & Engineering Chemistry Process Design & Development,* vol. 21, pp. 650-653, 1982.

[9] P. L. Lagus and K. H. Lie, "A numerical investigation of the constant tracer flow technique," in *Air Change Rate and Airtightness in Buildings*, M. H. Sherman, Ed., Philadelphia, ASTM, 1990, pp. 21-30.

Refinery PHAs – Revalidation or Re-Dos? Do They Ensure Safe Operations?

Kumar (Chris) Israni
Principal Consultant
ERM
840 W Sam Houston Pkwy N, Suite 600, Houston, TX
chris.israni@erm.com

Joshtin Cooper
Senior Consultant
ERM
3838 Causeway Blvd, Metairie, LA
joshtin.cooper@erm.com

Radhika Bompelly
Consultant
ERM
840 W Sam Houston Pkwy N, Suite 600, Houston, TX
radhika.bompelly@erm.com

Prepared for Presentation at
American Institute of Chemical Engineers
2017 Spring Meeting and 13th Global Congress on Process Safety
San Antonio, TX
March 26 – 29, 2017

Refinery PHAs – Revalidation or Re-Dos? Do They Ensure Safe Operations?

Kumar (Chris) Israni
Principal Consultant
ERM
840 W Sam Houston Pkwy N, Suite 600, Houston, TX
Chris.israni@erm.com

Joshtin Cooper
Senior Consultant
joshtin.cooper@erm.com

Radhika Bompelly
Consultant
radhika.bompelly@erm.com

Keywords: PHA, OSHA, Refinery, Revalidation, Redo

Abstract

OSHA 1910.119 Process Safety Management (PSM) has been the industry standard for over 20 years. Its elements have been clearly rolled out and adopted by refineries throughout the USA. However, the most recent grassroots refineries in the country were built over 45 years ago. There have been various initiatives from regulators and operating companies by way of compliance audits, emphasis programs and due diligence, to ensure that the process safety management aspects are adhered too, while following the path of compliance and conformance to the 14 elements, one of these being the every five year occurring Process Hazard Analysis (PHA) cycle.

The success of these PHAs rely on various factors – choice of methodology, scope inclusive of changes from the last few years, addressing new and earlier unidentified scenarios, tracking of previous actions and incorporation of findings from incidents and lessons learned. While this should be the scope of the PHA, it is often misrepresented with a check the boxes type of approach. The key is ensuring compliance as well as satisfying the requirements of a revalidation or redo – a choice made by a few operating companies.

It is commonly observed that while sites have completed this five year compliance requirement, some of effectiveness of the PHA is lost due to non-adherence to scope. This paper highlights the experiences of the authors while conducting PHAs using its various methodologies in the revalidation and re-do approaches.

Satisfying the every five year requirement, addresses the compliance need. The effectiveness of the PHA (irrespective of approach – revalidation / re-do) is depends on adherence to the claims made in the PHA. The operating risk stays low, when the identified scenarios and their associated safety mechanisms (claims) are continuously implemented and practiced.

1 Introduction

Petroleum refining has evolved continuously in the US for different products. Currently, there are 141 refineries in the US with a total capacity of more than 18 million barrels per day [1, 2, 10]. Apart from transportation fuels, petroleum refineries also produce liquids that are used by the petrochemical industry to make a variety of chemicals and plastics. Refineries have to ensure compliance with government regulations such as Occupational Health and Safety Administration (OSHA) 1910.119 [2] and Environmental Protection Agency (EPA) 40 CFR Part 68 [4].

2 Process Hazard Analysis and Methodology

Center for Chemical Process Safety (CCPS) Guidelines for Hazard Evaluation Procedures defines hazard as "an inherent physical or chemical characteristic that has the potential to do harm (to people, property, or the environment)" [4]. A Process Hazard Analysis (PHA) is defined as a systematic effort designed to identify and analyze hazards associated with the processing or handling of highly hazardous materials and a method to provide information which will help workers and employers in making decisions that will improve safety. The OSHA 1910.119 regulation [3], amongst its other requirements, requires the sites to perform a PHA on each process.

The method selected for PHA will be influenced by many factors such as the type of process, complexity of the process, time required, etc. The process hazard analysis methodology selected must be appropriate to the complexity of the process and must identify, evaluate, and control the hazards involved in the process. Proper selection of the methodology is one of the most basic factors that can aid in making a good PHA. The most commonly used techniques of PHA are What-If, Checklist, What-If/Checklist, Hazard and Operability Study (HAZOP), Failure Mode and Effect Analysis (FMEA) and Fault Tree Analysis (FTA) [5]. All PHA methodologies are subject to certain limitations [9]. So regardless of the method used, the PHA shall address all of the following issues:

- The hazards of the process;
- Previous incidents which could have been catastrophic;
- Engineering and administrative controls;
- The consequences of failure of engineering and administrative controls;
- Facility siting;

- Human factors; and
- The range of possible safety and health effects caused by the failure of controls.

The process hazard analysis shall be performed by a team with expertise in engineering and process operations, and the team shall include at least one employee who has experience and knowledge specific to the process being evaluated. Without adequate knowledge and experience specific to the process, a PHA team may not identify unique hazards or appropriately assess potential causes, consequences, or the effectiveness of safeguards to control process hazards [5, 6, 9].

3 PHA Revalidation

As mentioned, at least every five years after the completion of the initial PHA, the companies are required to update or revalidate their PHA as per the OSHA process safety management (PSM) regulation, 29 CFR 1910.119. Revalidation is a process of examining the initial PHA to ensure that it adequately addresses process hazards as they are currently identified. This includes modifying the previous PHA accordingly to reflect the necessary changes in the facility since the previous analysis. Also, current understanding of the process hazards might differ from time to time – the initial PHA may have a certain type of understand, but due to changes in technology, regulation and personnel, the current group may have a different understanding.

There could be many factors that would have introduced new hazards such as process changes, changes in occupancy patterns either on-site or off-site, knowledge in better understanding the potential hazards, incidents occurred and changes in the safeguards identified in previous PHA. So the PHA revalidation must address the following items [6]:

- Changes subsequent to previous PHA
- Identification and correction of PHA omissions and deficiencies
- Ensure that Process Safety Information (PSI) is complete, current and accurate and all the training procedures are up to date
- Review the new process technology or information since the previous PHA
- Update the PHA according to new or modified regulations (OSHA and EPA)

One of the advantages of revalidation is time effectiveness. Conducting an initial PHA requires considerable amount of time and effort. The revalidation process significantly reduces the time and effort as the PHA scope is less since revalidation only reviews the modified sections. The prior PHA serves as the starting point for the revalidation process. The new content is added to the PHA and results will be documented as a new PHA.

PHA Revalidation Approach

PHA revalidation is an essential and critical activity. The first step in any PHA revalidation is to consider the issues before determining the appropriate revalidation approach. The most appropriate approach depends upon several factors including the quality of previous PHAs, information available from the previous PHA, the complexity of the changes, the number of undocumented changes and the quality of Management of Change (MOC). The next step would be to define the scope of the revalidation; the scope should include all the affected changes but not be limited or dictated by the scope of prior PHA.

The procedure for PHA revalidation study is similar to that of the initial PHA and should follow the steps listed below:

- Confirm on the scope
- Select the study team and the leader
- Select the appropriate methodology
- Assemble the required information such as previous PHA data, updated P&IDs, equipment data, incident investigation reports, MOC and revised operating procedures
- Review of actions from previous PHA
- Prepare a list of items to be reviewed during revalidation
- Conduct sessions for revalidation
- Establish new recommendations and action plan
- Issue revalidated PHA report

Selection of team members for revalidation is as critical as selection of team members for an initial PHA. A team comprising a mix of both new and previous team members would be efficient as this helps in providing consistency of approach, implementing lessons learned from previous PHA and providing fresh insight and objectivity. Just as conducting an initial PHA, the team should be trained in advance on the objective of the revalidation to make the sessions effective. Training may include providing the process overview, purpose of PHA revalidation, methodology, etc.

Revalidation requires more time to be spent on collecting the information, reviewing the previous and preparing the information to be reviewed during the sessions. Selection of an efficient study leader results in proper documentation of the revalidation process and generating a clear, concise report for implementation of the new findings and retention of record for future use.

Essential Tasks for PHA Revalidation

In order to optimize the PHA revalidation and to avoid pitfalls it is important to follow essential tasks as described below [7]:

1. Review of modifications: involves reviewing all the MOCs, audit reports, comparing current P&IDs with those used for previous PHA, operating procedures, etc.
2. Review of previous incidents: involves gathering incident investigation reports and reviewing with the full team to address any potential issues that are likely to cause catastrophic consequences.

3. Review previous PHA recommendations: involves reviewing the previous PHA recommendations to check the status of closure and to address unresolved actions.
4. Update facility siting/human factors: involves preparing a checklist to address the modifications in the facility siting or human factor analysis that could affect the risk scenarios.
5. Evaluate non-routine operations: involves evaluation and addressing the hazards such as start-up, shutdown, maintenance, etc.
6. Evaluate the standards and regulations: involves reviewing of the new regulations and comparing with those used in previous PHA to determine the additional information to be added in the new PHA to ensure that the new PHA is compliant with the new regulations.

4 PHA Re-do

A PHA redo can be defined as a process to perform a complete, new PHA using one or more recognized techniques as described in the sections above that meet all the regulatory, industry and company requirements. A complete redo would be considered a feasible and cost effective option depending on various factors.

A PHA redo may be a most appropriate choice in situations where [7]:

- Gaps or deficiencies exist in the manner in which the prior PHA was conducted indicating that significant hazards might have been overlooked or not controlled appropriately as a result of improper use of process safety information.
- Uncontrolled changes were occurred in the process or equipment since the previous PHA and the complexity of the process may require a redo.
- Inappropriate use of technique or methodology in the previous PHA.
- Occurrence of various incidents since the previous PHA
- Previous PHA documentation do not meet the current standards of regulatory requirements due to new interpretations of codes and standards

A redo is a time and resource effective option if the facility has undergone a large number of MOCs since the previous PHA as reviewing all MOC documents may require more time than redoing a PHA to ensure that all the hazards were assessed adequately. A redo also provides an opportunity to conduct a PHA using an appropriate method if a different method was used in the previous one. A redo is also considered a better option after a certain number of revalidations in order to ensure that no potential hazards were overlooked in the revalidating process over the period of time. Also document retention after an indefinite period of time becomes unmanageable with the addition of large number of MOCs to the PHA revalidation reports.

5 Comparison between Revalidation and Redo

The primary aim to either revalidate or redo a PHA is to identify, evaluate, and control the hazards involved in the process. The decision to revalidate or redo a PHA for the facility depends on the number of factors. Both revalidation and redo has its own limitations as well as benefits. The information in the paragraphs below are based on the authors experiences with facilitating PHAs in refineries, using either of the two approaches. These were based on their interactions with refinery sites, at a national level, and hence covered the various regions within the United States.

In the case of PHA revalidations, the limitations observed were:

- Inadequate documentation of the previous PHA(s) results in inefficient revalidation process,
- Limitations in defining the scope as it might be dictated by the previous PHA,
- The same team that performed the prior PHA may not be available for revalidation,
- Improper or poor documentation of the MOCs may result in inadequate evaluation of the hazards,
- Incorrect or missing PSI for revalidation,
- Unavailability of P&IDs, node information, operating procedures used in the prior PHA to compare with the new changes,
- Missing information on the closed out actions which forms the basis for revalidation,
- Noncompliance of the previous PHA with the regulations and standards,
- Amount of time spent in collecting and reviewing the previous PHA could be more than actual revalidation session,
- May require more time than an actual PHA to incorporate each MOC when there are a number of changes since the prior PHA,
- Data could become unmanageable after several revalidations, and
- Some hazards may be overlooked over a period of time.

At the same time, the benefits gained were:

- The approach was a huge time saver,
- Lessons from previous PHA can be implemented during revalidation,
- Experience and on-the-job training of the team aids in identifying the new issues that were not evaluated in the previous PHA,
- The team with a mix of new and old team members provide consistency and new insight for the session, and
- Revalidation offers an opportunity to pull multiple PHAs together.

In the case of PHA redos, the limitations observed were:

- Time required to redo a PHA is more compared to that of revalidation,
- Redo might just be a repetition of previous PHA in case of small changes,
- The team may not be familiar with the previous PHA, and
- The method selected for redo may not be effective in evaluating certain hazards identified in the initial PHA.

At the same time, the benefits gained were:

- Filling the gaps or deficiencies identified in the prior PHA,
- Selection of an appropriate method if the method used in the previous PHA was inefficient for the process,
- Defining the scope according to the new changes in the process, and
- Incorporating the current standards and regulations.

6 Conclusion

The two approaches have their benefits, and neither of them replace the other.

The revalidation approach is preferred due to time savings implying budgetary savings. But this is only successful if:

- The previous PHA is of a good quality and is used in the revalidation,
- Updates to the action items from the previous PHA are available,
- Incident reports and MOCs since the previous PHAs are discussed in the revalidation,
- The study still uses the most current set of PSI,
- Incorporates any regulatory changes, and
- During the revalidation, thought is given to changes to the process and the team also contributes towards any missed scenarios.

The re-do approach, essentially means beginning the PHA from scratch and will be successful:

- As long as the drawings are current, and brainstorming is followed with an experienced team, the study will cover all aspects and
- Even if the previous PHA has any deficiencies, they are covered by the complete re-do approach.

It is our observation, that however time and labor intensive the re-do process is, many companies are choosing it on their ten or fifteen year cycles. That covers for a lot of gaps otherwise noticed in:

- The previous PHAs,
- Any skipped or missed scenarios, and
- Changes in personnel, thereby becoming a training tool for new team members.

Hence, the choice is up to the operating company, on the approach they take, the key is that irrespective of route taken, the PHA must always be on the current state of the process, and provide adequate information on hazards identified, to minimize risks to personnel, environment and the business, during operations.

7 References

[1] U.S. Energy Information Administration, http://www.eia.gov/energyexplained/index.cfm?page=oil_refining#tab1Accessed on February 17, 2017.

[2] U.S. Energy Information Administration, https://www.eia.gov/tools/faqs/faq.cfm?id=29&t=6, Accessed on February 17, 2017.

[3] OSHA 3132 Process Safety Management Informational Booklet, 2000.

[4] American Petroleum Institute, API Recommended Practice 75, Recommended Practice for Development of a Safety and Environment Management Program for Offshore Operations and Facilities, 3rd edition, Washington, DC: API, May 2004.

[5] CCPS Guidelines for Hazard Evaluation Procedures, Center for Chemical Process Safety, American Institute of Chemical Engineers, New York, NY, 1992.

[6] Nigel Hyatt, Guidelines for Process Hazard Analysis, Hazards Identification & Risk Analysis, Dyadem, 1st Edition, 2004

[7] Walter L. Frank and David K. Whittle, CCPS Revalidating Process Hazard Analysis, Center for Chemical Process Safety, American Institute of Chemical Engineers, New York, NY, 2001

[8] Kevin E. Smith and David K. Whittle, Six Steps to Effectively Update and Revalidate PHAs, Chemical Engineering Progress, 2001

[9] Israni, Chris and Cooper, Joshtin, "Technical Risk – It's Not a One-Size-Fits-All Proposition", paper presented at AICHE 2013 Spring Meeting, 9th Global Congress on Process Safety, San Antonio, Texas, April 28 – May 1, 2013

[10] US energy Information Administration https://www.eia.gov/dnav/pet/pet_pnp_cap1_dcu_nus_a.htm, last accessed March 6, 2017

How Small of a Cloud Do We Want to Detect with our Gas Detectors?

Vinicius Simoes
Genesis
11750 Katy Freeway Suite 100
Houston, TX, 77079
Vinicius.Simoes@genesisoilandgas.com

Massimiliano Kolbe, Sarah Chambliss
Genesis
11750 Katy Freeway Suite 100
Houston, TX, 77079
MKolbe@genesisoilandgas.com, Sarah.Chambliss@genesisoilandgas.com

Prepared for Presentation at
American Institute of Chemical Engineers
2017 Spring Meeting and 13th Global Congress on Process Safety
San Antonio, TX
March 26 – 29, 2017

How Small of a Cloud Do We Want to Detect with our Gas Detectors?

Vinicius Simoes
Genesis North America
11750 Katy Freeway Suite 100
Houston, TX, 77079
Vinicius.Simoes@genesisoilandgas.com

Massimiliano Kolbe, Sarah Chambliss
Genesis North America

Keywords: Gas Detectors, Cloud Dispersion, Explosion, VCE

Abstract

There is a big discussion in the industry about how small is really small. The objective of this paper is to cover a series of cloud sizes and their impact to determine how small of a cloud we should really be detecting in order to avoid personnel injury and structural damage. There are a lot of different discussions in the industry about what clouds we should be detecting.

1 Introduction

The purpose of this paper is to determine a cloud size or release rate that one should focus on when defining their gas detector placement. It is well known there are several different methodologies when it comes to cloud detection or release detections. There are several criteria which are applied by different companies or even industries (e.g., chemical, pharmaceutical). Those criteria could be based on dispersion, cloud volume, or separation distances. It is of paramount importance to determine the target for our analysis, e.g., flammable or toxic materials. Those materials will have very distinct thresholds which would directly impact the study. This paper focuses on the flammable releases, and more importantly on the explosion effects coming from those clouds that could be generated from such releases.

The criteria could be, for example, based on release rate ranges. This could be using very small releases (0.1 kg/s) up to small releases (5 kg/s), depending on what criteria are applied. There are a series of questions that could still be unanswered after the gas detector layout has been designed. For example, have we optimized the number and location of our gas detectors? Did we detect all the clouds needed? The idea of the study was to identify if we really need to detect clouds being formed from very small releases or if there is a certain range that one should focus on when it comes to overpressure analysis resulting from vapor cloud explosions (VCE).

2 Modeling Approach

2.1 Scenario 1

The scenario was conducted using a series of releases of Ethane under 120 °F and 500 psi. The modeling of the discharge, dispersion and explosion was done using Safeti (Software for the Assessment of Flammable, Explosive and Toxic Impact) Version 6.7. Safeti is well-recognized and a very comprehensive quantitative tool available for assessing process plant risks. It is designed to perform all the data processing, generation of results and presentation elements of a Quantitative Risk Assessment (QRA) within a structured framework. Safeti went through an extensive validation process.

The Safeti QRA software is designed to comply with the regulatory requirements of many countries. For example, specific modules have been included to ensure compliance with the Dutch Yellow Book, the Reference Manual Bevi Risk Assessments version 3.1 and UK HSE regulations.

The explosion model adopted for this analysis was TNO Multi-Energy. A set of TNO strengths were used for the assessment, ranging from 3 to 7. It is understandable that detonations are possible, but for the purpose of this exercise this was not included in the assessment. The study focuses on the far-field assessment of the explosions. TNO is well accepted on the far-field but it is important to be cautious when applying it to the near-field. For a near-field exercise the authors do believe that a more sophisticated model, such as Computational Fluid Dynamics (CFD) should be applied. Based on the tool and approach used, it is important to highlight that the focus of this paper has been the onshore industry.

The study used a sample layout, which can be seen in Figure 1. This layout was created to define several congested regions, structures and buildings. In total, five congested regions and three monitor points were defined in the model, and two buildings were assessed for overpressure. The east wall of the two buildings and the roof of the control room building were monitored.

Several ignition sources were placed along the congested regions to ensure proper coverage of the different cloud volumes and ignition locations.

Two weather conditions were used for this study, including 2 m/s with a class F for the Pasquill Stability and 5 m/s wind speed with a class D for the Pasquill Stability. This was done to ensure we properly covered both stable and unstable conditions. Table 1 presents the other atmospheric parameters used for the study.

Table 1: Atmospheric Parameters

Parameter	Value
Atmospheric Temperature	68 °F

Parameter	Value
Atmospheric Pressure	14.7 psi
Surface Roughness	183.2 mm
Humidity	70 %
Solar Radiation Flux	250 Btu/ft^2·hr

In addition, four release locations were assumed and are indicated in Figure 1 by the red marks. For the purpose of this study, the authors covered a series of locations to ensure proper representation of the possible explosive cloud locations.

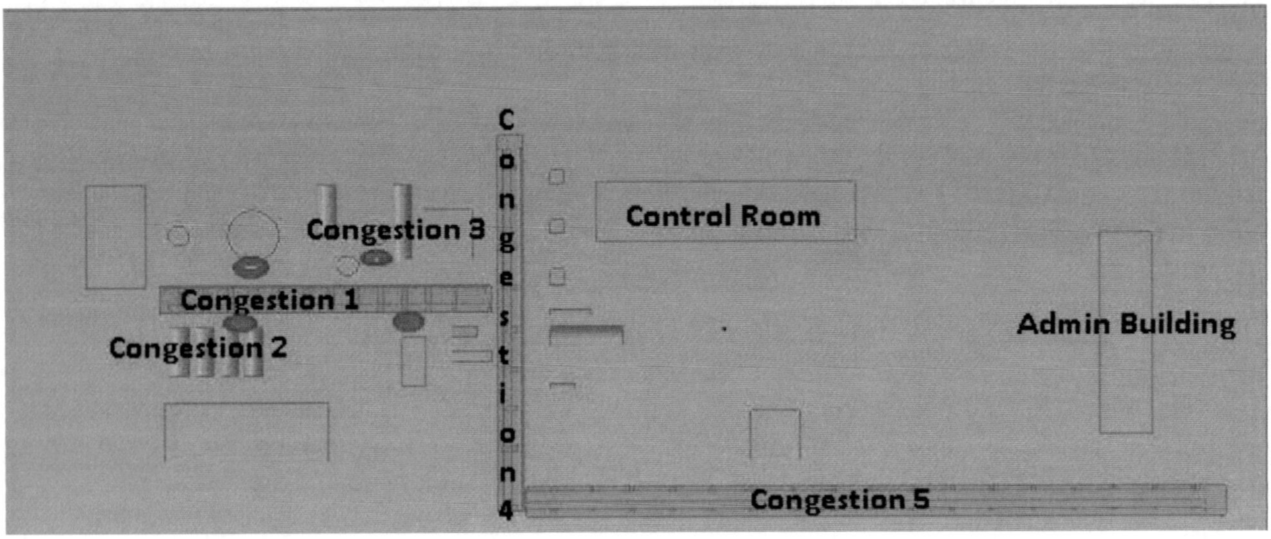

Figure 1: Facility Layout

Table 2 shows the list of Ethane release scenarios and their parameters.

Table 2: Ethane - Scenarios Parameters

Release Rate (kg/s)	Location	Pressure (psi)	Temperature (°F)
0.1	Location 1,2,3,4	500	120
0.5	Location 1,2,3,4	500	120
1	Location 1,2,3,4	500	120
2	Location 1,2,3,4	500	120
3	Location 1,2,3,4	500	120
4	Location 1,2,3,4	500	120
5	Location 1,2,3,4	500	120
7	Location 1,2,3,4	500	120
9	Location 1,2,3,4	500	120
10	Location 1,2,3,4	500	120
15	Location 1,2,3,4	500	120

Release Rate (kg/s)	Location	Pressure (psi)	Temperature (°F)
20	Location 1,2,3,4	500	120

2.2 Scenario 2

The scenario was conducted using a series of releases of Propane under 160 °F and 250 psi. The scenario used the same sample layout, congested regions, ignitions and weather stability class as described in section 2.1. Table 3 presents the other atmospheric parameters used for the study.

Table 3. Atmospheric Parameters

Parameter	Value
Atmospheric Temperature	68 °F
Atmospheric Pressure	14.7 psi
Surface Roughness	183.2 mm
Humidity	70 %
Solar Radiation Flux	250 Btu/ft2·hr

In addition, four release locations were assumed, and are indicated in Figure 1 by the red marks. For the purpose of this study, the authors covered a series of locations to ensure proper representation of the possible explosive cloud locations.

Table 4 provides the list of propane release scenarios and their parameters.

Table 4: Propane - Scenarios Parameters

Release Rate (kg/s)	Location	Pressure (psi)	Temperature (°F)
0.1	Location 1,2,3,4	250	160
0.5	Location 1,2,3,4	250	160
1	Location 1,2,3,4	250	160
2	Location 1,2,3,4	250	160
3	Location 1,2,3,4	250	160
4	Location 1,2,3,4	250	160
5	Location 1,2,3,4	250	160
7	Location 1,2,3,4	250	160
9	Location 1,2,3,4	250	160
10	Location 1,2,3,4	250	160
15	Location 1,2,3,4	250	160
20	Location 1,2,3,4	250	160

3 Results

3.1 Scenario 1 – Ethane Releases

A series of Ethane simulations were conducted as described in Section 2.1. Those simulations were conducted with different release rates that led to different cloud volumes and explosion locations.

From these simulations, explosions were modeled in the congested regions, and the flammable mass that could reach the congested volumes was determined. The analysis was conducted explicitly to determine how much of the release could be found in the congested region and contribute to the explosion energy and lead to overpressure that were then evaluated for each of the monitor points. The main findings are listed in Table 5.

Table 5: Results Summary for Ethane Scenarios

Release Rate (kg/s)	Monitor Points	Overpressure (psi)	Overpressure (bar)	Impulse ($N \cdot s/m^2$)	Flammable Cloud Volume (m^3)
0.1	East Wall CR	NA	NA	NA	0.07
0.1	Center CR Roof	NA	NA	NA	0.07
0.1	Admin Building	NA	NA	NA	0.07
0.5	East Wall CR	NA	NA	NA	0.35
0.5	Center CR Roof	NA	NA	NA	0.35
0.5	Admin Building	NA	NA	NA	0.35
1	East Wall CR	NA	NA	NA	4
1	Center CR Roof	NA	NA	NA	4
1	Admin Building	NA	NA	NA	4
2	East Wall CR	NA	NA	NA	6
2	Center CR Roof	0.51	0.04	10.12	6
2	Admin Building	NA	NA	NA	6
3	East Wall CR	0.65	0.05	15.56	12
3	Center CR Roof	0.35	0.02	8.98	12
3	Admin Building	NA	NA	NA	12
4	East Wall CR	0.74	0.05	19.56	21
4	Center CR Roof	0.41	0.03	11.23	21
4	Admin Building	NA	NA	NA	21
5	East Wall CR	0.90	0.06	22.69	32
5	Center CR Roof	0.45	0.03	12.98	32
5	Admin Building	NA	NA	NA	32
7	East Wall CR	1.36	0.09	40.47	60
7	Center CR Roof	0.59	0.04	19.39	60
7	Admin Building	NA	NA	NA	60

Release Rate (kg/s)	Monitor Points	Overpressure (psi)	Overpressure (bar)	Impulse (N·s/m²)	Flammable Cloud Volume (m³)
9	East Wall CR	1.75	0.12	58.21	92
9	Center CR Roof	0.71	0.05	27.36	92
9	Admin Building	0.25	0.02	12.3	92
10	East Wall CR	1.90	0.13	65.64	108
10	Center CR Roof	0.77	0.05	30.6	108
10	Admin Building	0.28	0.02	13.74	108
15	East Wall CR	2.44	0.17	92.5	221
15	Center CR Roof	0.81	0.06	42.8	221
15	Admin Building	0.33	0.02	19.0	221
20	East Wall CR	3.42	0.24	155.2	361
20	Center CR Roof	1.17	0.08	62.48	361
20	Admin Building	0.38	0.03	22.82	361

Another important part of the analysis was defining generic overpressure contours generated from specific leak rates in order to define the threshold for cloud size generated for the overall facility rather than for a specific building.

Figures 2 through 9 provide an overall guidance on the vicinity of the congested regions and were used to determine the cloud size that would impact the area around the explosion source. The figures provide overpressure contours for the simulations conducted.

Figure 2 presents the results for the 0.1 kg/s release rate and demonstrates that rates this small impact only the areas directly around the release and around congested regions. This simulation did not generate overpressure above 2 psi.

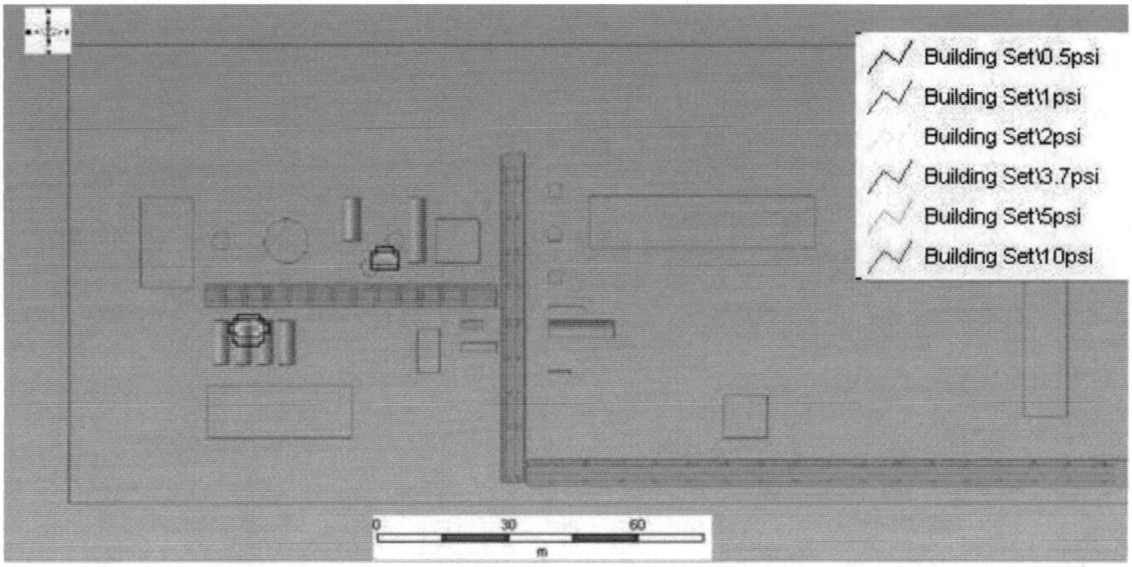

Figure 2: 0.1 kg/s release Ethane

Figure 3 presents the results for the 1.0 kg/s release rate and demonstrates that rates as small as those impact the areas around the explosion region. Higher overpressures were registered for this simulation, but the impact was still only around the explosion source, which would likely require a more refined model to produce robust results.

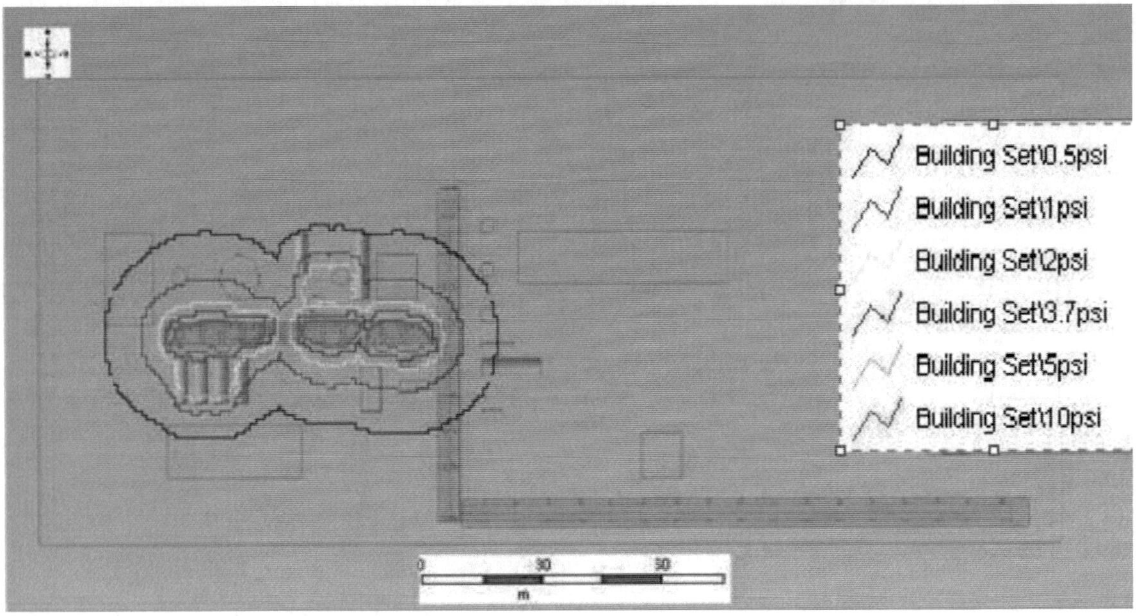

Figure 3: 1kg/s release Ethane

Figure 4 presents the results for the 3.0 kg/s release rate and demonstrates that those rates start to produce some overpressure on the building of concern, but the overpressure and associated impulse is not significant enough to cause damage to the structure and compromise the personnel located inside.

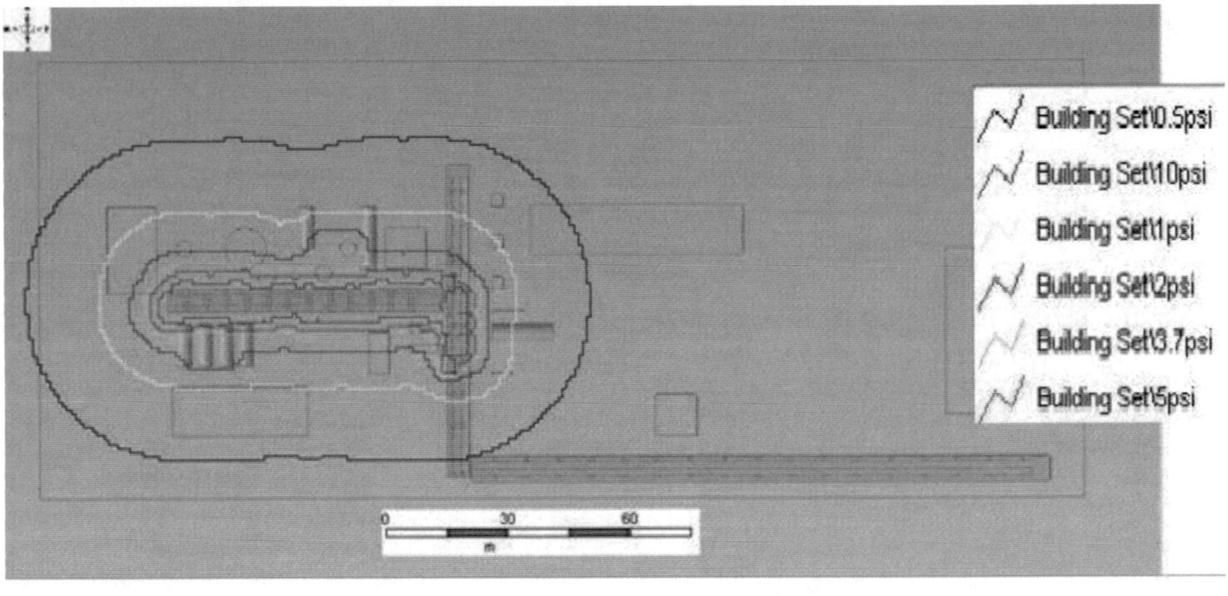

Figure 4: 3 kg/s release Ethane

Figure 5 presents the results for the 5.0 kg/s release rate and demonstrates that those rates start to produce overpressures above 1 psi on the building of concern. These values are still not an issue for the occupants, but the building would need to have appropriate windows, frames and other reinforcements.

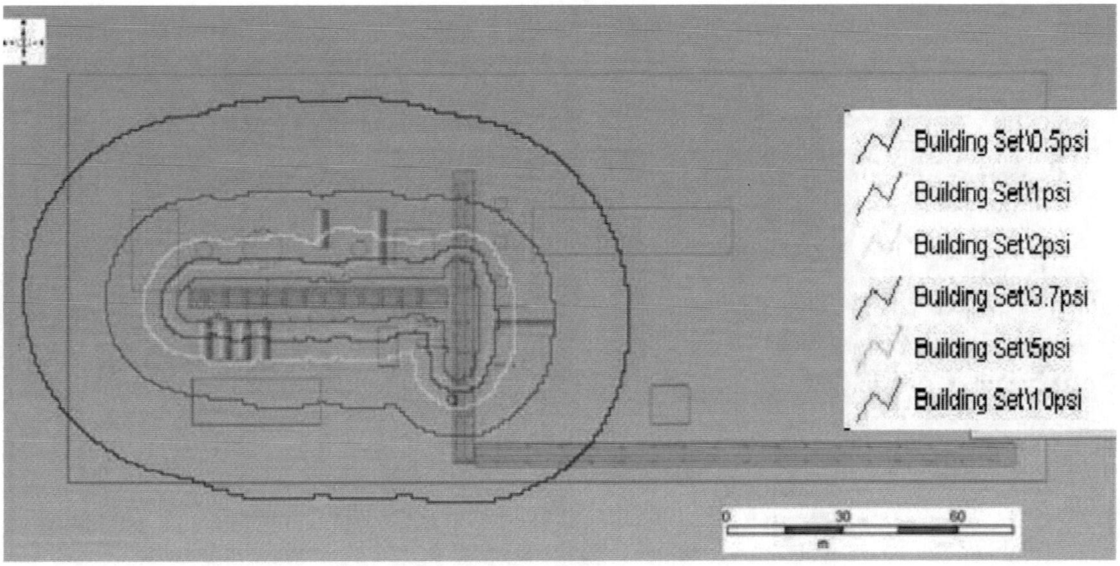

Figure 5: 5 kg/s release Ethane

Figure 6 presents the results for the 7.0 kg/s release rate and demonstrates that those rates start to produce overpressures around 2 psi on the building of interest. These buildings would require reinforcement, at a significant cost, in order to sustain overpressures generated from those scenarios.

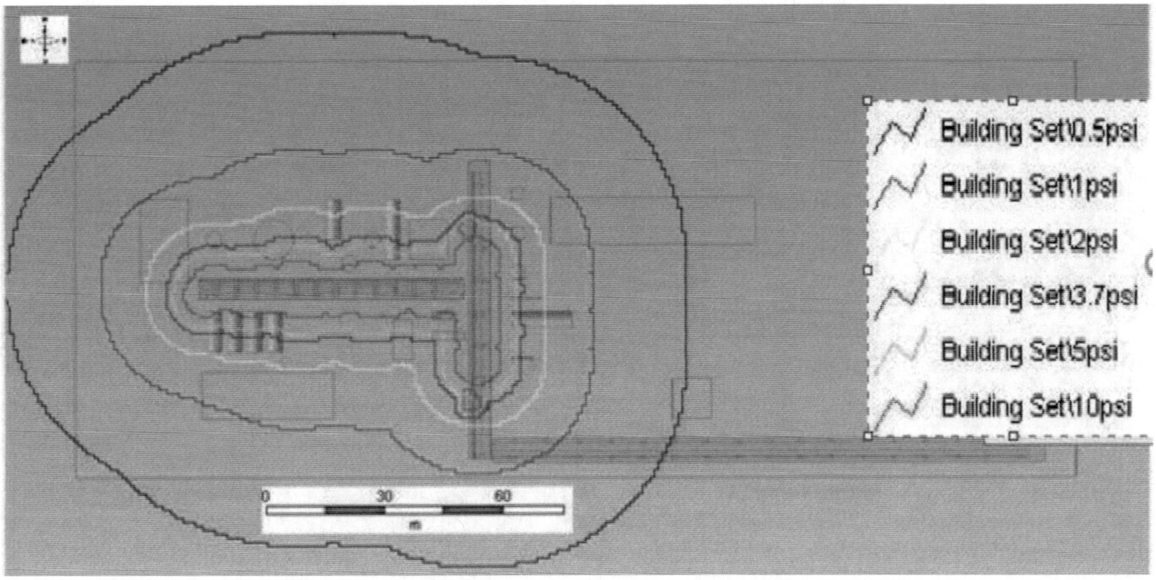

Figure 6: 7 kg/s release Ethane

Figure 7 presents the results for the 9.0 kg/s release rate and demonstrates that those rates also produce overpressures around 2 psi on the building of concern. Again, these buildings would require reinforcement, at a significant cost, in order to sustain overpressures generated from those scenarios.

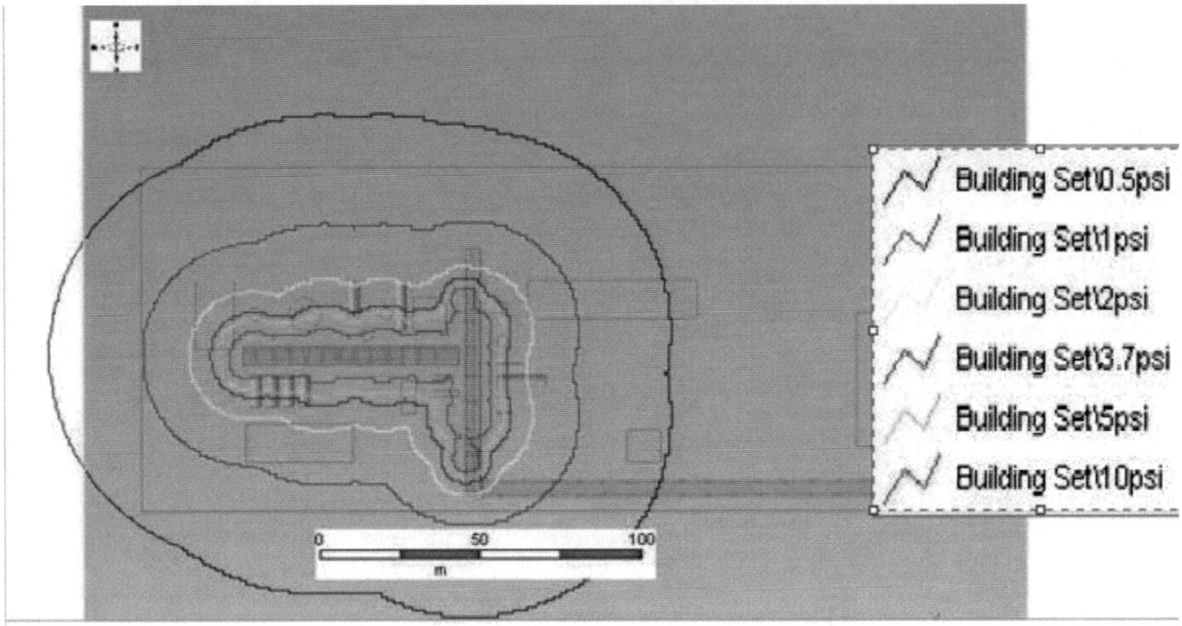

Figure 7: 9 kg/s release Ethane

Figures 8 and 9 present results for release rates of 10 kg/s and 15 kg/s, respectively. These demonstrate severe impact to the buildings and a proper retrofit would be required. These releases would produce overpressure and impulse significant enough to cause severe damage, dependent on the building type used in the facility.

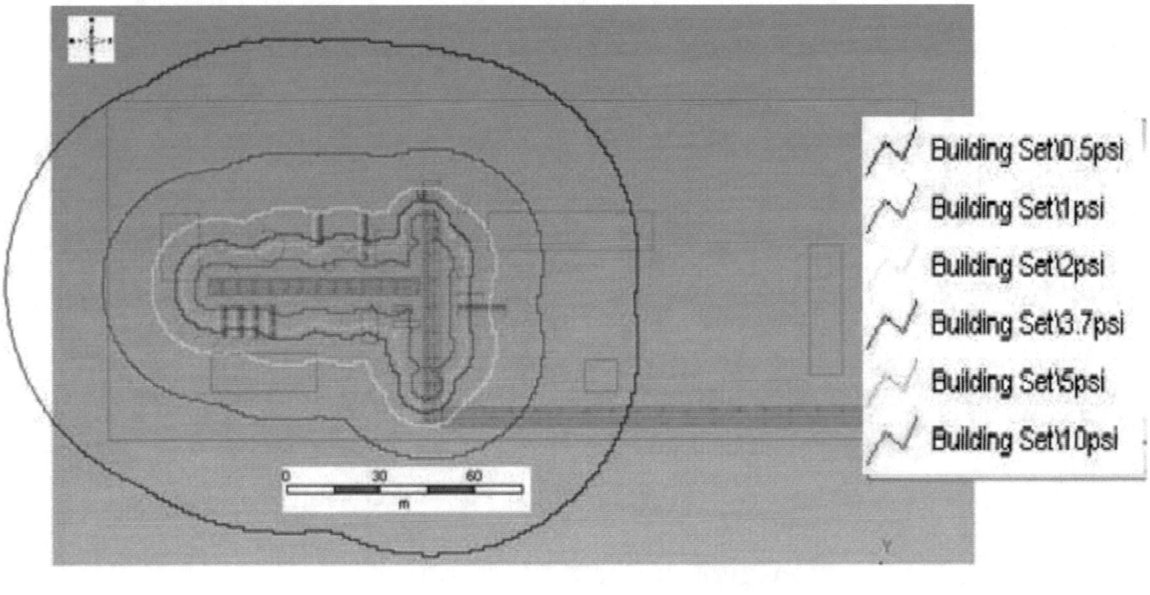

Figure 8: 10 kg/s release of Ethane

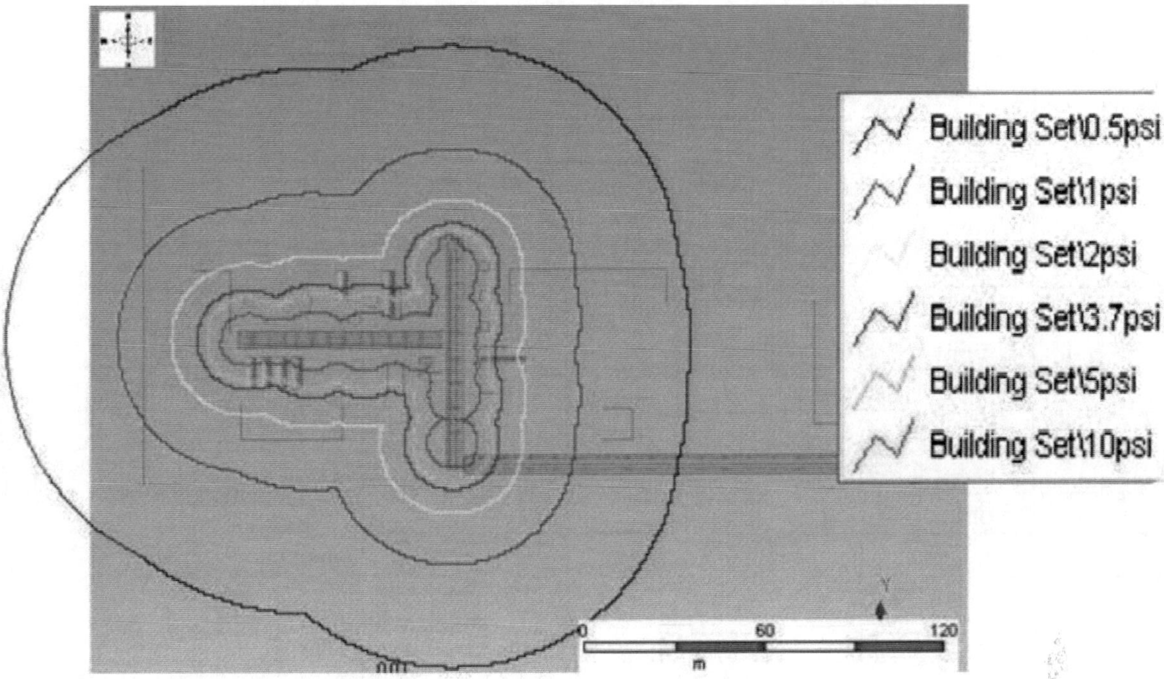

Figure 9: 15 kg/s release Ethane

3.2 Scenario 2 – Propane Releases

A series of Propane simulations were conducted as described in Section 2.2 with different release rates that led to different cloud volumes and explosion locations.

From these simulations, explosions were modeled in the congested regions and the flammable mass that could reach the congested volumes was determined.

The analysis was conducted explicitly to determine how much of the release could be found in the congested region and contribute to the explosion energy and lead to overpressure. The main findings are listed in Table 6 for each monitor point.

Table 6: Results Summary for Propane Scenarios

Release Rate (kg/s)	Monitor Points	Overpressure (psi)	Overpressure (bar)	Impulse (N·s/m²)	Flammable Cloud Volume (m³)
0.1	East Wall CR	NA	NA	NA	0.04
0.1	Center CR Roof	NA	NA	NA	0.04
0.1	Admin Building	NA	NA	NA	0.04
0.5	East Wall CR	NA	NA	NA	0.2
0.5	Center CR Roof	NA	NA	NA	0.2
0.5	Admin Building	NA	NA	NA	0.2
1	East Wall CR	NA	NA	NA	3
1	Center CR Roof	NA	NA	NA	3

Release Rate (kg/s)	Monitor Points	Overpressure (psi)	Overpressure (bar)	Impulse (N·s/m^2)	Flammable Cloud Volume (m^3)
1	Admin Building	NA	NA	NA	3
2	East Wall CR	0.55	0.04	11.59	6
2	Center CR Roof	0.42	0.03	6.75	6
2	Admin Building	NA	NA	NA	6
3	East Wall CR	0.70	0.05	17.2	12
3	Center CR Roof	0.38	0.03	9.9	12
3	Admin Building	NA	NA	NA	12
4	East Wall CR	0.81	0.06	21.2	21
4	Center CR Roof	0.42	0.03	12.15	21
4	Admin Building	NA	NA	NA	21
5	East Wall CR	1.05	0.07	26.95	33
5	Center CR Roof	0.49	0.03	14.52	33
5	Admin Building	NA	NA	NA	33
7	East Wall CR	1.60	0.11	49.64	61
7	Center CR Roof	0.65	0.05	23.03	61
7	Admin Building	0.23	0.02	10.43	61
9	East Wall CR	1.91	0.13	66.2	94
9	Center CR Roof	0.77	0.05	31.02	94
9	Admin Building	0.28	0.02	13.91	94
10	East Wall CR	2.03	0.14	72.87	110
10	Center CR Roof	0.81	0.06	34.1	110
10	Admin Building	0.29	0.02	15.26	110
15	East Wall CR	2.84	0.20	117.6	234
15	Center CR Roof	1.02	0.07	48.8	234
15	Admin Building	0.33	0.02	20.5	234
20	East Wall CR	3.77	0.26	178.8	387
20	Center CR Roof	1.26	0.09	70.87	387
20	Admin Building	0.39	0.03	25.29	387

Another important part of the analysis was defining generic overpressure contours generated from specific leak rates in order to define the threshold for cloud size generated for the overall facility rather than for a specific building.

Figures 10 to 17 provide an overall guidance on the vicinity of the congested regions and were used to determine the cloud size that would impact the area around the explosion source. The figures will provide overpressure contours for the simulations conducted.

Figure 10 presents the results for the 0.1 kg/s release rate and demonstrates that rates this small impact only the areas around the release and around congested regions. This simulation did not generate overpressure above 2 psi.

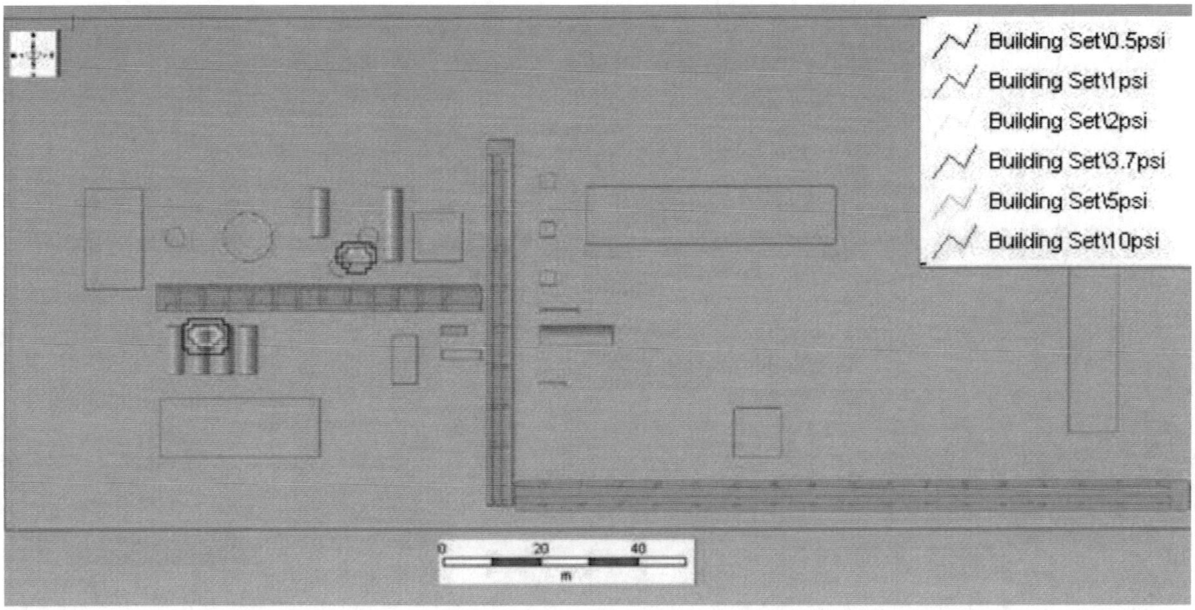

Figure 10: 0.1 kg/s release Propane

Figure 11 presents the results for the 1.0 kg/s release rate and demonstrates that rates as small as those are impacting the areas around the explosion region. Higher overpressures were registered for this simulation, but the impact was still only around the explosion source, which as previously described would likely require a more refined model in order to produce robust results.

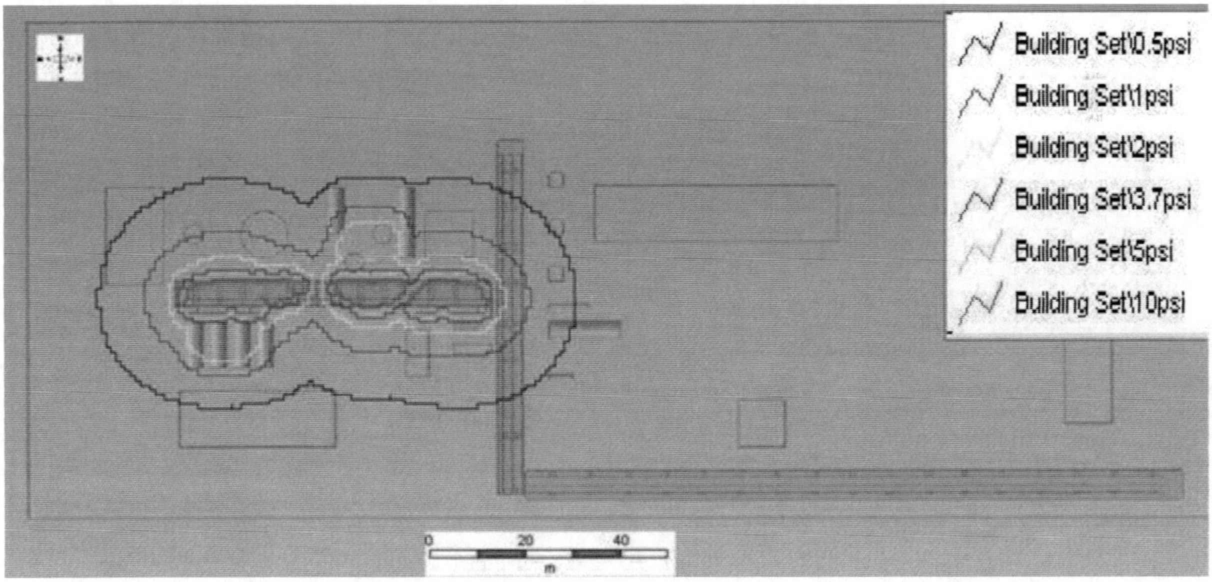

Figure 11: 1kg/s release Propane

Figure 12 presents the results for the 3.0 kg/s release rate and demonstrates that those rates start to produce some overpressure on the building of concern, but the overpressure and associated impulse is not significant enough to cause damage to the structure and compromise the personnel located inside.

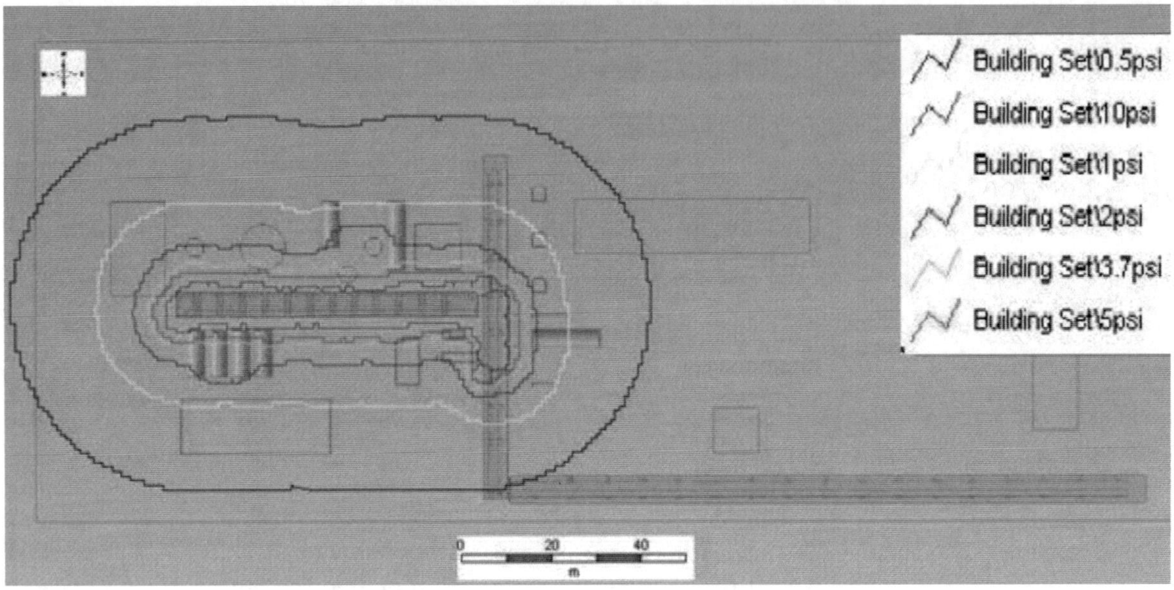

Figure 12: 3 kg/s release Propane

Figure 13 presents the results for the 5.0 kg/s release rate and demonstrates that those rates start to produce overpressures above 1 psi on the building of concern. These values are still not an issue for the occupants, but the building would need to have appropriate windows, frames and other reinforcements.

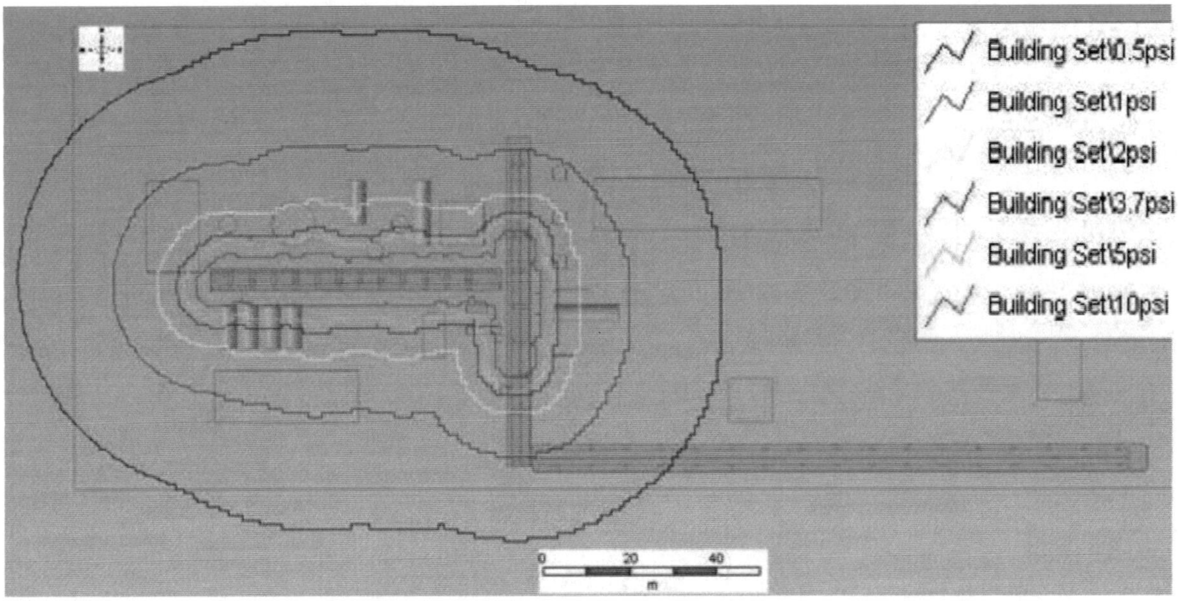

Figure 13: 5 kg/s release Propane

Figure 14 presents the results for the 7.0 kg/s release rate and demonstrates that those rates start to produce overpressures around 2 psi on the building of concern. These buildings would require reinforcement, at a significant cost, in order to sustain overpressures generated from those scenarios.

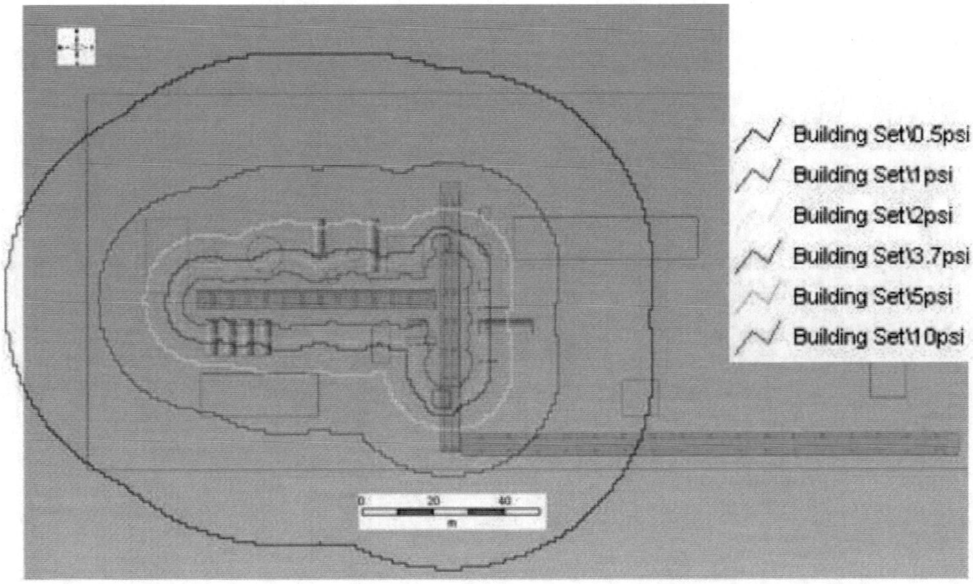

Figure 14: 7 kg/s release Propane

Figure 15 presents the results for the 9.0 kg/s release rate and demonstrates that those rates also produce overpressures around 2 psi on the building of concern. Again, these buildings would require reinforcement, at a significant cost, in order to sustain overpressures generated from those scenarios.

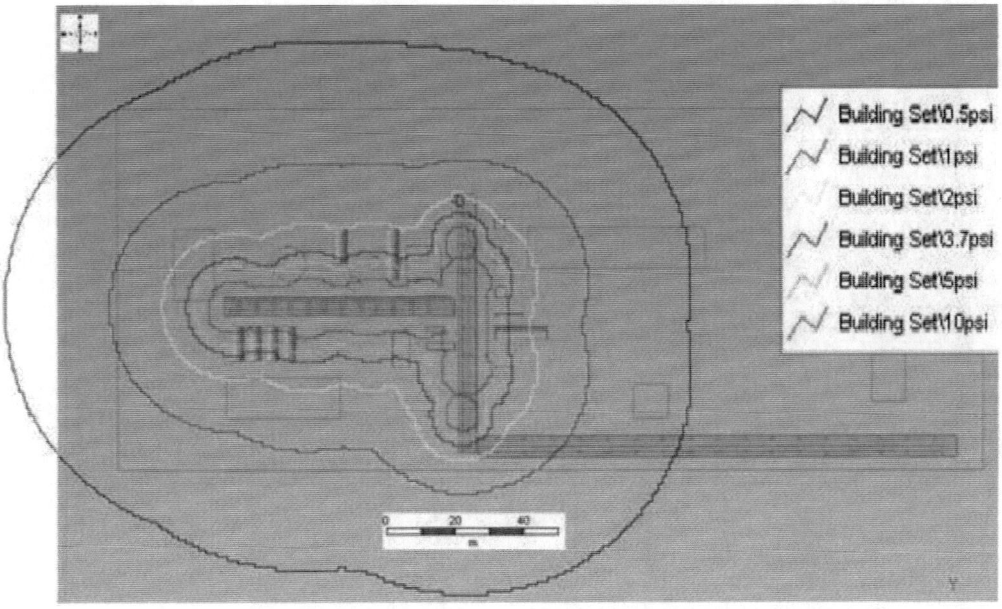

Figure 15: 9 kg/s release Propane

Figures 16 and 17 present results for the release rates of 10 kg/s and 15 kg/s, respectively. Those demonstrate severe impact to the buildings and a proper retrofit would be required. Those releases would produce overpressure and impulse significant enough to cause severe damage dependent on the building type used on the facility.

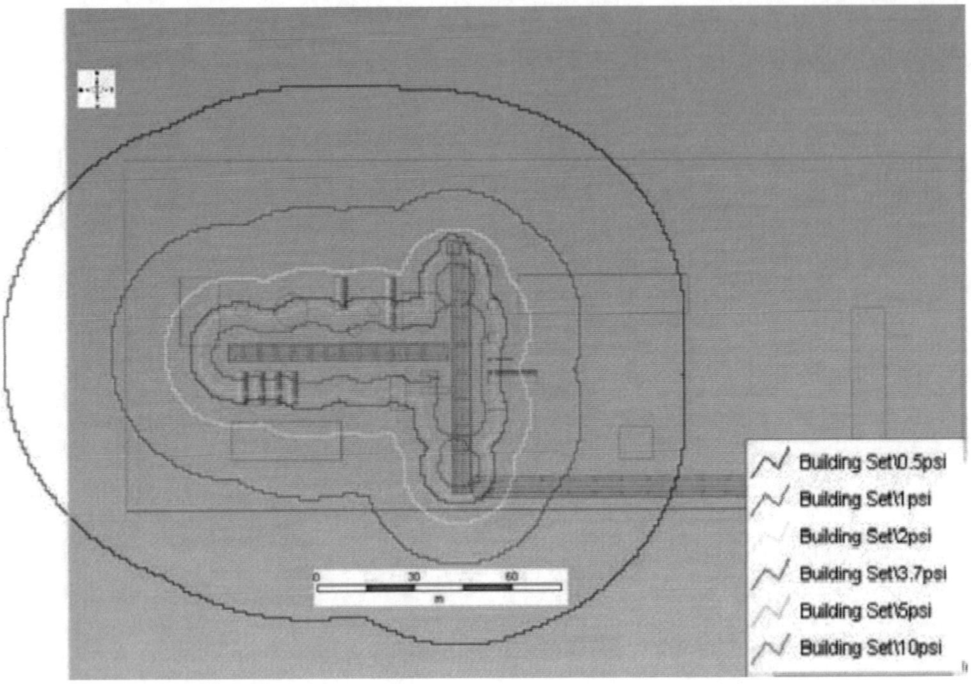

Figure 16: 10 kg/s release of Propane

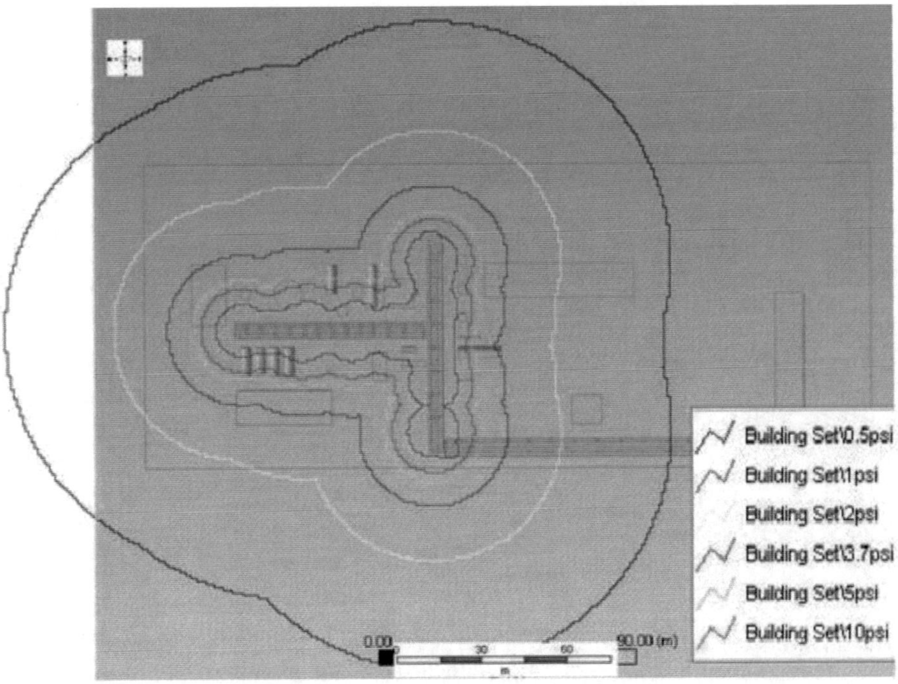

Figure 17: 15 kg/s release Propane

4 Discussion

Several conclusions could be drawn from the study. The first conclusion is that the material being released will be of importance. Propane releases seem to be forming larger clouds and lead to larger impacts to the occupied buildings studied, as expected. Another important conclusion is that there seems to be a threshold for the cloud sizes that we should be focusing on for our detection system when it comes to overpressure impact on buildings. The major conclusion that can be drawn from this study is that releases under 5 kg/s or releases which will form flammable clouds with volumes below 35 m^3 will have a local impact on the congested regions and not the areas outside of the explosion source. The releases below 5 kg/s will produce overpressure below 1 psi and impulses below 27 $N{\cdot}s/m^2$. Most of the regular constructions currently used in the onshore facilities will be able to withstand this load without significant damage to property or personnel. Furthermore, the releases that would lead to cloud sizes above the 35 m^3 threshold are produced from releases of 5 kg/s and above.

To determine the load that would damage the occupied buildings, the authors have assessed two typical building construction types widely used in the Unites States. This consists of Wood Frame and Unreinforced Concrete Masonry Unit (CMU). The authors have plotted the results from the simulations, described on Sections 3.1 and 3.2, using Pressure and Impulse (P-I) curves for each of those buildings types. The results can be found in Figures 18-21 for both types of buildings and releases.

The authors have used the Building Damage Level (BDL) 1 curve to determine the damage threshold of interest, which means the buildings exposed to this level will not present significant damage and will not require reinforcement. The definitions used in Table 7 are taken from a previous publication – Vulnerability Model for Occupants of Blast Buildings [1].

Table 7: BDL Definitions

Building Damage Level (BDL)	Damage Description
1	Onset of visible damage to reflected wall of building.
2.0 (2A)	Reflected wall components sustain permanent damage requiring replacement. Other walls and roof have visible damage that is generally repairable.
2.5 (2B)	Reflected wall components are collapsed or very severely damaged. Other walls and roof have permanent damage requiring replacement.
3	Reflected wall has collapsed. Other walls and roof have substantial plastic deformation that may be approaching incipient collapse.
4	Complete failure of the building roof and a substantial area of walls.

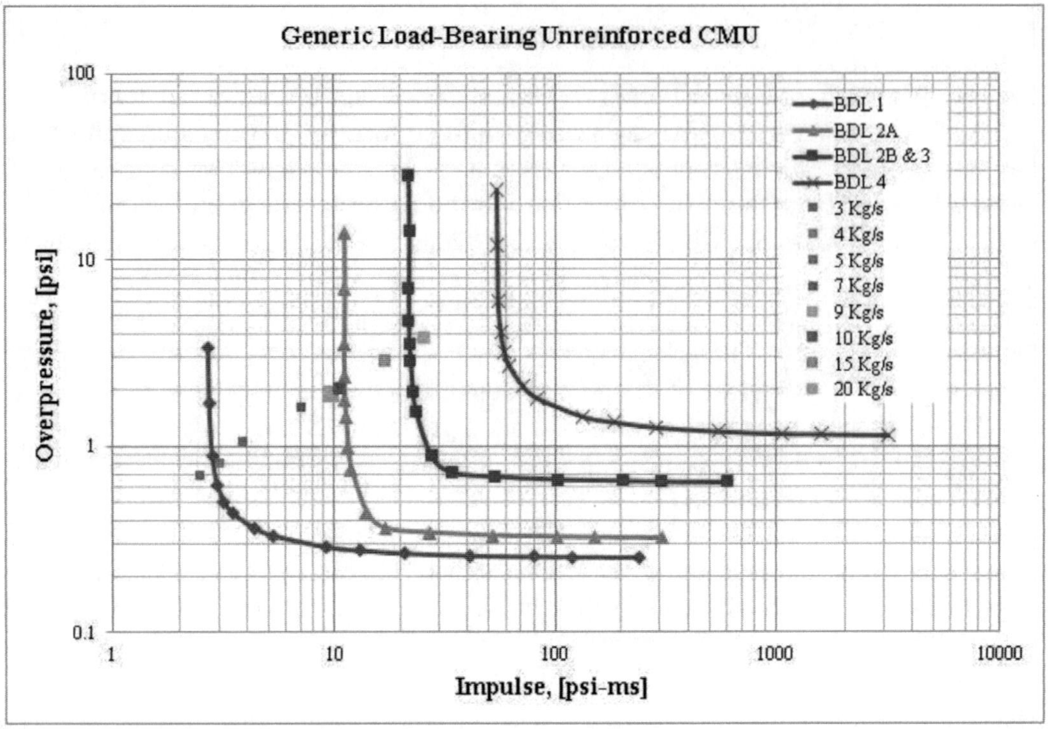

Figure 18: P-I Results for Propane – Unreinforced CMU

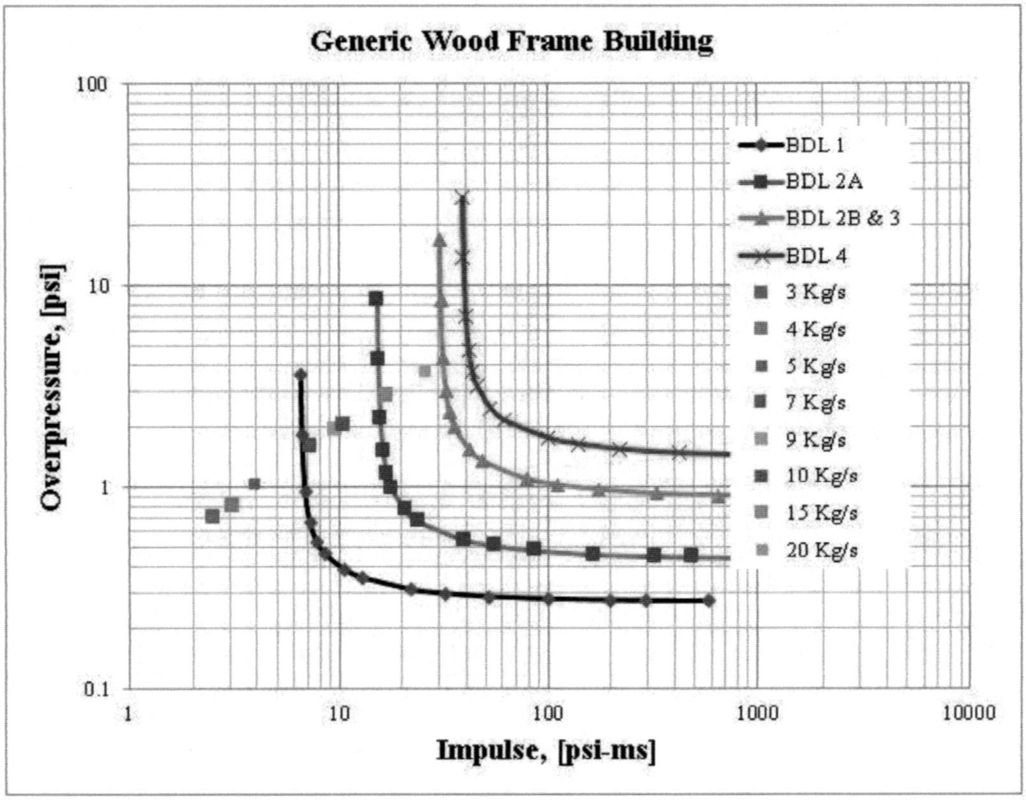

Figure 19: P-I Results for Propane – Wood Frame

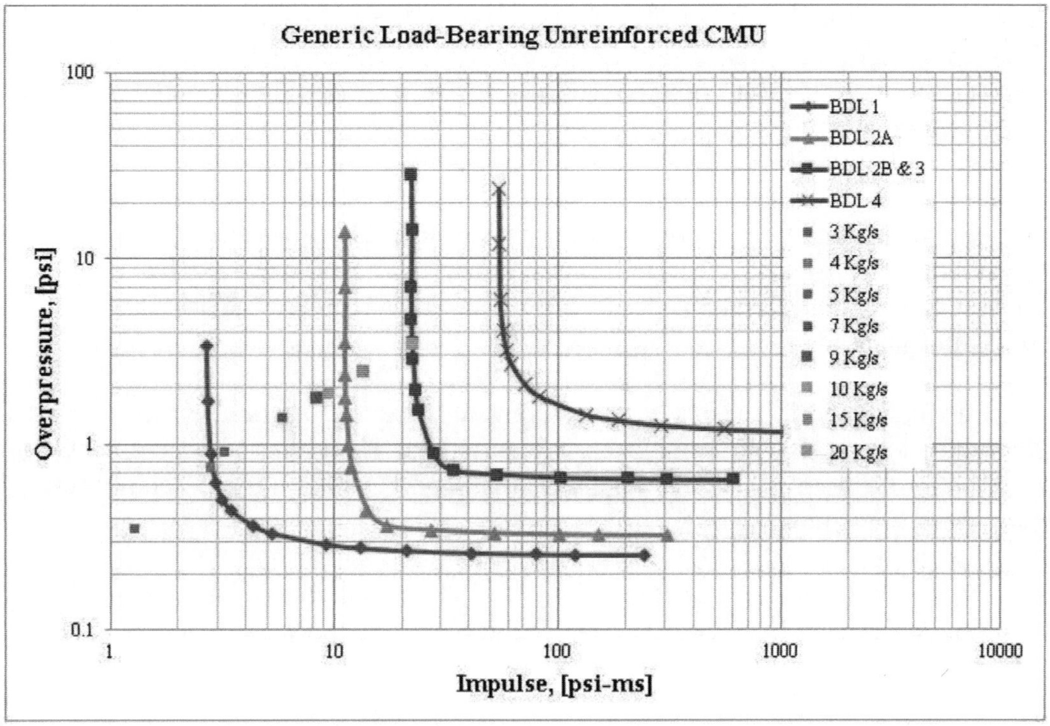

Figure 20: Results for Ethane – CMU

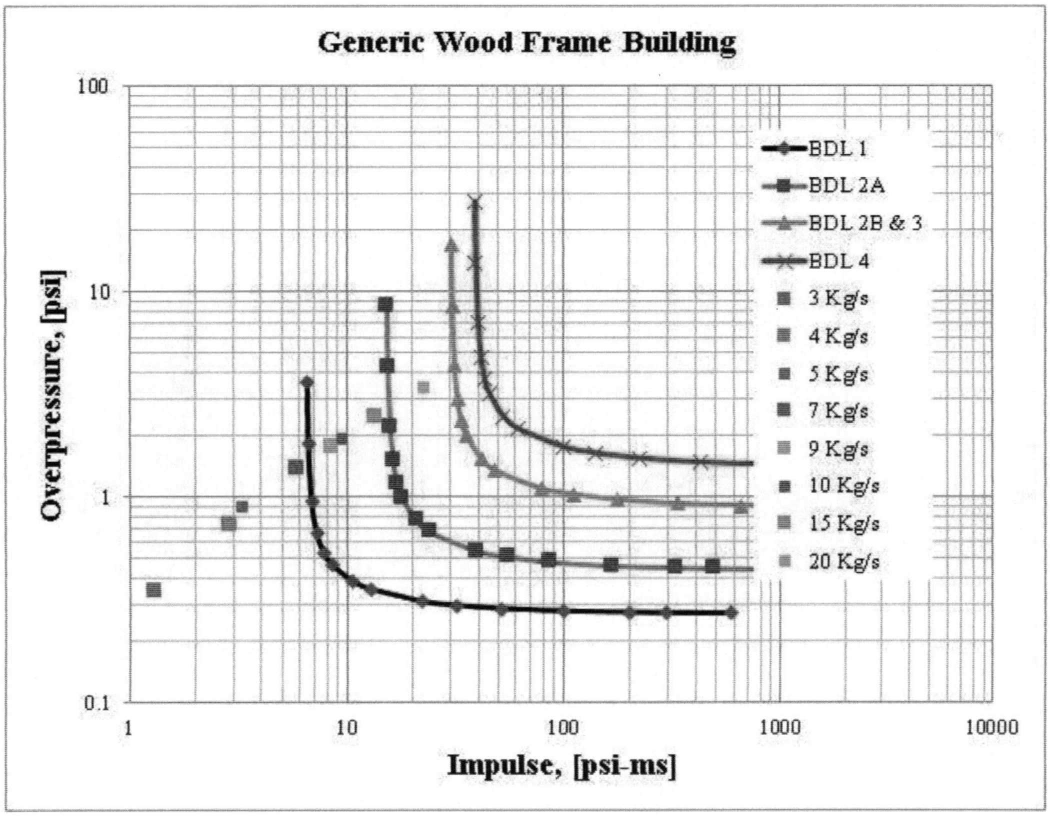

Figure 21: Results for Ethane – Wood Frame

Using the defined building damage levels for building assessments, the conclusion is that any cloud smaller than 35 m^3 would not produce damage to the typical construction types normally used for onshore facilities and would not present a concern to the building integrity and personnel inside the building itself.

An important point to be highlighted is that if one is concerned with damage to equipment within or very close to the explosion source, the study should be reevaluated with a more robust and refined model. The authors also conclude that the reinforcement of the buildings will need to be drawn from the Facility Siting Study, and this could allow, depending on the facility, a higher load on their occupied buildings.

As stated initially, the models used are normally reasonably accurate for far-field explosion and dispersions, but they should be used with caution on the near-field. The authors understand the model limitations for near-field explosions and highlight the importance of a more accurate model to be used for those predictions.

The authors would also like to highlight two very important factors regarding Computational Fluid Dynamic (CFD) cloud sizes when compared with Gaussian Dispersion Models and Gas Detector Placements.

First the authors would like to emphasize their understanding that a more refined simulation could lead to smaller flammable clouds. This is discussed in further detail in a previous publication titled "Evaluation beyond a consequence-only approach: When is it really needed?" [2]. In this publication the authors demonstrated that CFD modeling reduces the size of the clouds being estimated in the Gaussian models, even without significant obstacles. A similar facility was simulated and the modeling was carried out and explained in the paper. Figure 22 presents the difference in the cloud volumes obtained when comparing CFD to Phast results.

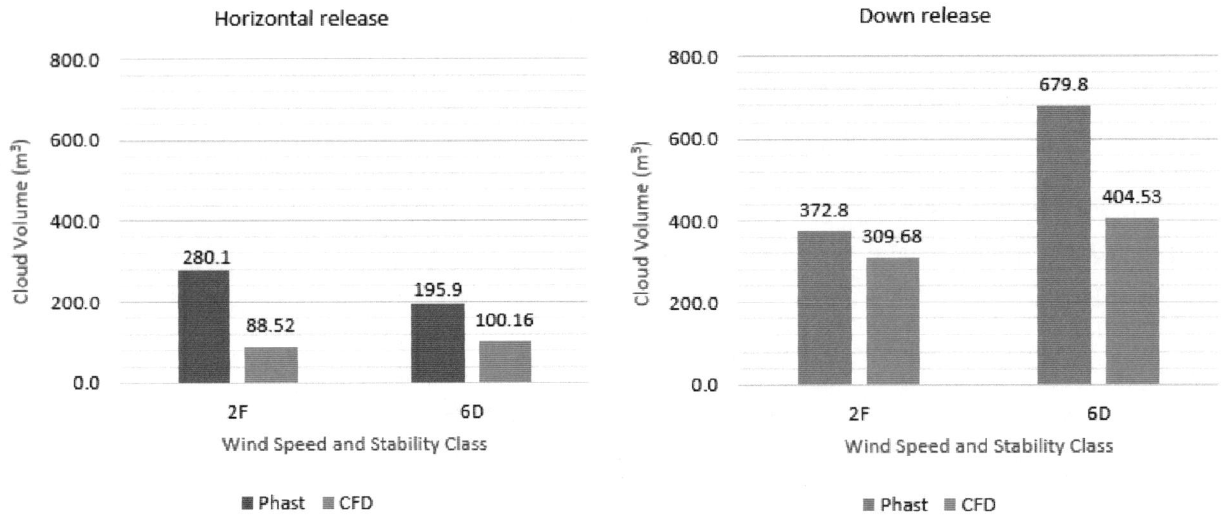

Figure 22: Cloud Volumes Phast vs CFD

This could lead others to question the need to detect even 5 kg/s releases, as the clouds would probably be reduced using CFD models. This topic is addressed in a separate publication, "Optimi – A Novel 3D Computational Tool for Gas Detector Optimization" [3]. In this separate publication, the authors ran a series of CFD simulations with 5 kg/s release rates which led to leak locations and cloud sizes to determine the characteristic length (or grid cells size) that should be applied for the gas detector optimization. For this approach, the authors have proposed that from the range of clouds generated we want to ensure detection of the smaller cloud threshold for the series of clouds being generated.

As the authors have highlighted in the Gas Detector Optimization publication [3], there were a series of CFD simulations with several wind and release directions that were modeled. The series of releases provide a wide range of cloud volumes ranging from 9 m^3 to 59 m^3, depending on the wind and release direction. For this analysis, it is critical to highlight that a very large recirculation zone was created behind the process area which causes some of the clouds to be accumulated depending on the direction of the release and wind conditions. Based on the results from the Gas Detector Optimization publication [3], the placement of detectors using a 5 kg/s release rate seems to provide the optimal point of detection, which aligns with the proposed approach of this paper based on the cloud volumes for the explosion assessment. In the next couple of figures, the layout and cloud volumes from the Gas Detector Optimization paper [3] are presented.

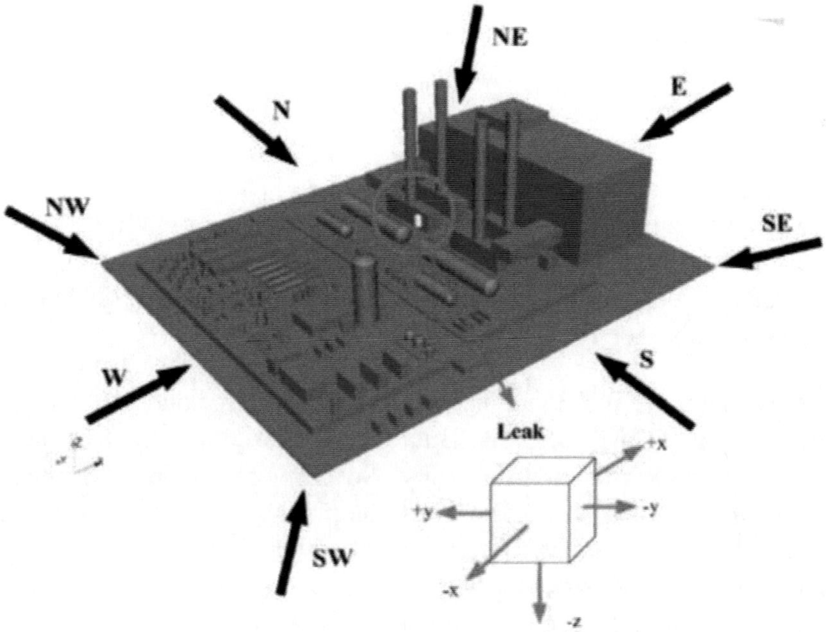

Figure 23: Sample Facility Layout

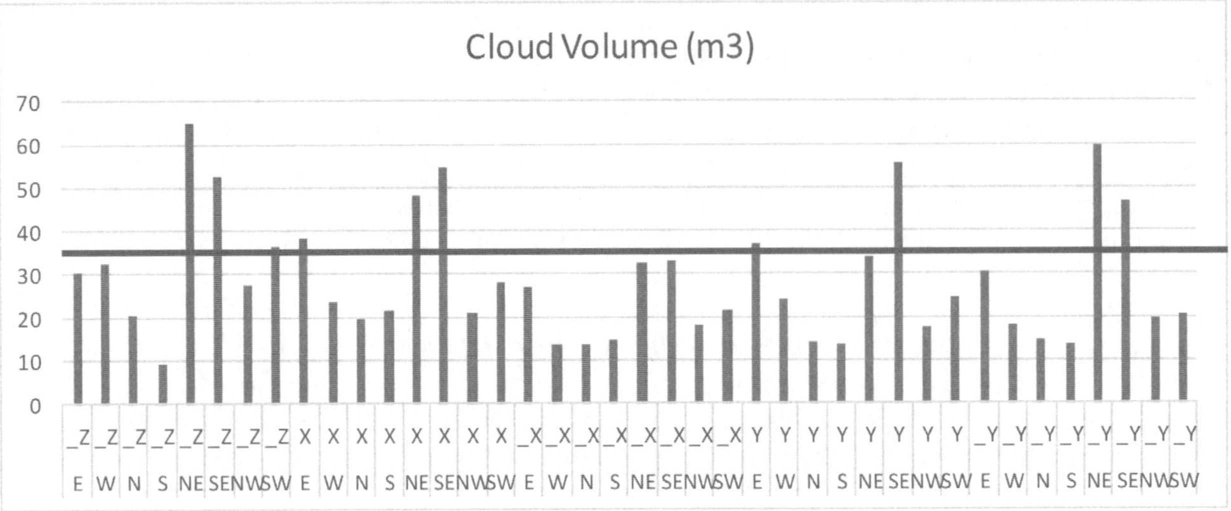

Figure 24: 5 kg/s CFD Cloud Volume

The red line on the graph above indicates the threshold proposed for this paper of 35 m^3. As indicated in Figure 23 and 24, some of the releases would have clouds above the suggested threshold of 35 m^3. It should be noted these releases are upwind releases, which were not modeled by the authors for this analysis due to limitations in the model. In addition, the cloud would be concentrated around the area of the release and upwind of the release rather than towards the congested regions.

5 Conclusion

From the simulations that were carried out, it's reasonable to conclude that there is certainly a threshold that one should focus on in order to place their gas detectors based on cloud volumes that produce significant damage to structures and buildings. The release rates of 5kg/s, or more specifically the cloud volume of 35 m^3, were determined to be the minimum threshold that would cause damage from a vapor cloud explosion, which would compromise facility structures and possibly lead to casualties. It is concluded that this minimal threshold should also be used to determine the optimal location for the gas detector.

It's expected that other materials will have similar or even larger cloud size thresholds based on their average or low reactivity, e.g. Methane, Butane. It's also important to note that the simulations conducted were specific to this facility. So it's very important to highlight that the authors are suggesting that caution should be used when applying a spacing criteria, and the main idea was not to determine a single point to be used but rather focus on a specific range to which one should look into. The conclusion is that very small releases will not significantly impact the structures and/or personnel, and the focus was exactly to try to determine a range in which one should focus their simulations on in order to achieve the proper spacing required for their detectors. The authors do believe a proper consequence or QRA model should be used in order to determine the appropriate separation for the gas detector spacing in each facility based on the importance of both the material being released and the configuration of the facility.

6 References

[1] Oswald, C. J., Baker, Q. A., "Vulnerability Model for Occupants of Blast Damaged Buildings", 34th Annual Loss Prevention Symposium, 2000.

[2] Vianna, S. S. V., Sofyan S., Chambliss S., Simoes V., Kolbe M., "Evaluation beyond a consequence-only approach: When is it really needed?", Mary Kay O'Connor Process Safety International Symposium, 2016.

[3] Vianna, S.S. V., Fiates, J., Simoes, V., "OPTIMI – A Novel 3D Computational Tool for Gas Detectors", Mary Kay O'Connor Process Safety International Symposium, 2015.

481423 Reinvention of Basf's Butadiene Process - It's All about Reliability

Abstract:

A reliable plant should at all time produce its products at desired quantity and quality. Looking into statistics, the root cause unplanned shut downs, reduced capacities or off-spec product are mainly problems with utilities, feedstock and operation. There are problems where process design cannot help, e.g. utility failures and lack of feedstock. Typical problems in butadiene service caused by low quality feedstock, insufficient protection from oxygen ingress into the process and operational upsets can be reduced by process design features. Though already having best features to ensure reliability, BASF has overhauled its process to further increase on stream times. The paper demonstrates, how by simple means as well as by sophisticated new process arrangements it was possible to control problems with poor feedstock, improve and simplify oxygen management, simplify operations and provide new type of alert system at increasing fouling risk.

Stephen Caskey
Jay Gorawara
Honeywell UOP

New High Capacity Adsorbents for Oxygenate, Nitrogen, and Sulfur Compound Removal from Olefinic C_4 Streams

Honeywell UOP

March 29, 2017 | 2017 AICHE Spring Meeting | San Antonio, TX

UOP 7828D-0

Agenda

- C_4 Sources
- Adsorbent Fundamentals
- Adsorbent Reactivity Concept
- Oxygenate and Contaminant Removal from C_4 Olefin Containing Streams
- Enhanced Contaminant (Sulfur Compound, Oxygenates) Removal from Paraffinic Hydrocarbons
- Nitrile Removal from C_4 Olefin Containing Streams
- Summary

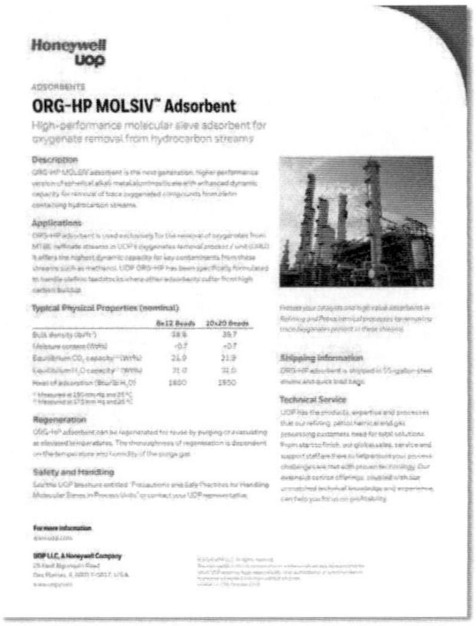
Honeywell UOP

ADSORBENTS

ORG-HP MOLSIV™ Adsorbent

High-performance molecular sieve adsorbent for oxygenate removal from hydrocarbon streams

Honeywell UOP

UOP 7828D-1

Refinery – Petrochemical Integration

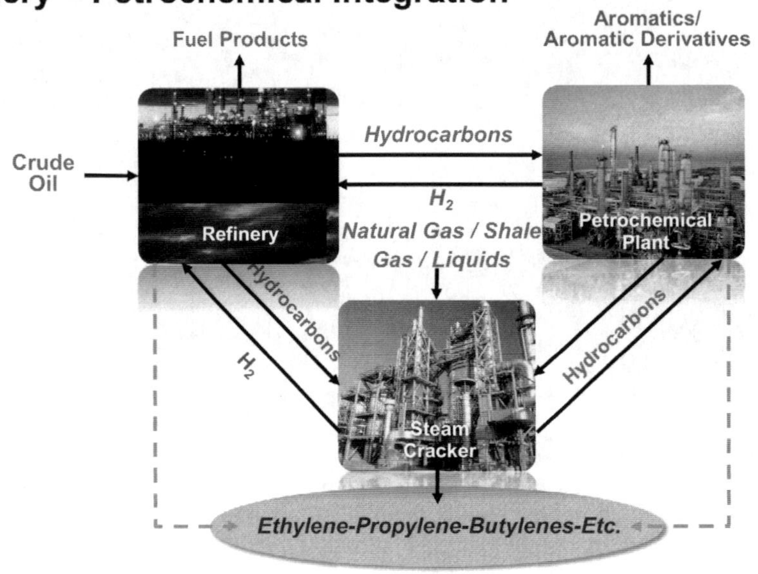

Honeywell
UOP

UOP 7828D-2

Global C_4 Olefins Volumes by Process Type

Crude C_4 yield, total 132 million ton / +4 % per annum (2014)

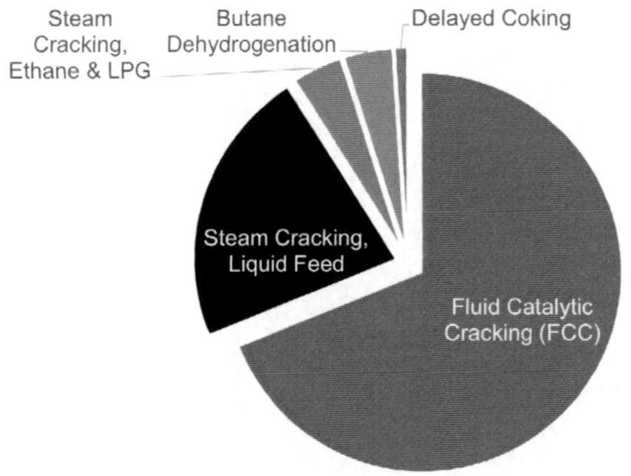

Bender, M. "An Overview of Industrial Processes for the Production of Olefins – C_4 Hydrocarbons," ChemBioEng Rev, 1, 2014, pp. 136 – 147.

FCC C_4s Are the Major Source of Crude C_4s

4

Relative Volumes of C_4 Olefins and Key Applications of 1-Butene

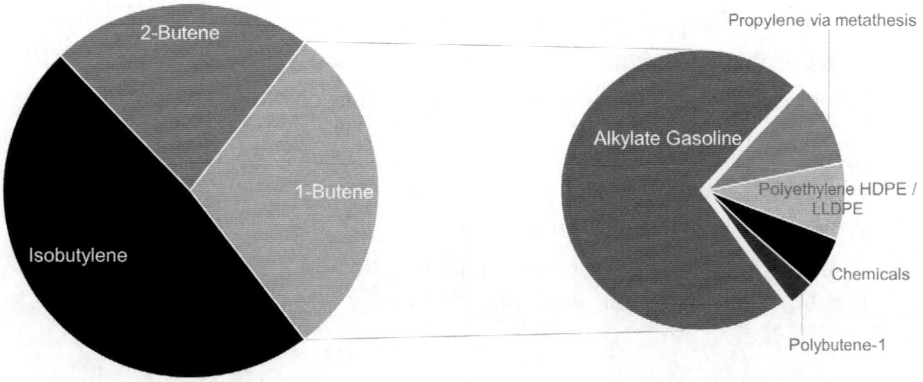

Relative Volumes of C_4 Olefins
Total = 62 million ton

Application Usage of 1-Butene by Volume

Bender, M. "An Overview of Industrial Processes for the Production of Olefins – C_4 Hydrocarbons," ChemBioEng Rev, 1, 2014, pp. 136 – 147.

Dominant Use of C_4s Remains Alkylate Gasoline but Petrochemical Applications Grow

5

Use of Adsorbents in Olefin Production

What tools do we have to do separations?

- Distillation
- Absorption
- Phase changes / cryogenic separations
- Membranes
- Adsorption

Benefits of using adsorbents

- Are a cost effective (and sometimes the only) way to remove trace levels of contaminants from process streams
- Improve product stream quality to meet specifications
- Protect valuable catalysts

Honeywell UOP

UOP 7820D-5

6

What is Adsorption?

- Adsorbed material concentrates on a solid surface
  - Can be physical adsorption – as in with molecular sieves
  - Or chemical interaction on surface (reversible or irreversible)
- Occurs spontaneously and evolves some heat
  - A process similar to condensation
  - Depending upon the adsorbent selected, this heat rise may need to be managed to prevent an unstable situation or a temperature run-away

Honeywell
UOP

UOP 7828D-6

7

A Complete Range of Contaminant Removal Solutions

Molecular Sieves

- Use of a 3A type zeolites enables exclusion of reactive species is very effective for dehydration
- Other contaminants may need a larger pore (such as a 4A, 5A, or 13X type) which may adsorb a reactive carrier stream thus a preload step may be required
- Careful selection of base material helps control high surface energy and active sites to meet adsorption requirements in reactive streams

Hybrid Adsorbents

- Engineered homogenous combination of a promoted alumina and a large pore zeolite
- Combines high selectivity for light acid gases with low olefin reactivity (low heat of adsorption of hydrocarbons) and high capacity for polar molecules makes these products ideally suited for purification of olefinic streams

Promoted Aluminas

- Activated aluminas are synthetic transitional phase aluminas with a tailored pore structure to facilitate rapid chemisorption, resulting in a short mass transfer zone
- Low affinity for reactive hydrocarbons, thus a preload step not needed
- Light acid gases such as CO_2, H_2S & COS are removed by chemisorption to very low concentrations

Metal Oxides & Sulfides

- Critical applications for hydrocarbon streams
- Trace arsine and phosphine removal from hydrocarbon streams to protect sensitive noble metal or polymerization catalysts
- Specialty applications - such as Hg & S compounds removal
- Trace CO, O_2 scavenging from C_2H_4 and N_2

Honeywell
UOP

UOP 7828D-7

Types of Adsorbents: Molecular Sieves

Molecular sieves are crystalline solids with microscopic pores.
(The pores sizes of each type of molecular sieve are identical).

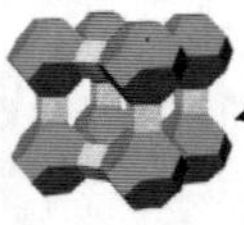

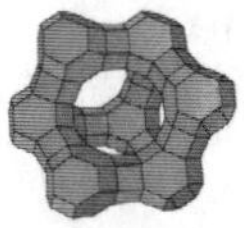

Selectivity

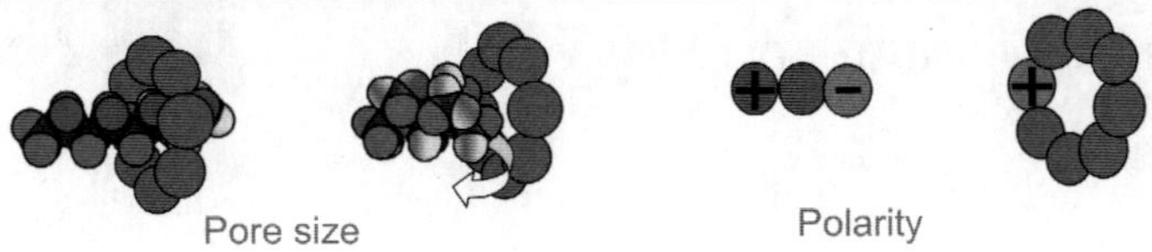

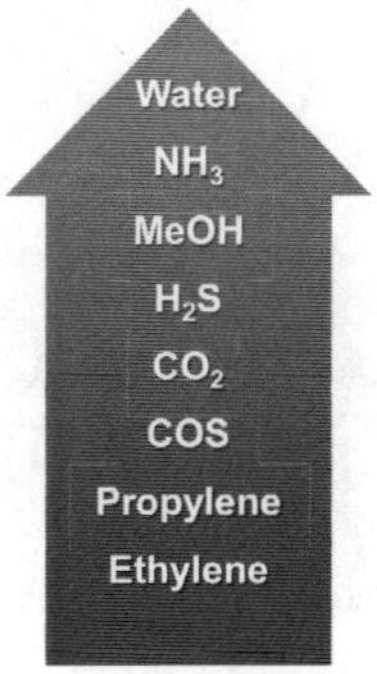

Honeywell
UOP

UOP 7828D-8

Adsorption Fundamentals

Typical Adsorption Cycle

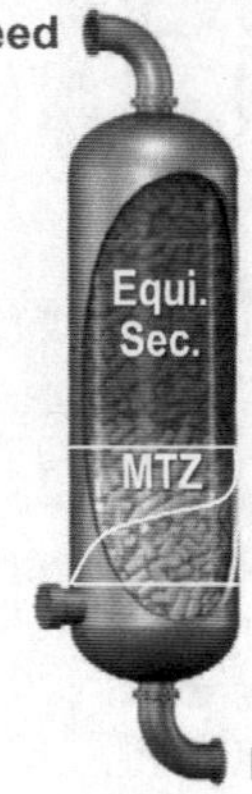

- The adsorption process moves like a wave through the adsorbent bed.
- The contaminated feed entering the adsorbent vessel becomes progressively cleaner as it flows through the bed toward the outlet.
- The inlet (equilibrium section) of the adsorbent bed eventually becomes completely saturated with contaminant.
- The lower (mass transfer section) of the adsorbent bed reduces the contaminant content of the feed in a stable front that advances through the bed.
- The product leaving the adsorbent vessel remains clean until "breakthrough" of the front occurs.

Honeywell
UOP

UOP 7828D-9

10

Multiple Adsorbates Example

- Order of Selectivity
  - H_2O
  - Oxygenates
  - Sulfur compounds
    - i.e. sulfur compounds will not displace H_2O
- Multiple Equilibrium and Mass Transfer Zones

Honeywell UOP

UOP 7828D-10

11

Typical Olefin Treating Operating Sequence

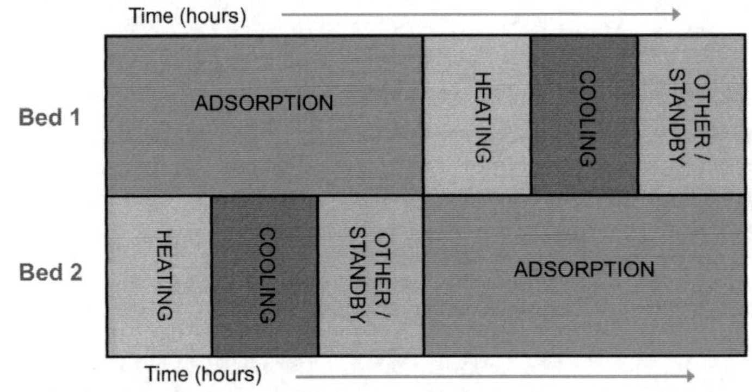

1. Adsorption process units are one of the key unsteady state processes in hydrocarbon processing
2. Continuous process operation is often achieved by using multiple adsorbers, operated in a cyclic mode; consecutive cycles of adsorption and regeneration

Honeywell UOP

UOP 7828D-11

UOP's Low Olefin Reactivity Adsorbents

Why low olefin reactivity is a "must have" for cutting edge adsorbents?

- Adsorbent's reactivity is often overlooked
  - Limited adsorbent reactivity data is available
- With undesirable reactivity, the full potential of the adsorbent may not be achieved due to the following possible events:
  - The adsorbent reacts with a component of the main stream
  - Side reactions compete with contaminant removal
  - New side products created from the main stream can further react with the adsorbent
  - Performance declines due to deactivation (coking for example)
- Safety concerns could be encountered due to undesirable exothermic side reactions or high heat of adsorption of main stream

Low Reactivity is a Key Feature of Advanced Adsorbents; Complete Inertness Towards the Main Stream is Desired

Low Reactivity Adsorbents – The Industry Standard

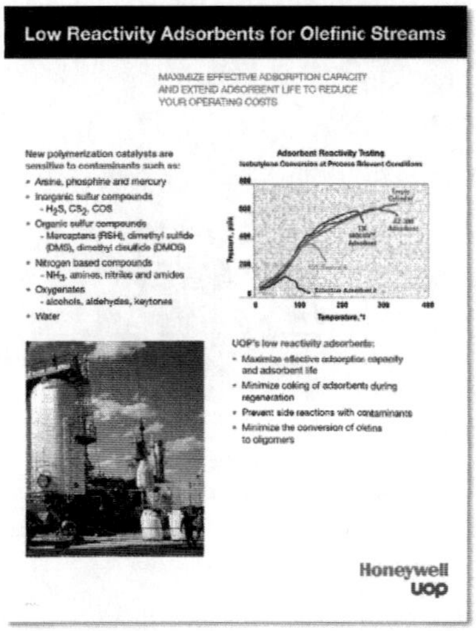

Low Reactivity Adsorbents for Olefinic Streams

MAXIMIZE EFFECTIVE ADSORPTION CAPACITY AND EXTEND ADSORBENT LIFE TO REDUCE YOUR OPERATING COSTS

New polymerization catalysts are sensitive to contaminants such as:

- Arsine, phosphine and mercury
- Inorganic sulfur compounds
  - H_2S, CS_2, COS
- Organic sulfur compounds
  - Mercaptans (RSH), dimethyl sulfide (DMS), dimethyl disulfide (DMDS)
- Nitrogen based compounds
  - NH_3, amines, nitriles and amides
- Oxygenates
  - alcohols, aldehydes, keytones
- Water

Adsorbent Reactivity Testing

UOP's low reactivity adsorbents:

- Maximize effective adsorption capacity and adsorbent life
- Minimize coking of adsorbents during regeneration
- Prevent side reactions with contaminants
- Minimize the conversion of olefins to oligomers

Honeywell UOP

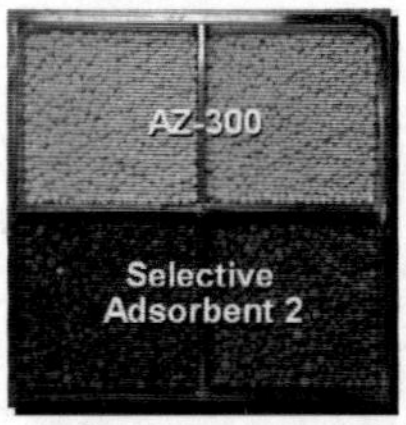

UOP Isobutylene Test - quantifies differences between products

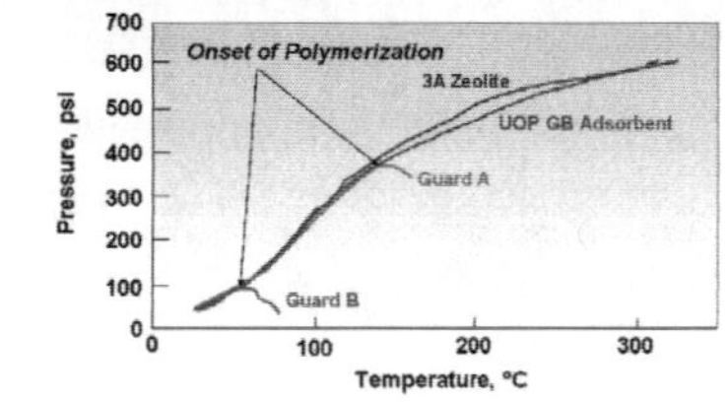

Example of a Simple Test to Probe the Reactivity of Adsorbents with Isobutylene

14

Accelerated Reactivity Testing Example

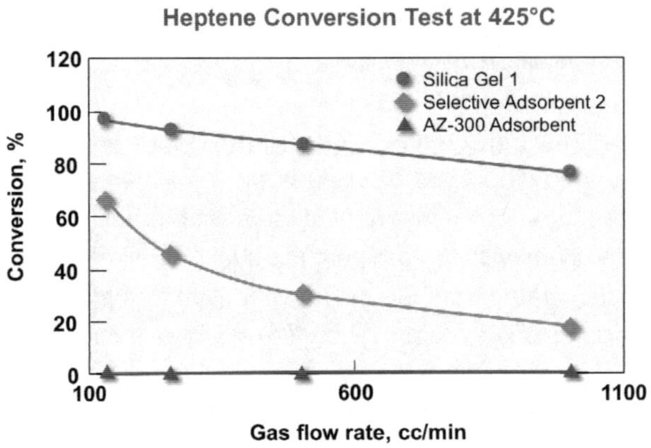

Kanazirev, V., Smith, D.L. "Adsorbent Reactivity: An Important Factor in Contaminant Removal," Proceedings of the 16th Ethylene Producers Conference, April 2004, 55d.

High Conversion is Indicative of Undesirable Reactivity

15

Adsorbent Reactivity – A Potential Safety Issue

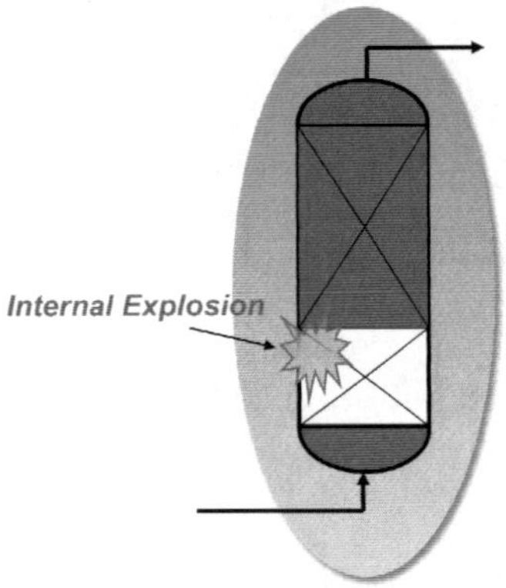

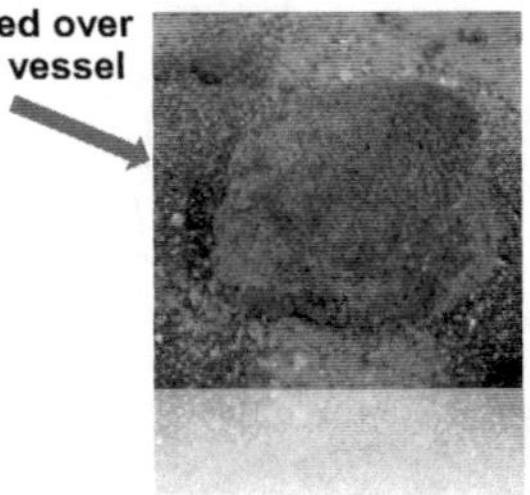

What could cause vessel rupture ?

Adsorbent Reactivity, Often Overlooked, is CRITICAL

16

Reactivity Summary

- Consider adsorbent reactivity along with capacity and other performance criteria when choosing adsorbents for contaminant removal
- Reactive adsorbents promote unwanted reactions that in some cases present safety concern
- UOP uses reactivity tests to:
  - Determine reactivity of existing adsorbents
  - Commercialize new low reactivity adsorbents

Honeywell UOP

UOP 7828D-16

17

Adsorbents in Olefin Plants

- UOP MOLSIV™ Adsorbents dehydrate olefin streams such as cracked gas and other ethylene, propylene and butylene process streams, but they offer much more:
  - Ability To Process Lower Cost Feedstock
  - Remove Additional Contaminants
  - Obtain Longer Adsorbent Life
  - Ability To Handle Process Upsets
- UOP Specialty Absorbents and Promoted Aluminas used to chemisorb trace contaminants to very low (ppb) levels necessary for Petrochemicals use

Our primary focus: Enable Reliability

UOP Adsorbents Innovation

- UOP continues to develop new technology solutions to meet industry demands
- Some of our newest adsorbent products:

Honeywell
UOP

New ORG-HP Key Features

- Oxygenate Removal Grade Product
- Enhanced adsorption capacity for contaminants for olefinic C_4 hydrocarbon feeds
- Improved diffusivity to allow ready access to contaminants
- Improved regenerability with key contaminants

ORG-HP: High Capacity and Regenerability with Key Contaminants

Basic Petrochemicals

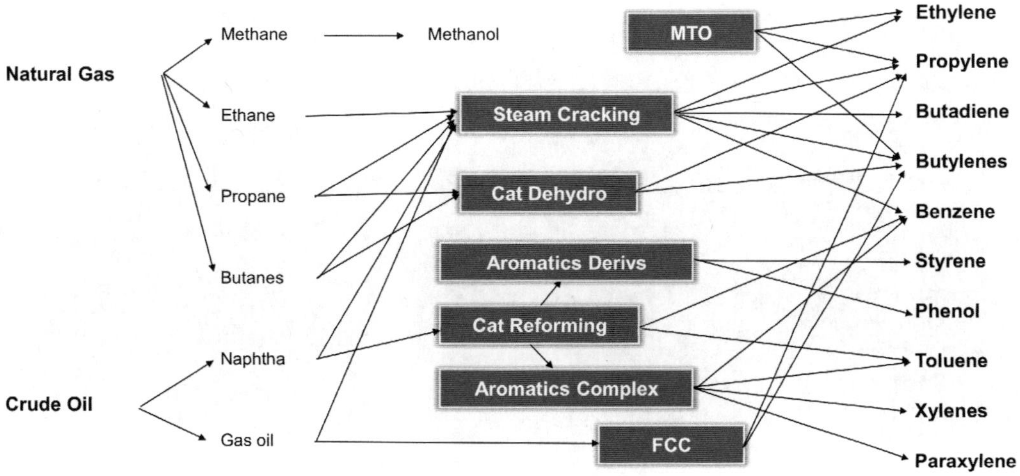

Honeywell UOP

UOP 7626D-20

Block Flow Diagram for MTBE from FCC C_4 Olefins

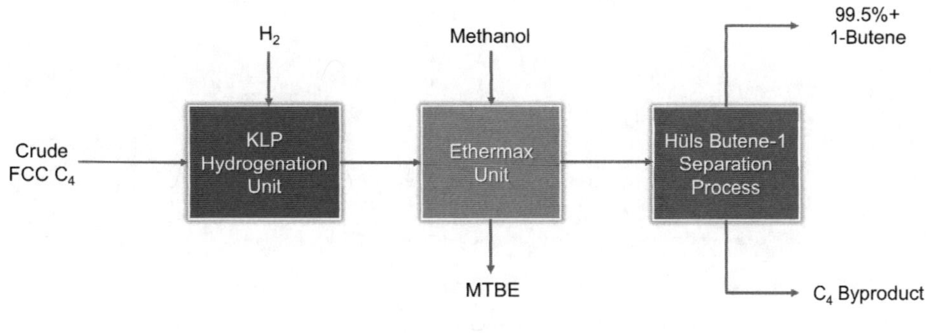

Purification is Integral to Utilization of FCC C_4s Olefins

22

Typical C_4 Fraction Composition and Contaminant Profile

Component	Unit	FCC	Steam Cracking
Isobutane	%	37	2
n-Butane	%	12	9.8
Isobutene	%	24	26
1-Butene	%	15	13.6
2-Butenes (cis and trans)	%	11	12
1,3-Butadiene	%	0.2	36
Others	%	balance	balance
Acetone	wtppm	800	50
Aldehydes	wtppm	20	5
Methyl Mercaptan	wtppm	200	5-10
Ethyl Mercaptan	wtppm	100	5-10
Disulfides	wtppm	5-10	2-5
Sulfides	wtppm	2-5	2-5
Acetonitrile	wtppm	50	5

Schulze, J. and M. Homann. "C_4 – Hydrocarbons and Derivatives," Springer-Verlag, Berlin, 1989.

FCC C_4 Feeds: Particularly Difficult to Purify Limiting Petrochemical Application

23

Petrochemical Utilization of FCC C_4s

Press release

EVONIK INDUSTRIES

July 1, 2015

Successful start of production: Evonik expands leading market positions in C_4 products

Essen/Antwerp. Evonik Industries has strengthened its leading position in C_4-based products and has successfully, and on time, put into operation new production plants in Antwerp (Belgium). Also in the Marl Chemical Park (Germany), the C_4 capacities are being increased. For this, the company has invested a total amount in the three-digit million range (Euro) in the two sites.

With long-term supply contracts, Evonik has sustainably ensured the raw material supply for the operation of the new plants. As a technology leader, Evonik has also for the first time made FCC- C_4 material flows from refineries usable. This demonstrates Evonik's technological excellence on C_4 and is an important contribution to a sustainable production.

Evonik Press Release. "Successful start of production: Evonik expands leading market positions in C_4 products," July 1, 2015.

Increasing Use of FCC C_4s for Petrochemical Applications

Potential Olefinic C_4 Purification Process

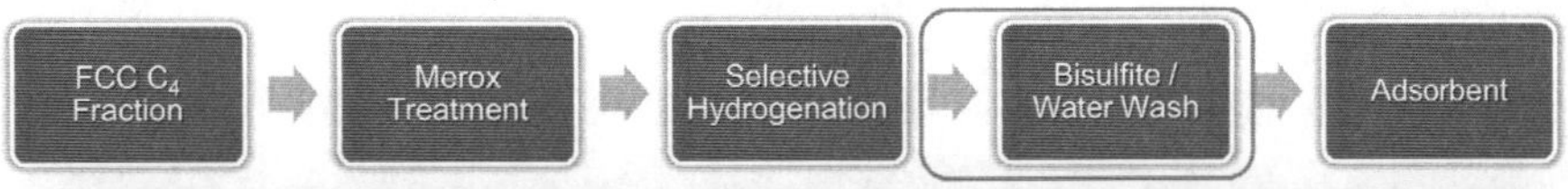

	Example % Removal for Common C_4 Purification Processes			
	Merox Treatment	**SHP**	**Bisulfite / Water Wash**	**Adsorbent**
Ketones	0%	0%	>50%	+99%
Aldehydes	0%	0%	>90%	+99%
Mercaptans	98%	100%	0%	*
Disulfides	0%	0%	0%	*
Sulfide	0%	0%	0%	*
Nitriles	0%	0%	>50%	+99%
1,3-butadiene	0%	99+%	0%	NA

*unit performance depends on design and operation

Multiple Purification Strategies Needed to Address the Challenge

New ORG-HP Exhibits Low Reactivity with Olefins

Conversion of 1-Hexene at 280°C for Prolonged Time

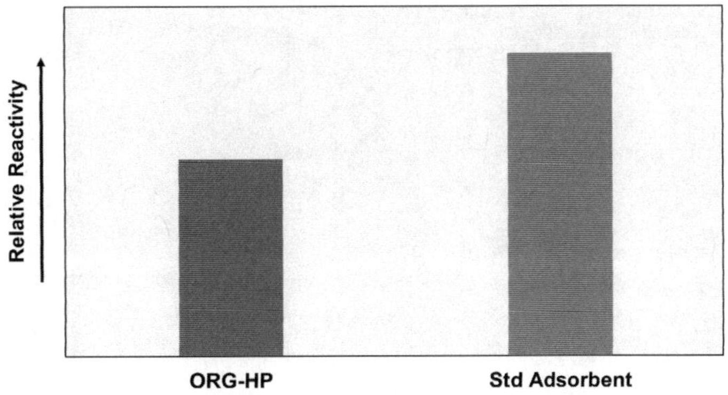

Lower conversion is a desirable indicator of less acid catalyzed reactivity with olefins

ORG-HP vs. ORG-E (Alcohol Capacity and Regenerability)

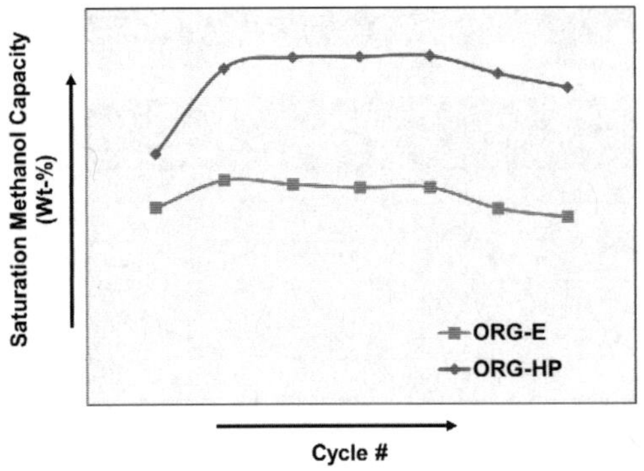

- Highly reactive, olefinic stream
- Modeling projects significant life enhancement for ORG-HP

Based on dynamic lab scale testing with a C_4 feed containing 2000 ppm MAPD & 1% isobutylene

Oxygenate (Acetone) Removal with ORG-HP

Equilibrium Data

Simulated Feed Prepared with High Olefinic Content

	% wt	ppmw
Isobutane	37	
n-Butane	12	
Isobutene	24	
1-Butene	15	
2-Butene (cis/trans)	11.8	
Isoprene (2 Me-1,3 C4==)	0.2	
Acetone		500

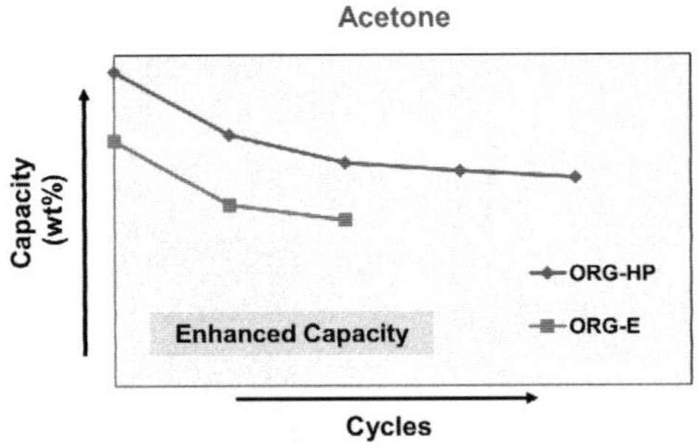

- Adsorbent used: ORG-HP
- Activated at 350°C under N_2.
- Adsorption testing at 40°C
- Regeneration temp of 283 °C with N_2 as regenerant

ORG-HP Successfully Tested Under Harsh Simulated C_4 Feed

28

New Platform of C_4 Hydrocarbon Treating Adsorbents

- ORG-HP
  - oxygenate and sulfur compound removal from olefinic C_4s
- **HPG-300**
  - **oxygenates, sulfur compound removal from paraffinic C_4s**
- ACN-300
  - nitrogen compound removal from olefinic C_4s

Honeywell
UOP

UOP 7828D-28

29

New HPG-300 Key Features

- High Performance Grade Product
- Optimized for oxygenate and sulfur compound removal
- Added pore volume for contaminants
- Enhanced pore volume to withstand harsh coking conditions

HPG-300: Optimized Pore Volume for High Oxygenate & Sulfur Capacity

UOP Oleflex/MTBE Complex Block Flow Diagram

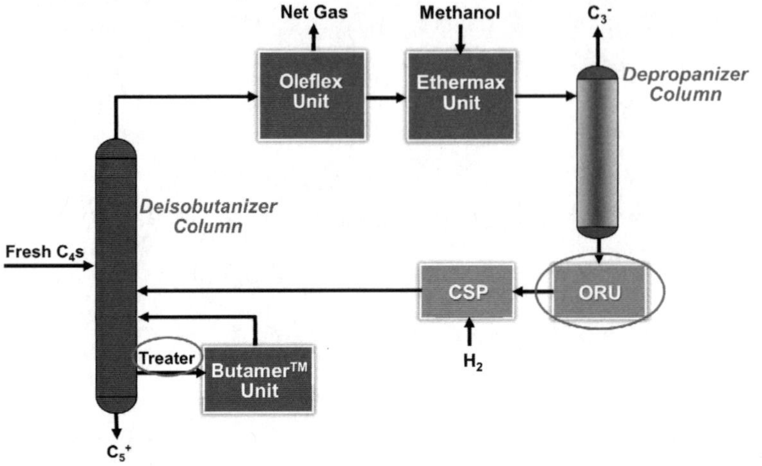

Myers, D, Funk, G, Vora, B. "Shale Gas Monetization – How to Get Into the Action," American Fuel & Petrochemical Manufacturers Annual Meeting, March 2013, AM-13-53.

C_4 Treatment Plays Key Role in Enabling Technology

UOP's Butamer Process

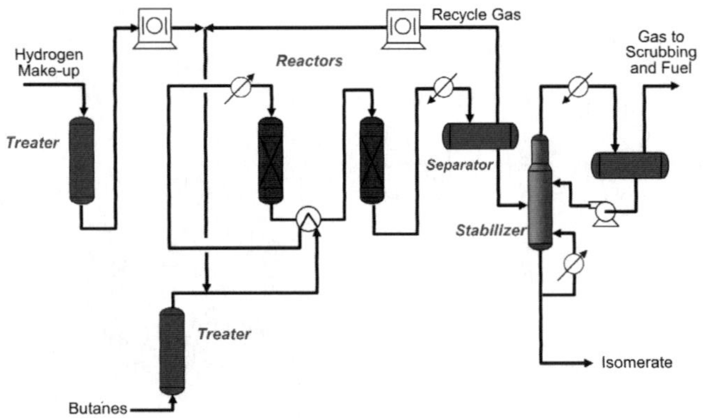

- Oxygenates are permanent poisons and sulfur compounds are a temporary poisons for Isom catalyst
- Sulfur compounds breakthrough first and are key indicator of adsorbent performance

Feed Pre-treatment Protects Isomerization Catalyst from Deactivation

Sulfur Removal with New HPG-300

	% wt	ppmw
C4 hydrocarbons	100	
Dimethyl Sulfide		20
Ethyl Mercaptan		20

- Adsorbent used: **HPG-300**
- Activated at 350°C under N_2
- Adsorption testing at 40°C
- Regeneration temp of 280°C with N_2 as regenerant

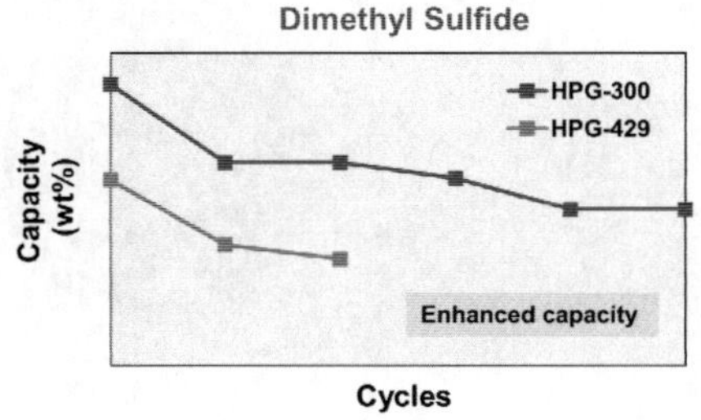

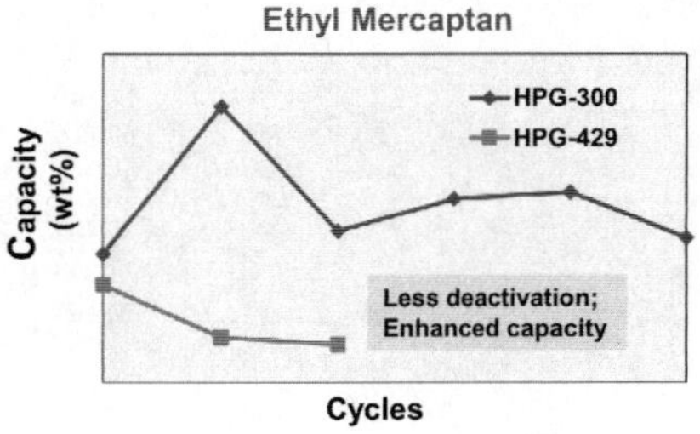

HPG-300: Excellent Sulfur Compound Removal for Catalyst Protection

New Platform of C_4 Hydrocarbon Treating Adsorbents

- ORG-HP
  - oxygenate and sulfur compound removal from olefinic C_4s
- HPG-300
  - oxygenates, sulfur compound removal from paraffinic C_4s
- **ACN-300**
  - nitrogen compound removal from olefinic C_4s

Honeywell UOP

UOP 7926D-33

34

New ACN-300 Key Features

- Nitrile Removal Adsorbent
- Optimized adsorption sites for high capacity
- Balanced acidity / basicity for capacity and regenerability enhancement

ACN-300: Improved Nitrile Removal Adsorbent

35

Block Flow Diagram for Etherification Unit

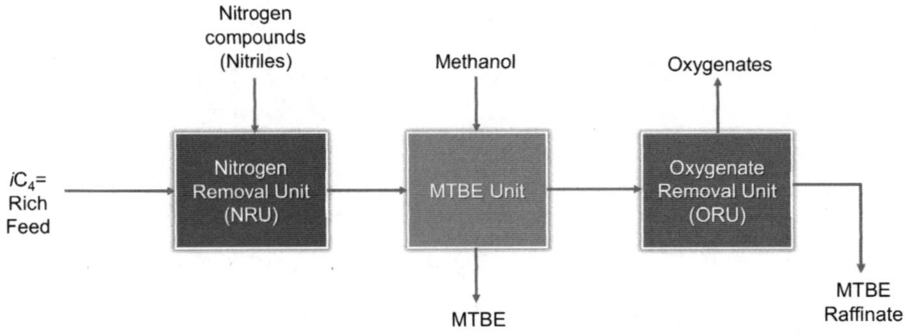

Nitriles & Basic Nitrogen: Key Contaminants for Etherification Process

Nitrile Removal with New Nitrile Removal with New ACN-300

Equilibrium Data

Simulated Feed Prepared with High Olefinic Content

	% wt	ppmw
Isobutane	37	
n-Butane	12	
Isobutene	24	
1-Butene	15	
2-Butene (cis/trans)	11.8	
Isoprene (2 Me-1,3 C4==)	0.2	
Acetonitrile		60

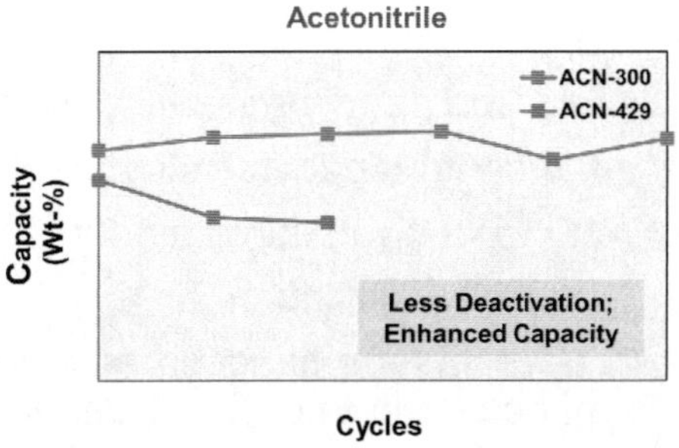

- Adsorbent used: ACN-300
- Activated at 350°C under N_2.
- Adsorption testing at 40°C
- Regeneration temp of 283°C with N_2 as regenerant

ACN-300 Provides Enhanced Nitrile Removal Capability

Economic Benefit of New Adsorbents

Higher capacity provides several benefits

- New units – *decrease vessel & equipment sizes*
  - Lower CAPEX
- Existing units – *increase cycle time*
  - Lower OPEX
  - Decreasing regeneration frequency preserving adsorbent life
  - Ability to handle unexpected upsets in the feed
  - Option to increase throughput or select lower cost feedstock

Extending adsorbent life

- *Reduces annualized turnaround and disposal costs*

UOP 7828D-37

Summary

- Recovery of increasingly valuable olefinic C_4s, especially for petrochemical use, is placing high demands on purifications processes
- Adsorption is a key purification option in maximizing value for C_4 refinery producers and petrochemical consumers
- Key adsorbent properties include:
- **Capacity** for a given contaminant
- **Life** / reactivity / coking
- UOP's new platform of adsorbents have enhanced performance for both capacity and longer adsorbent life

Call us to help you select the proper adsorbent and design the right contaminant solution for your next project

Q&A

Honeywell
UOP

480073 Converting Propane and Butane (LPG) in to High-Value Products

Abstract:

Shale oil and gas operations in North America generate abundant and low-cost supplies of LPG. At the same time, high production levels of shale oil and gas have created market niche opportunities for high-octane gasoline and low-temperature diesel blendstocks.

The industry has long contemplated the combinations of technologies to convert components of LPG, propane and butane into gasoline and diesel blendstocks. However, prior to the advent of high differentials between oil, gas and LPG, these combinations did not make economic sense.

These processes should be updated with advancements in the individual technologies and advancements in the process intensification, as well as adapted to the market requirements by simplifying separations. The resulting processes represent additional high-value and high-volume outlets for low-cost North American propane and butane.

AUTHOR INDEX

AUTHOR INDEX

AIChE
120 Wall Street, FL 23
New York, NY 10005-4020

ISBN 978-1-5108-4147-5